The Second Workshop on Grand Unification

University of Michigan, Ann Arbor
April 24-26, 1981

Jacques P. Leveille,
Lawrence R. Sulak, and
David G. Unger, editors

1981

Birkhäuser
Boston · Basel · Stuttgart

Editors:

Jacques P. Leveille
Lawrence R. Sulak
David G. Unger

The Harrison M. Randall Laboratory of Physics
The University of Michigan
Ann Arbor, Michigan 48109

Library of Congress Cataloging in Publication Data

Workshop on Grand Unification (2nd: 1981: University of Michigan)
The Second Workshop on Grand Unification.
1. Grand unified theories (Nuclear physics) --Congresses.
I. Leveille, Jacques P., 1951-
II. Sulak, Lawrence R., 1944-
III. Ungar, David G., 1951- IV. Title.
QC794.6.G7W67 1981 539.7'54 81-38522
ISBN 3-7643-3055-4 (Basel) AACR2

CIP-Kurztitelaufnahme der Deutschen Bibliothek

Workshop on Grand Unification 02, 1981, Ann Arbor, Mich.:
The Second Workshop on Grand Unification / ed. by:
Jacques P. Leveille ...- Boston; Basel; Stuttgart:
Birkhäuser, 1981
ISBN 3-7643-3055-4
NE: Leveille, Jacques P. (Hrsg.); HST

ISBN: 3-7643-3055-4
Printed in USA

CONTENTS

FOREWORD

Recently there has been rapid progress towards understanding the separate theories of the strong, weak and electromagnetic interactions within the framework of the standard SU(3) x SU(2) x U(1) model. The purpose of the Second Workshop on Grand Unification was to discuss the physics beyond the standard model and the major topic was grand unified theories which unify the strong, weak and electromagnetic sectors. Grand unified theories are presently being used to calculate experimentally accessible quantities such as the proton lifetime and nucleon decay branching ratios. Meanwhile, experiments are currently being performed, and new, dedicated experiments mounted, to measure these quantities. Reports on these experimental and theoretical activities occupied much of the workshop. Furthermore, since grand unified theories allow one to extrapolate the behavior of the universe back to the first instants after the big bang, their cosmological implications and the constraints on these theories from cosmology were of great interest at the workshop.

The conference opened with a keynote address by S. L. Glashow in which he discussed among other topics baryon minus lepton number conservation, neutrino masses and a neutrino-free universe. To maximize the interplay between theorists and experimentalists, theoretical and experimental talks were interleaved. An experimental highlight of the workshop was the presentation by S. Miyake of three candidate events for proton decay. The Indian-Japanese group working in the Kolar Gold Fields has established that the inherent neutrino background is as low as initially predicted and has measured a new lower limit on the lifetime of the proton of 3×10^{30} years. The current limits from the on-going experiments of the Minnesota group at the Soudan Mine and of the Pennsylvania group at the Homestake

Mine were presented by M. A. Shupe and R. I. Steinberg respectively. The experiments of the Irvine-Michigan-Brookhaven, Harvard-Purdue-Wisconsin, French, and Italian groups, which are all in the proposal or construction phases, were detailed by D. Sinclair, C. Wilson, R. Barloutaud and W. Fiorini respectively. These detectors can potentially observe nucleon decay if the lifetime is less than 10^{32} -10^{33} years. To extend the limit of observation to a lifetime of 10^{34} years, J. G. Learned proposed a deep ocean detector. Decay mode independent searches have been made with "passive" detection schemes. C. L. Bennett presented a new, lower bound on the lifetime of 2×10^{27} years from his passive technique. Neutron-antineutron mixing is another possible consequence of some theoretical frameworks, including several grand unified models. K. Green reviewed the phenomenology of n-n mixing and the experiments designed to observe its consequences.

On the theoretical side, P. Langacker surveyed the theoretical expectations for baryon number violation within the context of grand unified theories. W. J. Marciano presented the confrontation between predictions of the SU(5) model and current data. There is presently good agreement on the value of $\sin^2\theta_W$. Y.-P. Yao discussed the low energy effective theories and parameters that result from unified theories with widely separated mass scales. C. T. Hill and M. Machacek pointed out possible new consequences of grand unified theories on the low mass fermion spectrum. Schemes beyond the standard model, but which do have new mass scales between 10^2 and 10^{14} GeV, were presented by M. Veltman, S. Raby and J. Pati. The novel idea of combining supersymmetry with technicolor to provide answers to current difficulties, such as naturally explaining the existence of an extremely large gauge hierarchy, was discussed by M. Srednicki, H. Georgi, and S. Dimopoulos. P.Q. Hung outlined alternative models in which the weak and electromagnetic interactions have a single gauge coupling constant.

The consequences of grand unified theories for fundamental problems in classical cosmology and vice versa were reviewed by M. S. Turner. Within the unification framework the observed baryon asymmetry may be understandable. Alternatively, the observed baryon asymmetry can be used to constrain grand unified theories as explained by M. Yoshimura.

S. Weinberg concluded the workshop by placing the discussions of the three days into perspective and by challenging the participants with new issues in this exciting field.

In addition to the lectures and informal discussions, conviviality was assured by an evening of imbibing, French cuisine and excellent wines in an environment that included the Hall of Dinosaurs at the Natural History Museum on the University of Michigan campus. The down-to-earth nature of spontaneously broken symmetry was demonstrated by the soap bubbles of L. Sander, and the deep-under-earth nature of the experimental aspects of the field was highlighted by a visual trip to the IMB salt mine, courtesy of Michigan Media TV.

The financial support for the workshop was provided by the National Science Foundation, the Department of Energy, and the Physics Department of the University of Michigan.

We want to thank the lecturers for their cooperation in preparing this volume. All members of the local organizing committee must be thanked for their hard work; in particular Gordy Kane for his unfailing enthusiasm. We are grateful to Lois Kane for producing the art used on the cover of this book. A special debt of gratitude is due to Barbara Nieman whose efforts and patience in the months prior to and during the workshop made this meeting the success that it was.

June 1981

J. P. Leveille

L. R. Sulak

D. G. Unger

A NEUTRINO-FREE UNIVERSE[†]

Sheldon Lee Glashow[*]

Center for Theoretical Physics
Massachusetts Institute of Technology
Cambridge, Massachusetts 02139

The work I shall discuss was done in collaboration with H. Georgi and S. Nussinov. References to the literature may be found in our recent Harvard preprint, HUTP-81/A026.

(1) Everything is allowed that is not explicitly forbidden. This is not a Scandinavian sex law, but a perceived truth of elementary particle physics. Hence, parity-violation, CP violation, and (soon) baryon number violation. Nobody, convincingly, argues that neutrino masses *must* vanish. In SU(5), when it is naively implemented, neutrino masses do vanish. On the other hand, O(10) (again naively implemented) *requires* non-vanishing neutrino masses. In general, it is argued that neutrino masses are small by the same factor that the unifying mass is large. Grand unified theories suggest (but do not demand) neutrinos with masses from electron volts to microvolts.

(2) Hints of neutrino mass. Recent papers address the possibility that neutrinos have measurable masses, and indeed, that effects due to these masses have already been seen. Here is a sampling of the literature.

(3) Neutrinoless double beta decay. This process is induced by the existence of a Majorana mass of the electron neutrino. Anciently, Primakoff and Rosen argued that the experimental data excluded such a mass if it were larger than 800 eV. More recently, stronger limits have been deduced. From an analysis of data concerning ^{76}Ge and ^{82}Se decays, Haxton, *et al*. conclude that the Majorana mass of the electron

[†]This research is supported in part by the National Science Foundation under Contract No. PHY77-22864 and the D.O.E. Contract DE-AC02-76ERO-3069.

[*]On leave from Harvard University.

neutrino must be less than 15 eV. In an independent study of the ββ decays of ^{128}Te and ^{130}Te, Doi, <u>et al</u>. conclude that the Majorana mass of the electron neutrino <u>is</u> 41 ± 1 eV. These two studies are only in apparent contradiction. The model of neutrino masses we shall discuss is compatible with both results.

(4) <u>Endpoint spectra</u>. A finite neutrino mass will affect the shape of the electron spectrum in beta decay near its endpoint. In a recent Soviet study of tritium decay, evidence for a non-zero neutrino mass has been claimed: $14\text{eV} < m < 46\text{eV}$. The interpretation of this experiment is clouded by uncertainty due to atomic and molecular corrections which are comparable to the alleged effect. Clearly, further studies of tritium are required, and in a simpler atomic environment than as a constituent of valine. Another promising endpoint study that may reveal the neutrino mass involves inner bremsstrahlung in electron capture. Such experiments have been recently proposed, discussed, and initiated by De Rújula and his collaborators.

(5) <u>Neutrino oscillations</u>. Massive neutrinos may (but need not) display the phenomenon of neutrino oscillations, wherein one kind of neutrino becomes another. These phenomena may be parameterized in terms of (three or more) neutrino masses and (four or more) neutrino mixing parameters. Large mixing angles and large masses have been ruled out by experiment. However, there are two weak experimental indications that neutrinos do oscillate. One is the well-known solar neutrino deficit. This <u>could</u> be explained by small neutrino masses (at least microvolt) and large mixing angles. With three neutrino species, the maximum suppression of solar neutrinos is by a factor of 0.33, and a random choice of angles gives a mean suppression of 0.60. The Gallium experiment, as proposed by Bahcall and as initiated by Davis, can reveal whether the observed solar neutrino deficit is a problem for solar physics or for particle physics. The second positive indication for neutrino oscillations, a reactor experiment, was announced by Reines <u>et al</u>. a year ago. Confirmation of this result, as yet, is lacking. Several very sensitive searches for neutrino oscillations are in various stages of planning or implementation at present. It is a subject of primary interest and importance.

(6) <u>Galactic Haloes</u>. At various levels, there are indications of the existence of substantial amounts of non-luminous mass in the universe. For example, the rotation curves of galaxies show (almost without exception) the existence of massive invisible galactic haloes.

The mass in the diffuse non-luminous halo exceeds the mass in the localized luminous galaxy by a factor of 2 to 20. The non-luminous component may be in the form of extinct or sub-luminous stars. Another possibility is that it consists of massive neutrinos, so long as the neutrinos weigh at least ~20 eV. The radiative decay of these neutrinos may produce a significant flux of UV photons. If these are softer than 13.6 eV, they may be directly observed. If they are harder, they can affect the ionization state of the intergalactic medium. Dennis Sciama and Adrian Melott recently suggested that neutrino masses of order 100 eV and radiative lifetimes of order 10^{27} seconds could explain certain observational peculiarities. If correct, this analysis has two interesting side effects. Their neutrino mass is just right to give the universe a critical mass density, so that it is neither open nor closed but flat. Secondly, their required radiative lifetime is too short to be compatible with the standard theory with three fermion families. If there were a fourth family lurking at or nearby the Z^0 mass, the short lifetime they need can easily be obtained as De Rújula and I have argued. I suggested last year, jokingly, that galactic haloes made of neutrinos or of "bricks" could be distinguished by a determination of the resolving power of gravitational lenses. Apparently, this suggestion is not entirely absurd.

(7) Galactic clusters. The observed behavior of clusters of galaxies suggests the presence of missing mass at a larger distance scale. Steigman and Schramm have pointed out that this missing mass can be explained in terms of neutrinos of mass 6±3 eV. Their essay, "A Neutrino Dominated Universe", won first prize in the Gravity Foundation competition.

(8) Each tub on its own bottom. The evidence for neutrino masses, scanty though it is, is of four types:

(a) Oscillations

(b) Endpoint effects

(c) Gravitational effects

(d) Neutrinoless double beta decay.

Each of these effects may, or may not, be present in nature at an observable level. There are evidently sixteen distinct possible experimental scenarios. Any theorist worth his salt can construct an acceptable model corresponding to any of these possibilities: all effects, any three, two or one, or no effects at all. For example, oscillations may be prohibited by a discrete symmetry; neutrinoless double beta decay may be prohibited by lepton-number conserving Dirac neutrino masses; endpoint effects do not occur if only the electron neutrino is massless; and massive neutrinos are cosmologically irrelevant if they have already decayed or annihilated one another. We need independent experimental data concerning each of the neutrino mass effects.

(9) B-L symmetry plays a special role in the standard G = SU(3) x SU(2) x U(1) model, and in simple grand unifications based upon SU(5) or O(10). The fifteen dimensional family of G admits precisely one additional current which is linearly independent of the generators of G and is free of gauge-gauge-current anomalies. This current--baryon number minus lepton number-- is moreover an exact global symmetry of the standard theory. Models of neutrino mass may be classified in terms of what has become of B-L.

(10) B-L as a local symmetry. In order to accomplish this, the fermion representation must be changed to eliminate $(\text{gauge})^3$ anomalies. Most simply, this is done through the addition of right-handed neutrinos. Then, B-L may be gauged. It cannot be an exact gauge symmetry, because there is no second variety of massless photon. It must be spontaneously broken. The corresponding vector boson must be much heavier than Z^0 lest neutral-current phenomena are drastically changed. A large VeV of a Higgs singlet may give the B-L boson its mass. It may also give large Majorana masses to the right-handed neutrinos. An interplay of the Higgs doublet with the Higgs singlet results in small Majorana masses for the left-handed neutrinos, of order m^2/M where M is the mass of the right-handed neutrino, and m is typical of charged fermion masses. This possibility is realized within a grand unified theory based on O(10). With a single heavy mass scale, neutrino masses are expected with values between volts and microvolts. In general, all the phenomenological effects of ¶8 are produced. This is the most conventional picture of neutrino masses today. Though it is a viable possibility, it is not the possibility that we shall pursue here.

(11) B-L may be an exact global symmetry. This is the case for the standard theory and for its minimal extension to SU(5), in which the neutrino masses quite naturally vanish. It is possible to retain exact B-L symmetry, and yet, to give neutrinos mass. This is done by introducing right-handed neutrinos, and appropriate Yukawa couplings of the Higgs doublet. Neutrino masses are set to be very small at the

beginning: an awkward but acceptable procedure. In this case, neutrino masses are necessarily Dirac masses. Neutrinoless double beta decay is rigorously forbidden.

(12) B-L symmetry may not be a symmetry of the Lagrangian at all. This is an ugly technical possibility which may be implemented in various ways, and which is devoid of specific experimental predictions. We shall not consider it.

(13) B-L may be a spontaneously broken global symmetry. This possibility necessarily entails the existence of a massless Goldstone boson, the Majoron. It will be necessary to demonstrate that such a particle is phenomenologically innocuous. Two recent papers propose models in which B-L is spontaneously broken. In one case the B-L breaking VeV is very large, in the other case it is very small.

(14) The Chicasige, Mohapatra, Peccei Model. The standard theory is supplemented by right-handed neutrinos. A complex Higgs singlet ϕ, bearing B-L quantum number of 2, is given Yukawa couplings to the right-handed neutrinos. B-L is spontaneously broken and ϕ develops a large VeV. Right-handed neutrinos develop large masses M, and left-handed neutrinos develop small Majorana masses of order m^2/M. B-L symmetry has been broken at a very large mass scale, and the Majoron is seen to be harmless.

(15) The Gelmini-Roncadelli Model. In this case, the authors limit themselves to 15-dimensional families without introducing right-handed neutrinos. (Of course, this is not a necessary restriction.) What manner of Higgs mesons satisfy the following two conditions?

(a) They enjoy Yukawa couplings to the fermions.

(b) They may develop non-zero VeVs which are compatible with exact charge and color symmetry.

There are just two possibilities. One is the familiar Higgs doublet which may give masses to all the charged fermions. The other is a complex Higgs triplet whose VeV gives Majorana masses only to the neutrinos. The VeV of the doublet breaks SU(2) x U(1) but leaves B-L intact. The VeV of the triplet breaks B-L as well as SU(2) x U(1). Call the doublet VeV u and the triplet VeV v. In view of the successes of the standard theory in describing neutral current phenomena, we may conclude that $v \ll u \sim 250$ GeV. This is the gist of the G.R. model. B-L symmetry, in this case, is broken at a much smaller energy than is the electroweak theory.

(16) Doubly Charged Mesons. The Higgs doublet and the new Higgs

triplet include neutral states, singly charged states, and doubly charged states. We shall denote the mass of the doubly charged meson χ^{--} by M, and it is of order $u\sqrt{\lambda}$ where u ~ 250 GeV is the Salam-Weinberg VeV and λ is a meson self-coupling. We may write 15 GeV $\lesssim$ M $\lesssim$ 250 GeV where the lower limit is experimental and the upper limit follows from self-consistency of the theory. The existence of this particle is a key prediction of the G.R. scheme. Electron-positron annihilation can produce pairs of these particles with a cross section of one R-unit. Electron-electron collisions can produce χ^{--} singly and resonantly, but we have shown that the decay width for $\chi^{--} \to 2e^-$ can be no larger than several KeV. More interestingly, if M is sufficiently small, χ^{--} may be produced in the decays of W^- and Z^0. In general, χ^{--} may mediate a new class of weak interactions of leptons, generating decays like

(a) $\mu^- \to 2e^- + e^+$

(b) $\tau^- \to 2\mu^- + \mu^+$.

Severe experimental limits may be placed on (a). However, there is no difficulty in choosing the parameters of the G.R. model such as to provide the observed suppression. On the other hand, further searches for these decay modes are clearly useful.

(17) Singly-Charged Mesons. Of the two singly-charged fields in the theory, one linear combination corresponds to the Goldstone mode and is eaten by $W^\pm$. The remaining combination is predominantly composed of the Higgs triplet, with a doublet admixture of order v/u. Its mass is given by $M(\chi^-) = M/\sqrt{2} + O(v^2/u)$. This is another crucial prediction of the G.R. model. Where else are new particles predicted to have a mass ratio of $\sqrt{2}$? Both the χ^{--} and the χ^- will have two kinds of decay scheme: weak and purely leptonic, and will have decay widths in the range of several MeV.

(18) Neutral Mesons. There are four neutral fields in the theory. One Goldstone mode, predominantly doublet, is eaten by Z^0. The other, predominantly triplet, survives as a truly massless Goldstone boson. It enjoys bilinear couplings to neutrinos which are flavor-diagonal in tree approximation. It has pseudoscalar couplings to charged fermions which are also flavor-diagonal, and are of order $m\,v/u^2$ where m is the fermion mass. In addition, there are two scalar Higgs bosons. One, predominantly doublet, is the familiar Higgs boson with a mass $\lesssim$ u. The other, predominantly triplet, has a small mass $\lesssim$ v. We shall see

subsequently that $v \lesssim 100$ KeV, so that this scalar Higgs is very light indeed. Neither the Goldstone particle nor the light Higgs present phenomenological problems. Decay modes like $K^+ \to \pi^+\chi^0$, $\pi^+ \to e^+\nu\chi^0$, $\mu \to e\chi^0$ are predicted to take place, but at rates well below established limits.

(19) Decays of W and Z. The following table shows the possible decay modes of W and Z into the various χ states, the kinematically allowed domain, and the predicted partial widths. (In these estimates, M has been taken much smaller than its kinematical limit, so that p-wave suppression may be neglected.)

Decay Mode	Partial Width	Mass Range
$Z^0 \to \chi^{++}\chi^{--}$	110 MeV	M < 47 GeV
$W^+ \to \chi^{++}\chi^-$	240 MeV	M < 49 GeV
$Z^0 \to \chi^+\chi^-$	20 MeV	M < 66 GeV
$W^+ \to \chi^+\chi^0$	240 MeV	M < 117 GeV
$Z^0 \to \chi^0\overline{\chi^0}$	340 MeV	for any M

These decay modes present a possibility to discover χ^- and χ^{--} both in e^+e^- collisions at the Z^0 mass; and by W,Z production in hadron-hadron colliders. Of particular interest is the decay of Z^0 into $\chi^0\overline{\chi^0}$, or more precisely, into the Goldstone particle plus the light Higgs. This decay must take place with a determined rate whatever the value of M. It is a "dark" decay mode, like the 2-neutrino mode, for neither of the light mesons is observable. The width for this decay is precisely twice the value of the width for Z^0 to decay into each neutrino mode. A Z^0 factory will measure the dark width by comparing the actual width with the smaller "visible" width deduced from the e^+e^- partial width and apparent branching ratio. This experiment must reveal at least five "neutrinos" (a dark width ~1 GeV) if the G.R. model is correct.

(20) Stellar Evolution. In this section, we obtain a strong upper limit upon the magnitude of v, the triplet VeV. The argument is an adaptation of an axion limit deduced by Dicus *et al*. Consider the core of a red supergiant star, with a radius $R \sim 3 \times 10^9$ cm, a temperature $T \sim 10$ KeV, and a density $\sim 10^4$ g/cm^3. The photon mean free path is

$L \sim 10^{-4}$ cm. In order to escape, the photon undergoes a random walk suffering $N = (R/L)^2 \sim 10^{26}$ collisions with electrons. In each of these collisions, there is a small probability P that the photon is converted into the Goldstone boson M: $\gamma + e \to M + e$. This probability is given by the ratio of the cross section for this process to the Compton cross section: $P = 8T^2v^2(3\pi\alpha u^4)^{-1} \sim 10^{-13}(v/u)^2$. We shall assume that the product NP is less than unity. That is to say, we require that the star emits no more energy in the form of Majorons than it does in photons. This yields the constraint $(v/u)^2 \lesssim 10^{-13}$, which is to say, $v \lesssim 100$ KeV. This argument shows that the triplet VeV is truly small compared to the doublet VeV. Thinking in terms of grand unification, we find that the G.R. model has a second, and lesser, gauge hierarchy problem. It is the fact that v is so small that guarantees that Majorons are not produced at detectable rates in most elementary particle decays. An exception to this is the case of neutrinoless double beta decay.

(21) Neutrinoless Double Beta Decay. In the G.R. scheme, there are three distinct modalities of double beta decay:

$$(Z,A) \to (Z+2,A) + \begin{cases} 2e^- + 2\bar{\nu} \\ 2e^- \\ 2e^- + \chi^0 . \end{cases}$$

The third modality involves no neutrinos, but the emission of a Goldstone boson or a light Higgs. Existing analyses of double beta decay data must be reconsidered in the context of the G.R. model. Unsuccessful searches for the neutrinoless mode in ^{76}Ge and in ^{82}Se were calorimetric, and were thus insensitive to the Goldstone mode. The conclusion of Haxton et al., that the Majorana mass of the electron neutrino is less than 15 eV remains valid. The clearest "observation" of double beta decay may be the study of Xenon isotopic abundances in old Tellurium ores. A recent analysis by Doi et al. concludes that ^{128}Te double beta decay is dominated by the neutrinoless mode. In a conventional model with Majorana neutrino masses, they conclude that the electron neutrino weighs 30 eV, in agreement with the Russian endpoint experiment, but in disagreement with Haxton et al. (We choose to ignore any nuclear physics uncertainties in the Doi analysis, as well as the experimental uncertainties in the Hennecke data on Tellurium.) Our reinterpretation of the Doi analysis in terms of the G.R. model yields a quite different result. We obtain

$$\left(\frac{m}{30\mathrm{eV}}\right)^2 + 10^6 g^2 \approx 1$$

where m is the Majorana mass of the electron neutrino, and g is its Yukawa coupling to Majorons. There is clearly no conflict between this result and the Haxton result: the decay of ^{128}Te is dominated by the neutrinoless Goldstone mode. In passing, we have developed the important constraint that $g < 10^{-3}$. This allows us to bound the partial width for $\chi^{--} \rightarrow 2e^-$, and goes much of the way to produce the observed suppression of $\mu \rightarrow 3e$. Further data on double beta decay would be of immense interest.

(22) Neutrino Cosmology. It is often said that cosmology puts an upper bound on neutrino masses; that the sum of the neutrino masses cannot exceed 100 eV. This conclusion must be reconsidered in the G.R. model wherein neutrinos have additional interactions. In particular, we might imagine that heavy neutrinos will decay into light neutrinos plus a Majoron. This does not take place in tree approximation, but it is induced at one loop. We obtain $\tau(\nu_H \rightarrow \nu_L + M) = [g^{-3}(100\ \mathrm{KeV}/v)^5]\ 10^{10}$ years, where we know that the expression in brackets must exceed unity. It follows that this mode of neutrino decay is not cosmologically relevant. Another reaction to which our neutrinos are subject is annihilation, $\nu + \nu \rightarrow 2M$, into a state of two massless majorons. In the evolution of the universe, as soon as the temperature falls to the neutrino mass, this process will serve to deplete the relict neutrinos. This depletion will be effective providing that the inequality

$$m_\nu < 0.03\ \mathrm{eV} \left(\frac{v}{100\ \mathrm{KeV}}\right)^{1.6}$$

is satisfied. Since we know that v cannot exceed 100 KeV, we conclude that neutrinos more massive than .03 eV will have annihilated by the present epoch. If neutrinos are to play a role in galactic haloes, they must weigh tens of eV. If they play a role in clusters, they must weigh several eV. If they are to contribute significantly (at least 1%) to the critical mass density, they must weigh at least 1 eV. In the G.R. scheme, none of this is possible. Neutrinos may have eV masses or even KeV masses with no cosmological implications. There can be few such relict neutrinos about today: they have all annihilated into massless and quite invisible Goldstone bososn. There can be no neutrino dominated universe.

(23) The Moral. We have given an exegesis of the Gelmini-Roncadelli model, but we can give no compelling reason for it to be

Nature's way to give neutrinos their masses. It is as appealing as any other scheme, and more appealing than most. But, we have another point to make. We have become accustomed to the notion of a great desert in high-energy physics extending from the electroweak mass scale to the unifying mass scale. The exciting physics at 10^{14} GeV can never be explored directly. We can only see it darkly through the windows of proton decay and neonatal cosmology. The Gelmini-Roncadelli model (as well as certain recent considerations of "invisible axions" to solve the strong CP puzzle) present us with another kind of desert, one at very low energies. The theory has a genuine massless Goldstone boson. It produces no sensible long range forces. Its couplings are too weak to be relevant to most laboratory experiments, becoming important only in hot stars, supernovae, or over cosmological times. The double beta-decay window is at least as dirty as the proton-decay window. Nature's secrets can be hidden away in the low-energy weak-coupling domain just as effectively as in the ultra-high-energy domain.

THE KOLAR GOLD FIELD NUCLEON DECAY EXPERIMENT
A STATUS REPORT

M.R.Krishnaswamy,M.G.K.Menon,N.K.Mondal,V.S.Narasimham and
B.V.Sreekantan
Tata Institute of Fundamental Research,Bombay 400 005, India
N.Ito, S.Kawakami and Y. Hayashi
Osaka City University,Osaka,Japan
and
S.Miyake
Institute for Cosmic Ray Research,University of Tokyo

A status report on the experiments now in progress at Kolar Gold Fields (KGF) is presented here with a brief description of the candidate events suggestive of nucleon decay.

1. Experimental Details

A ; Detector at 2300 m

An dxperiment designed specifically to search for nucleon decay has been in progress since November 1980 in the Kolar Gold Mines at a depth of 2300 m equivalent to 7000 hg/cm^2 (Kolar rock). The detector (Figure 1) comprises 34 horizontal layers of iron plates and proportional counters with a total weight of 140 tons of iron. The proportional counters are made from square pipes of cross sectional area 10 x 10 cm^2 and thickness 2.3 mm and are arranged in an orthogonal configuration with alternate layers made up to 4 m and 6 m long counters. In between these layers, iron plates of thickness 12 mm are inserted ; these, as well as the counter walls provide the source of bound nucleons. The basic trigger is provided whenever there is a 5-fold coincidence of the OR-ed counter pulses in each layer ; in such cases the ionisation in each counter is also recorded. From the data obtained in the past 5 months we estimate that the efficiency to record single muons etc. is close to 100 %.

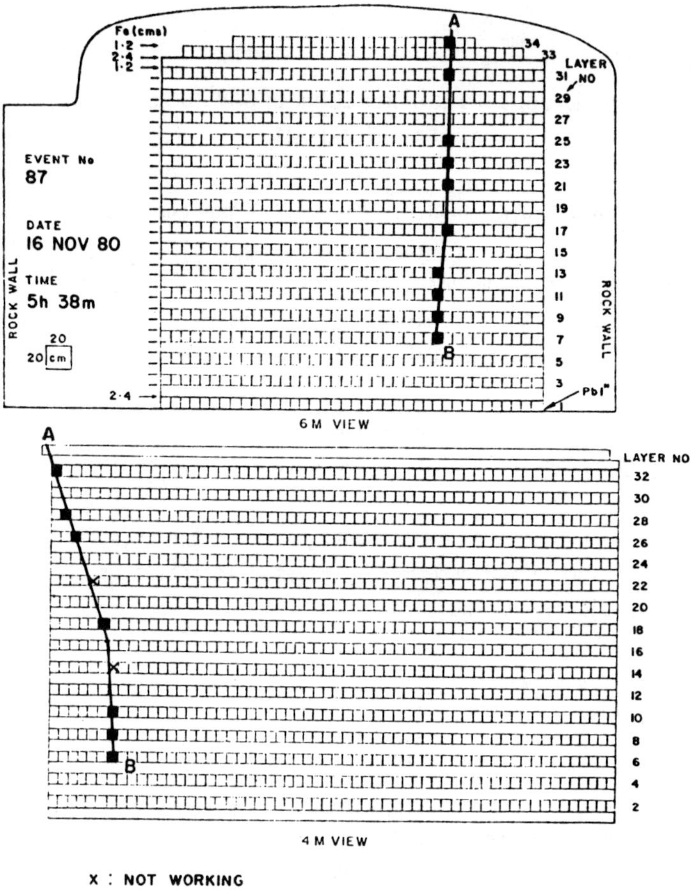

Figure 1. The two orthogonal views of the detector are shown here approximately to scale. The odd and even numbered layers correspond to the 6 m and 4 m long counters arranged in alternate layers except for layer No. 34 which is also made up of 6 m counters. Event 87 is shown in this figure ; the black squares represent the the counters that responded ; the 'X' mark indicates the two counters along the tracks A and B that were not operating.

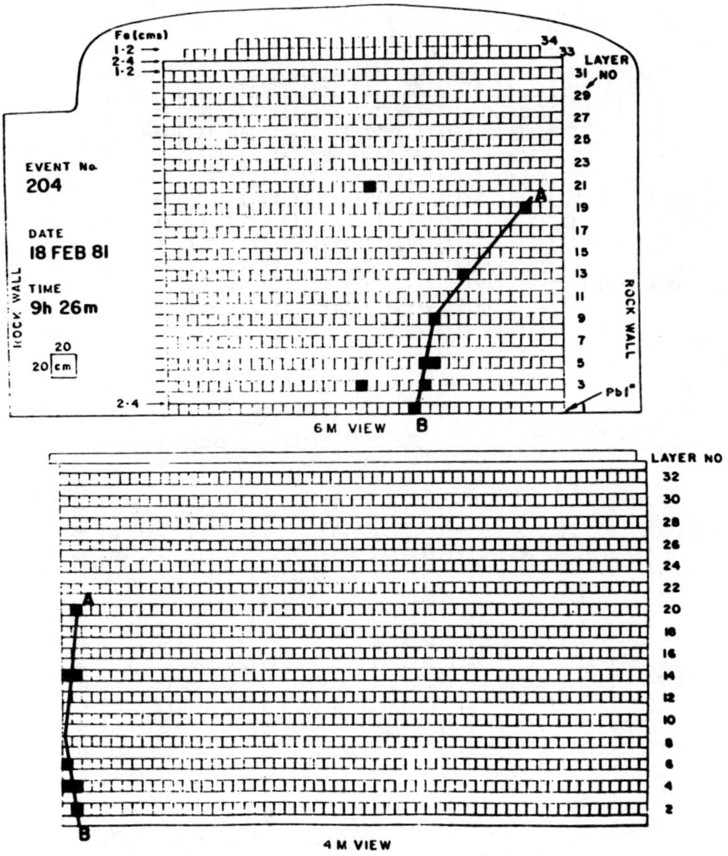

Figure 2. In this event the meeting point of tracks A and B is close to the edge of the detector. Track A has probably stopped inside the detector whereas track B has penetrated the bottom layer of 1" lead absorber. The total visible energy loss is ~ 0.46 GeV.

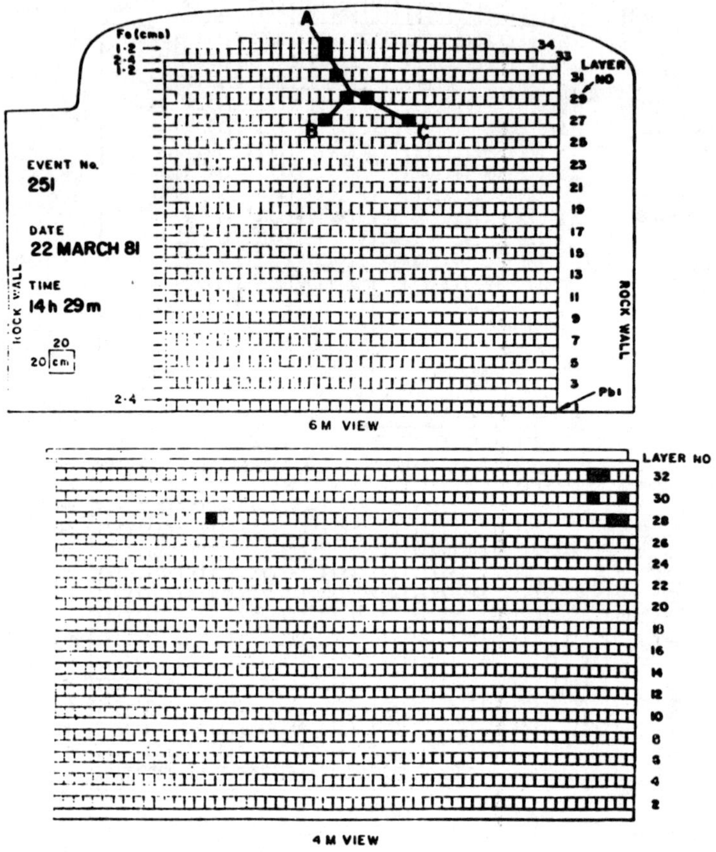

Figure 3. Among the 3 tracks in the 6 m view, the track A has penetrated the top layer. The projected angles between tracks A,B and B,C are 100° and 110° respectively. The total visible energy loss in this event is $\sim$ 0.4 GeV.

B. Telescope at 1200 m

This detector has been in operation since 1977 at a depth of 1200 m equivalent to 3375 hg/cm^2 (Kolar rock) to look for new particle events and has been described earlier. Briefly, this is arranged as a horizontal telescope with 9 columns of Neon flash tubes (2 cm diameter and 2 m long) suitably sandwiched with 4 walls of proportional counters and 4 walls of 7.5 - 10 cm thick iron absorbers as shown in Figure 4. By tilting alternate NFT walls by 4.5° with respect to the horizontal plane, 3-D reconstruction of tracks has been made possible. The overall dimensions of the detector are 4.2 m length, 3.2 m height and 3 m width. In view of the high track resolution, a fiducial central volume of the detector with about 11.5 tons of iron could be used to search for nucleon decay events.

2. Results from the new detector at 2300 m

The rate of event collection is about 1.8/day, and the data recorded in 131 days can be broadly classified as follows : (a) atmospheric muons penetrating and stopping, (b) neutrino interactions in rock leading mostly one muon with zenith angle $>55°$, (c) neutrino interactions inside the detector and (d) other processes. These results are summarised in Table 1 and compared with estimates based on the present knowledge of fluxes and cross sections of muons and neutrinos. The excellent agreement as seen in columns 2 and 3 of this table clearly indicates that all known backgrounds are fully accounted for.

Candidates for nucleon decay

We have so far recorded under the category (d) above 3 events (Events 87, 204 and 251) which are not understood in terms of any of the presently known backgrounds and have indeed the track configurations and visible energies consistent with nucleon decay. These events are shown separately in Figures 1,2 and 3 in the two orthogonal projections (called 4 m and 6 m views) of thc detector.

Events 87 and 204

These events (Figures 1 and 2) appear to have similar character with the following outstanding features :

(1) Approximately 50 % of the counters along their paths do not show measurable ionisation in the upper section (track A) of these events.

TABLE 1

Basic Data : Detector volume : 6 m x 4 m x 3.7 m
Total weight : 140 tons (Iron)
Live-time : 131 days
Total No. of Events : 243

Category of Events	Observed No.	Estimated No.
1. Single muons penetrating the detector (θ = 0 - 55°)	223	218 ± 30
2. Parallel Muons (E_μ > 15 TeV and multiplicity > 2)	1	1 - 2
3. Stopping muons	1	1 - 2
4. ν-interactions in rock *		
θ = 0 - 55°	Grouped under category 1.	4
θ = 56° - 90°	8	8 ± 2
5. ν-interactions inside the fiducial volume	3	4 - 5 **
6. Kolar Events (Multiprong,wide angle,air vertex events)	1 (?)	1
7. Unusual events (Nucleon decay candidates)	3	?

* Zenith angle range in above 4. includes below horizen.

**Estimated No. in above 5. are approximate and need correction due to trigger efficiencies.

(2) In the lower section (track B) the ionisation is continuous over 8 - 10 radiation lengths and (3) The angle between the two tracks A and B is 18° or 162° for event 87 and 32° or 148° for event 204 indicative of either a large deflection or decay with nearly back to back configuration at the meeting point.

The feature (1) cannot be due to the traversal of either muons or charged hadrons since it is inconsistent with the known detector efficiency. Thus we discount the possibility of scattering of a charged particle as being responsible for these events. In fact the Monte Carlo simulations show that it is attributable to a low energy ($\sim$1GeV) electron or photon. The track B could however be due to a μ, $\pi^{\pm}$ or an electron. The visible energy of these events is less than the nucleon mass. The only background that is relevant here is from low energy neutrino-interactions and the rate of such events mimicking the nucleon decay is estimated as $\sim$1 in 10 years. These events are therefore candidates for nucleon decay into 2-body modes such as e,π°; $e,\pi^{\pm}$ etc.

Event 251

This has 3 tracks with their vertex well inside the detector as seen in the 6 m projection (Figure 3). The 4 m view does not however permit unique reconstruction of the event. The important features here are (a) large opening angles of tracks as seen from the meeting point and (b) short range of particles B and C which stop inside the detector. It is very difficult to fit the event in a scheme involving μN-interaction with a secondary hadron either decaying or interacting at vertex. An interaction by low energy ν's can in principle produce such an event but as stated above their frequency is negligibly small. We thus consider this event also a candidate of nucleon decay into a multibody final state such as $e\rho$ or $e\omega$.

In all these events one track is touching the detector boundary and hence it is difficult to estimate the total energy. For two of these events the meeting point is well inside the detector volume so that one is sure that no other charged particle has escaped observation. The extremely low level of background, the observed track configurations as well as the fact that the visible energy loss is $< M_pc^2$ makes these events candidates for nucleon decay predicted by standard theories [2]. Assuming them as indeed due to nucleon decays, we estimate the lifetime of bound nucleons as $\sim 3.5 \times 10^{30}$ years.

3. Results from the experiment at 1200 m

Some of the important results on multiple muons, cascades and Kolar events were reported earlier [1]. We present here only the data relevant for discussion on nucleon decay. During the live-time of about 3.5 years,there are just 2 events which have characteristics consistent with nucleon decay. These are illustrated in Figures 4 and 5 by transposing the actual photographic records of events on the front view of the detector. Since in this detector the alternate walls are tilted by 4.5°, the splitting of tracks is a measure of the depth at which they are located along the length of the neon flash tubes. In the first event (Figure 4,dated 10/1/79) all 5 tracks seem to originate from the iron wall at a vertex situated at least 20 cms inside the detector. The absence of an energetic muon (which could otherwise have produced an interaction in iron) coupled with the near isotropic configuration of the tracks strongly indicates decay.

In the second event (Figure 5,dated 27/5/80) all tracks originate at a vertex at about 40 cm inside the iron absorber. The tracks 1 and 2 have in all probability stopped in the adjacent iron walls and the shower (3) appears to be electromagnetic in nature (either due to photons or electrons). The configuration of these three sets of tracks and visible energy loss of about 550-750 MeV (based on track counting method) indicates that this event is another candidate for nucleon decay into modes such as $P,N \rightarrow e(\rho,\omega)$ etc.

We have examined whether these events could be generated by the interaction of cosmic ray muons or neutrinos either in rock or the iron absorbers and find the probability very small even though the detector is operated at a rather shallow depth. If these events are considered as due to nucleon decay, the life-time can be estimated from the fiducial mass of 11.5 tons, (efficiency x branching ratio of 0.5) as $\sim 3 \times 10^{30}$ years.

Discussion

The candidate events reported from the two sets of experiments at depths of 1200 m and 2300 m are to be treated as suggestive of nucleon decay in the absence of any other plausible explanation. In the experiment at the greater depth, where 3 events were recorded in 131 days of operation, the contribution of $\nu,\tilde{\nu}$-interactions which can mimic them is estimated as < 0.1 per year ; this is based on the well known fluxes

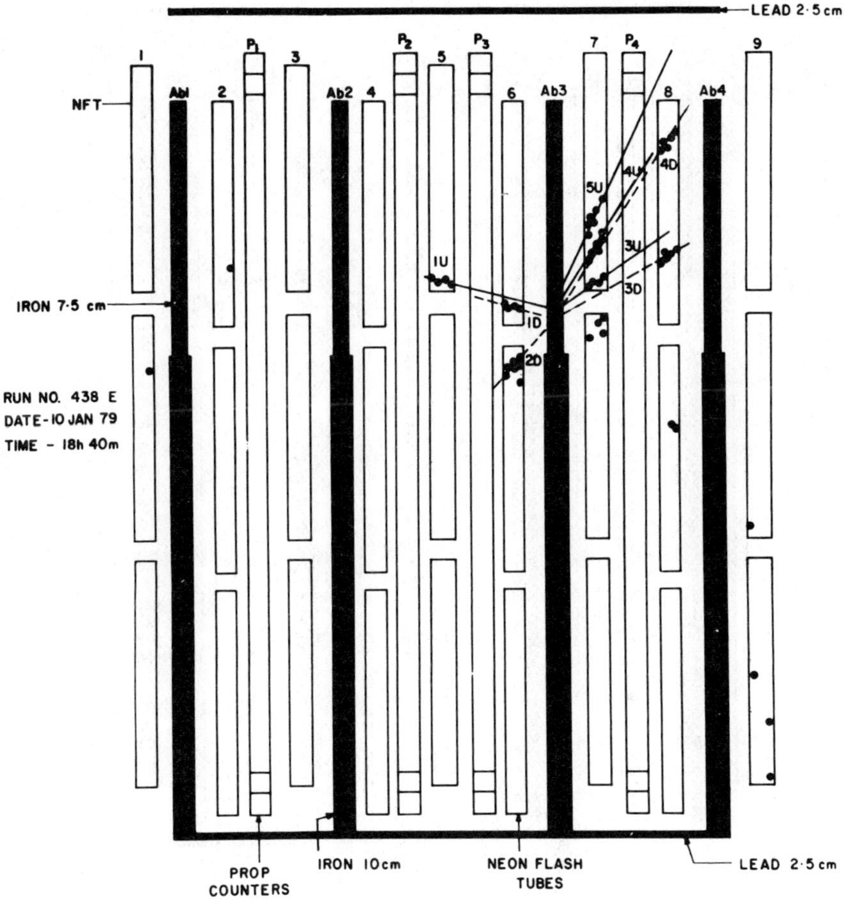

Figure 4. The 4 walls of the proportional counters used for trigger are labelled P_1,P_2,P_3 and P_4. The absorbers Ab 1 ...4 are of 10 cm thickness upto a height of 2 m and 7.5 cm thickness above. The NFT walls are numbered 1 - 9 with alternate walls tilted by 4.5°. The black dots in this diagram represent the NFT"s that have responded. The meeting point of the 5 tracks is determined by an extrapolation of tracks in the two sets of tilted NFT planes separately.

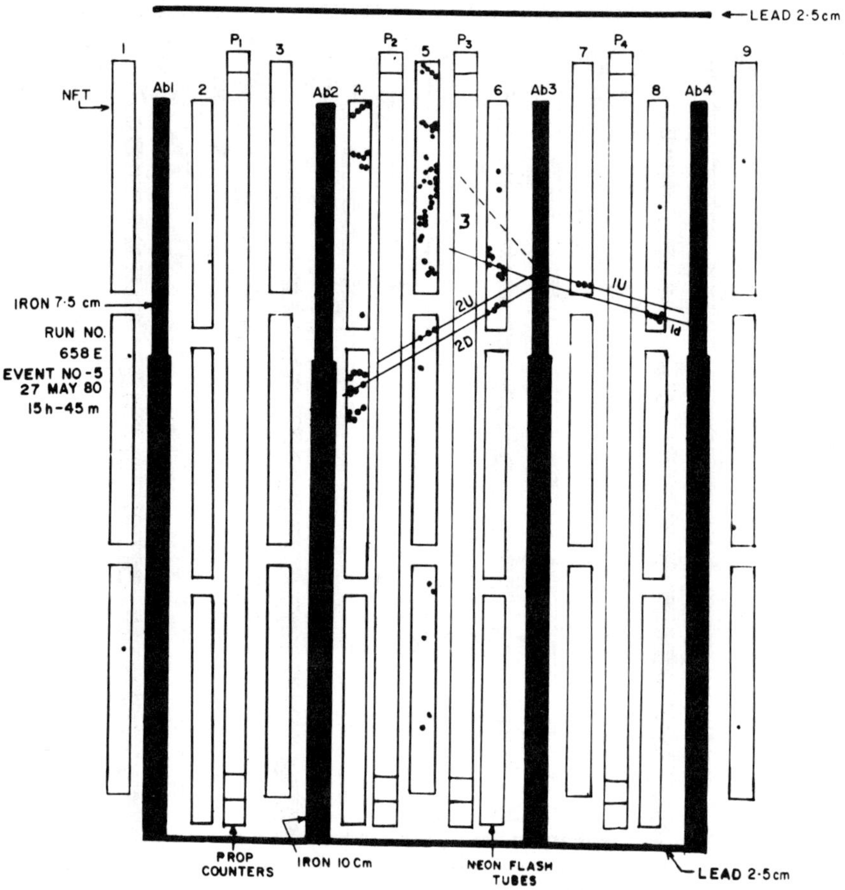

Figure 5. This event has 2 separate tracks (1 and 2) and a shower (3). The labels 'u' and 'd' distinguish the NFT planes tilted up and down. Particle '2' has produced a small shower in the proportional counter wall P_2 and is fully absorbed in the iron wall Ab 2. The shower (3) is of low energy ($\sim$ 350 MeV) and is spread over a cone of half angle of about 30°.

and cross-section of cosmic ray neutrinos and the angular distribution of their secondaries determined from both cosmic ray and accelerator experiments at the energies of the order of a few GeV. The cosmic ray muons at the depth of 2300 m have a mean energy of 400 GeV and their flux is so low that the contribution from any of the known types of secondaries produced in their interaction is negligible.

It is to be noted that, in view of the disposition of matter,there would be more decay events in the peripheral regions of the detector. Events with vertex well inside the detector and with fully confined tracks are likely to be recorded with a frequency of about 1 in 4 or 5 cases. The operation of the detector for a period of a year or two should therefore lead to a few central events.

Acknowledgements

We wish to thank Messers R.M.Wankar,V.M.Punekar, and S.Manjunath for their technical assistance in the construction of the detector and its maintenence. We are grateful to the officers and other staff of the Bharat Gold Mines Limited for their cooperation and help at all stages of the activity in the Mines. The partial financial support given by the Ministry of Education,Japan is gratefully acknowledged.

References

[1] Krishnaswamy,M.R.,Menon,M.G.K.,Mondal,N.K.,Narasimham,V.S. Hayashi,Y.,Ito,N., Kawakami, S. and Miyake, S. Proc. Cosmic Ray Conf.,Kyoto 13, p.14,p.24,p.378 (1979)

[2] Pati, J.C. and Salam, A., Phys. Rev. D 8, 1240 (1973)
Georgi,H. and Glashow, S.L. Phys. Rev. Lett., 32,438 (1974)
Georgi,H. , Quinn, H. and Weinberg, S., Phys. Rev. Lett., 33 451 (1974)

SEARCH FOR NUCLEON DECAY: THE DEEP UNDERGROUND WATER CERENKOV DETECTOR AND THE HOMESTAKE TRACKING SPECTROMETER*

M.L. Cherry, T. Daily, K. Lande, C.K. Lee, E. Marshall, R.I. Steinberg
Department of Physics, University of Pennsylvania
Philadelphia, Pennsylvania 19104

B. Cleveland
Department of Chemistry, Brookhaven National Laboratory
Upton, New York 11973

ABSTRACT

We report the initial result of a search for nucleon decay using a 300-ton water Cerenkov detector located 1480 m underground in the Homestake gold mine at Lead, South Dakota. We find $T_N > 2.4 \times 10^{31}$ B_μ y, where T_N is the nucleon total lifetime and where B_μ is the model-dependent probability of production and detection of a stopping and decaying muon following a nucleon decay event. In the SU(5) grand unified model, calculations of B_μ vary from approximately 5 to 10%, thereby yielding nucleon total lifetime limits ranging from 1 to 2×10^{30} y.

We also describe our plans for a conceptually new nucleon decay detector which will permit study of this process for lifetimes up to 3×10^{32} y. This detector will be a completely-vetoed completely-active 1000-ton liquid scintillation tracking spectrometer also located at a depth of 1480 m in the Homestake Gold Mine. This instrument will be sensitive to all nucleon decay modes; it will have a spatial resolution of 30 cm (0.6 radiation lengths), thereby permitting identification of the decay secondaries; it will measure the total ionization energy released by the nucleon decay with a resolution of 10%, thereby permitting study of the inclusive decay mode distribution by observing the full energy peak and neutrino escape peaks; and it will measure the charge of lepton secondaries as well as the polarization of secondary muons. By virtue of its excellent energy and spatial resolution and great depth underground, the spectrometer will be remarkably free of background and ambiguity.

I. SEARCH FOR NUCLEON DECAY WITH A DEEP UNDERGROUND WATER CERENKOV DETECTOR

The extraordinary success achieved in recent years by application of the renormalizable gauge theory formalism to the unification of the electromagnetic and weak interactions and to the introduction of a fundamental theory of the strong interaction has strongly motivated the search for an acceptable theory in which the presently separate electroweak and strong interactions would be explained as low energy manifestations of a single underlying unified gauge theory (the "grand unified gauge theory")[1,2]. Formulation of a successful grand unified theory would allow a deeper understanding of the nature of the elementary particles and their interactions and hopefully would eventually lead to a fundamental theory of the universe in which all physical phenomena (including gravity) would be understood.

In most of the grand unified models under present consideration,[3] a consequence of the unification would be the existence of new gauge bosons possessing both flavor and color. The interactions of these bosons would generally violate baryon number conservation and would therefore usually lead to nucleon decay. For consistency with the known approximate stability of nucleons, the new gauge interactions must be extremely weak at laboratory energies. This weakness implies that the gauge boson masses must be greater than approximately 10^{14} GeV. Because such energies are far beyond those directly attainable in the laboratory, the best evidence for the validity of grand unification is likely to be produced by extending the sensitivity of laboratory searches for nucleon decay.

The experimental status of nucleon stability has been reviewed recently.[4,5] The most sensitive directly measured limit on the nucleon total lifetime[6], $T_N > 2.2 \times 10^{26}$ y, was derived from an underground radiochemical search[7] for ^{37}Ar produced by nucleon decay in ^{39}K. The best limit on the nucleon partial lifetime for decay into specific modes was obtained by detection deep underground of stopping muons in a thin[8] liquid scintillation detector.[9] The result of this experiment was $T_N > 6 \times 10^{30}$ B_μ y, where B_μ is the probability for production of a stopping and decaying muon. The value of B_μ depends on the particular grand unified model being considered and on all details of the experiment which influence the probability of producing and detecting a muon that stops and decays (such as the particular nucleus in which the nucleon decay takes place, the experimental geometry, etc.). Estimates of B_μ for the experiment reported in reference 9 have recently been published.[10] For the SU(5) grand unified model[2], B_μ was estimated to be 15%, leading to a nucleon lifetime $T_N(SU(5)) > 10^{30}$ y.

The present Homestake nucleon decay experiment consists of a search for the decay of stopping muons which would be produced by decay of one of the nucleons of the water molecules comprising almost all of the mass of the detector. Assuming energy conservation, the only known hadrons to which nucleons can decay are the fourteen mesons lighter than 938 MeV. Eleven of these particles (all except π^0, π^-, and ρ^-) yield a substantial fraction of muon decays in a water detector. In addition, nucleon decays yielding direct positive or negative muons would also be detected efficiently. Our technique therefore is sensitive to a wide range of nucleon decay modes and thus avoids the difficulties attending a decay-mode-specific experiment in which the decay mode under study happened to be strongly suppressed or perhaps even forbidden.

The experiment is located in a 10 m x 10 m x 20 m chamber 1480 m below the surface of the earth (4400 meters water equivalent) on the 4850 foot level of the Homestake gold mine, Lead, South Dakota. The cosmic-ray muon flux, which is the primary source of background for any nucleon decay experiment at this or lesser depths, is approximately 1000 $m^{-2}y^{-1}$, about 10^7 times lower than that at the surface of the earth and about 10^3 times lower than that in shallow (1500-2000 m.w.e.) nucleon decay experiments. The results we report here[11] were obtained between January 1979 and February 1981 with a primary detector[12] consisting of 36 optically-separated rectangular water Cerenkov modules, each measuring approximately 2 x 2 x 1.2 m, together with an upper anticoincidence shield of 33 mineral oil-based liquid scintillation counters, each measuring 1.6 x 1.6 x .05 m (see fig. 1). For part of this period, a separate lower anticoincidence shield consisting of a single 150-ton water Cerenkov detector was in operation. The water Cerenkov modules of the primary detector were arranged in two groups of 6 x 3 modules covering opposite walls of the chamber, as shown in fig. 1. The 36 modules together contained a fiducial volume of 150 tons of water or 9×10^{31} nucleons.

Each Cerenkov module (fig. 2) consisted of highly reflective white plastic walls containing about 4 m^3 of filtered water to which 10 mg/l of Amino-G wavelength shifter had been added. Four 12.5 cm diameter hemispherical photomultiplier tubes collected the nearly isotropic Cerenkov light from each module. The size of the modules, together with the use of wavelength shifter, highly reflective walls and efficient large-area photomultiplier tubes resulted in a detection sensitivity of about 4 MeV per photoelectron. With the electronic threshold set at the single photoelectron level, the typical system trigger requirement of three phototubes firing within a 100 ns coincidence gate resulted in a detection threshold of about 10 MeV for electrons. This value was chosen as the optimum compromise between rejection of the large background of low energy electrons produced by natural radioactivity and the efficient detection of electrons from muon decay (mean energy 38 MeV).

The principal experimentally measured quantities available for the prompt and delayed pulses of each multiple-module event are the number and configuration of modules triggered, the visible energy deposited in each module, the presence or absence of anticoincidence signals, the time of occurrence of the event, the relative delays between prompt and subsequent pulses, and the direction of motion of the particle yielding the muon decay. The energy scale was calibrated by assigning an energy of 365 MeV to the pulse height produced by throughgoing vertical muons. The observed energy resolution for these throughgoing muons was typically 30%.

To aid in distinguishing true nucleon decay events from background (most of which is produced by cosmic-ray muons), we have performed detailed Monte Carlo simulations of the various nucleon decay modes suggested by the SU(5) model. These

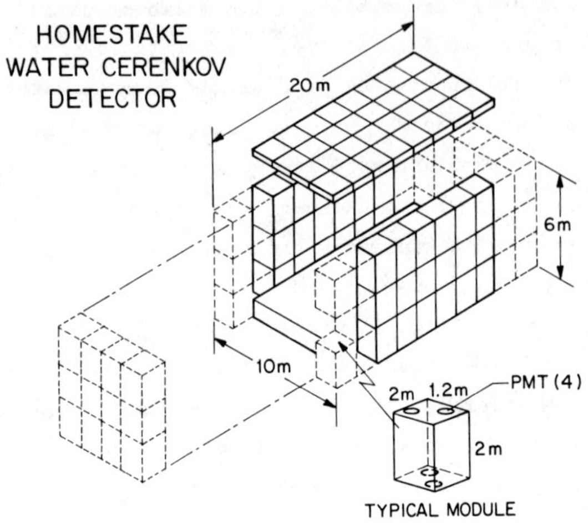

1. Overall view of the Homestake Water Cerenkov Nucleon Decay Detector showing a typical 4 m^3 module and the upper and lower anticoincidence shields. The modules shown with broken lines were not in operation for the present data.

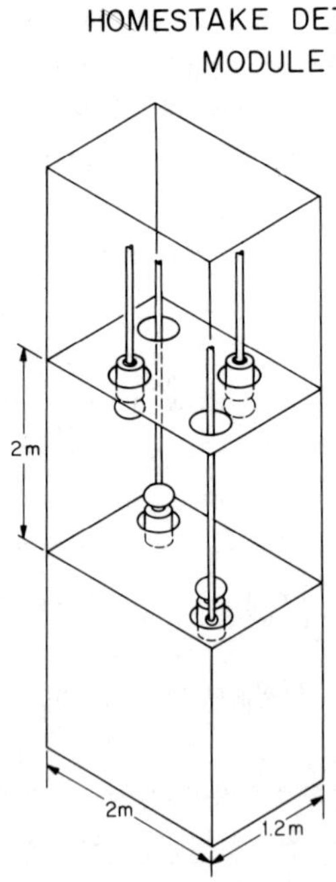

2. Isometric view of a typical water Cerenkov detector module showing the optics and photomultiplier tubes.

calculations also determine the detection efficiency of our apparatus for each decay mode. The Monte Carlo program includes detailed simulation of the propagation of electrons and photons[13] as well as muons and hadrons. For the latter, we have included the effects of intranuclear propagation as calculated by Sparrow[14]. Such intranuclear effects are important for all practical nucleon decay detectors, since free nucleons constitute a very small fraction of all materials other than pure hydrogen. Sparrow's calculations[14] indicate that, since nucleon decay pions have an energy close to that of the (3,3) pion-nucleon resonance, about 50% of all pions released inside an oxygen nucleus will either be absorbed or will charge exchange. The production rate of stopping muons from the pionic decay modes of those nucleons bound in heavy nuclei will therefore be reduced by a factor of two.

Based on our Monte Carlo simulations, we accept as valid nucleon decay candidates events with prompt visible Cerenkov energy between 50 and 600 MeV. The reduction of visible energy from that of the nucleon rest mass is due to not observing the pion rest mass and the energies of sub-relativistic particles as well as to incomplete containment of electromagnetic showers.

Since the Monte Carlo calculations indicate that less than 10% of nucleon decays would trigger 3 or more modules, we consider only one or two module events as potential nucleon decay candidates. Furthermore, to permit determination of the direction of motion of the nucleon decay product yielding the muon decay, we include only two-module events in the final data. By eliminating downward-moving particles we discriminate strongly against the cosmic-ray muon background at the expense of a calculated 22% loss of efficiency for detection of nucleon decays. The time window for accceptance of muon decay pulses extends from 800 ns (the system dead time) to 15 μs.

Full operation of the Homestake water Cerenkov nucleon decay detector began early in 1979. We present here the results of two runs totalling 407 days of operation. For the first run (255 days), the anticoincidence shield consisted of only the top layer of liquid scintillation counters, while for the second (152 days), both the top and bottom anticoincidence shields were in operation.

The results of the two runs are summarized in Table I.

A total of 235 multiple-module muon decays were observed in 188 separate events. For runs 1 and 2 there were respectively 27 and 15 upward or sideward muon decays without anticoincidence signal. The module number distributions for these events are shown in Table II, together with the corresponding distributions for events accompanied by a veto pulse. It is apparent that the anticoincidence efficiency for run 2, as would be expected, was substantially better than that for run 1. Applying the two-module requirement for nucleon decay candidates, there remain 3 events from run 1 and 2 events from run 2. In Table III we list the visible energies for these 5 events, together with other relevant information. Events 1 and 4 can be excluded because their visible energy is too large. Events 2, 3 and 5 remain as viable nucleon decay candidates.

To estimate the background, we examine the energy distribution of the corresponding vetoed events (see Table III). There is one such event (visible energy = 440 MeV) in the nucleon decay window. Since the overall anticoincidence efficiency is about 50%, the unvetoed and vetoed backgrounds should be approximately equal. We therefore estimate the background as one event, thereby yielding a net signal of $3 - 1 = 2 \pm 2$ events.

To calculate the effective exposure time for the two runs, we make use of the 235 observed muon decays, all of which may be assigned to cosmic-ray-muon produced background. At the depth of this experiment, cosmic-ray muons have a mean energy of 365 GeV[16], corresponding to a range of 1.8×10^5 cm of water. In one centimeter of water we therefore expect 5.5×10^{-6} muon stops per incident muon. Since the muon flux at this depth is 4×10^{-9} cm^{-2}sec^{-1}, we expect

$$\frac{4 \times 10^{-9} \times 5.5 \times 10^{-6}}{6 \times 10^{23}} = 3.7 \times 10^{-38} \frac{\text{muon decays}}{\text{nucleon sec}} ,$$

or one muon decay per 9×10^{29} nucleon y. In addition to these direct muon decays, there are also muon decays resulting from pions produced near the detector by muon photonuclear interactions. Muon decays resulting from this process increase the above rate to one muon decay per 2×10^{29} nucleon y.[15] Since 235 muon decays were observed, the effective exposure time was $235 \times 2 \times 10^{29} = 4.7 \times 10^{31}$ nucleon y.

An alternative calculation of the effective exposure time can be made by multiplying the number of nucleons in the detector (9×10^{31}) by the running time (1.1 y) and by the estimated muon decay detection efficiency (70% in accepted lifetime range x 75% above electronic threshold x 95% contained in active detector volume = 50%). This calculation gives 5×10^{31} nucleon y, in excellent agreement with the previous method.

The nucleon total lifetime is therefore

$$T_N > \frac{4.7 \times 10^{31} \text{nucl y} \times B_\mu}{2 \text{ nucleon decay candidates}} = 2.4 \times 10^{31} B_\mu \text{ y},$$

where B_μ is the model-dependent probability of production and detection of a stopping and decaying muon following a nucleon decay event.

We make use of the Monte Carlo calculations to determine B_μ. In Table IV we calculate the two-module total detection efficiencies for a number of two-body decay modes consistent with the SU(5) model. The efficiencies vary from 2% for the $n \to \nu K^0$ mode to 23% for the $p \to e^+\omega$ mode, with a typical value around 10%. To determine B_μ (see Table IV), we require predictions of the branching ratios for nucleon decay into the various two-body final states. For illustrative purposes, we show in Table IV the calculation of B_μ using branching ratios from the nonrelativistic model of Kane and Karl[17]. Our results for B_μ range from about 5% (using the branching ratios of ref. 17) to about 10% (using refs. 18 or 19).

Our result may therefore be stated as a nucleon total lifetime of

$$T_N > 1 - 2 \times 10^{30} \text{ y.}$$

Since the observed signal is not sufficiently well separated from the background, we interpret our result as representing a lower limit on the nucleon lifetime.

We plan to continue this experiment with a number of modifications which should allow significant improvement of the present results. As more running time is accumulated with the bottom counter operative, the anticoincidence inefficiency will decrease from the 58% value observed during run 1 to the 31% obtained during run 2. Instrumentation of the remaining side areas of the room will increase the detector mass by about 50% and will also help to improve further the anticoincidence performance. New electronic circuitry to be installed soon will improve the energy resolution and extend the dynamic range of our observations and will also reduce the electronic dead time following each nucleon decay event from 800 to about 300 ns. These modifications, together with one year additional running time, should allow improvement of the present nucleon lifetime limit by a factor of about 3.

II. THE HOMESTAKE TRACKING SPECTROMETER

There are now a number of nucleon decay experiments in operation, under construction, or in the design stages. These experiments fall into two categories: water Cerenkov detectors and calorimeters. The water Cerenkov detectors are well suited to reconstruct the nucleon mass for the $e^+\pi^0$ mode, and are capable of recognizing some other modes as well. The calorimeters give track information for most decay modes, but provide little information about the total energy released in a nucleon decay event.

The ideal nucleon decay detector should

1. be sensitive to all decay modes and be able to determine as many parameters of the decay process as possible;
2. have good position resolution and tracking capability in order to identify secondary tracks and recognize

multiple secondaries;

3. have excellent energy resolution and a low threshold in order to differentiate among decay modes and record the full energy emitted in a decay (including the ionization losses of subrelativistic particles, charged nuclear excitation fragments, and neutrons emitted by nuclear excitation);
4. be capable of recognizing decays of secondary particles (e.g. μ --> e);
5. be able to suppress the backgrounds produced by cosmic ray muons and neutrinos;
6. be sufficiently massive to contain the total energy released in the decay.

We are planning to construct a conceptually new instrument for nucleon decay which satisfies to a remarkable degree all the requirements of the ideal nucleon decay detector. Our plan is to locate this instrument -- a tracking spectrometer -- in the Homestake Gold Mine at Lead, South Dakota at a depth of 1480 m of rock, or 4400 meters water equivalent (m.w.e.). This location, unique in the United States, is ideally suited for such a detector because of the extremely low flux of cosmic ray muons sufficiently energetic to penetrate to such depths.

Our tracking spectrometer will

1. detect and recognize all nucleon decay modes for nucleon lifetimes $T < 3 \times 10^{32}$ years;
2. measure the total ionization energy released by the nucleon decay with resolution sufficient to discriminate against background induced by cosmic-ray muons and neutrinos, and natural radioactivity, and to identify and differentiate among the neutrino escape peaks associated with the various nucleon decay modes;
3. provide track information for nucleon decay secondaries;
4. measure the charge of lepton secondaries;
5. measure the polarization of nucleon-decay-produced muons.

Our approach is to construct a highly modular 1000-ton liquid scintillation detector viewed by an array of photomultiplier tubes. The detector will be housed in a new and dedicated room deep underground, totally enclosed by anticoincidence counters. The liquid scintillator will measure a signal corresponding to the total hadronic and charged leptonic ionization energy produced by the nucleon decay. From the patterns of energy deposited, detector modules fired, and delay times of secondary particle decays, we will be able to reconstruct many of the details of the nucleon decay process.

The tracking spectrometer will be a unique device in that it will combine a deep underground location, efficient background rejection, large mass, good energy resolution, and tracking capability. It will clearly reject background events due to charged particles entering from outside the detector volume and to neutrinos interacting inside, and will clearly identify different decay modes.

The detector will consist of long cylindrical counter elements filled with a liquid scintillator and viewed by two photomultipliers, one at each end. The counter elements are arranged in stacks of crossed X-Y arrays to provide tracking information for the nucleon decay secondaries. The detector will be housed in a new room at the Homestake mine, close to the existing Penn-Brookhaven water Cerenkov nucleon decay detector. The walls, ceiling, and floor of the new room will be completely covered with liquid scintillation anticoincidence counters.

Those nucleon decay channels with hadronic and charged leptonic secondaries only will deposit the entire 938 MeV nucleon rest mass in the spectrometer. For those decay modes in which energy is carried off by neutrinos, we will observe smaller, distinctive energy signals:

Decay Mode	Visible Energy
$\pi\nu$	479 MeV
$K\nu$	602 MeV
$\rho\nu$	785 MeV
$XX\nu$	continuum

Thus the various neutrino-associated decay channels can be distinguished by the visible energy deposited. For example, the $\pi\nu$ mode will consist of two subsets, $p \rightarrow \pi^+\nu$ and $n \rightarrow \pi^0\nu$.

The neutron decays will produce prompt pulses of 479 MeV, while the proton decays will be characterized by a prompt peak at 479 - 140 = 339 MeV followed by a muon decay 2.2 microseconds later. Similar considerations apply to the other decay channels. Therefore by combining time resolution sufficient to observe muon decays with energy resolution (dE/E = 10%) sufficient to resolve the neutrino escape peaks, the liquid scintillation detector elements make it possible to measure the total energy released in nucleon decay, including the nuclear excitation of the residual nucleus. This last point is particularly important for that subset of decays involving π and ρ secondaries, since about 50% of the pions are absorbed by the residual nucleus. The detector will also record the energy carried off by neutrons, including those emitted by the residual nucleus as well as those resulting from pion scattering from other nuclei. Our Monte Carlo simulations indicate that nucleon decays will typically trigger 10 to 18 counter elements, thus giving sufficient information to reconstruct the entire nucleon decay event.

The absolute calibration of our detector elements will be carried out with light pulsers, radioactive sources, through-going cosmic-ray muons, and electrons from muon decay. The cosmic-ray muons provide an excellent continuous system calibration, giving an energy deposit of 50 MeV per counter element.

The electronic readout of the photomultiplier information will be performed with circuitry identical to that which is now being installed for our water Cerenkov detector. This circuitry determines the time of photomultiplier firing to a precision of 2 ns and measures the pulse height of the photomultiplier with a 64-channel analog-to-digital flash converter. The readout information is transferred to a buffer memory capable of storing 16 events. The readout cycle is completed in 100 ns, and the circuitry can accept another pulse immediately thereafter. The stored memory information is then transferred to a dedicated minicomputer and combined with the information from other detector elements.

In order to maintain a high cycle rate, we plan to have each stack of detectors handled by its own microprocessor and then to use the minicomputer to read out the local microprocessor information. In this way only the stack or stacks that have stored data will be involved in the readout process.

At the depth of the Homestake laboratory, the angle-integrated flux of cosmic-ray muons is 4×10^{-5} $m^{-2}sec^{-1}$, or 10^{3} $m^{-2}yr^{-1}$. Over the full spectrometer area, this flux gives rise to 4×10^{5} muons per year. We expect that at least 99% of these muons will be vetoed by the top anticoincidence counter layer. Most of the muons entering the detector room will leave by either the side walls or the floor, where they will again encounter a veto counter layer. A veto efficiency of 99% on muon exit will result in only 40 unvetoed throughgoing muons per year. The tracking capability of this detector will easily permit us to recognize and reject these remaining throughgoing muons. We expect that about 1000 muons per year will either interact or stop and decay inside the detector. For these muons, only the top veto will apply, resulting in 10 unvetoed events per year. The timing of the events in each layer will permit us to identify these as downward-moving events. In addition, from our experience with our present nucleon decay detector, we know that the muon interactions will result in large multiparticle splashes that are very distinctive and easily recognized and rejected. The decaying muons will have downward trajectories and path lengths in the detector that will be uniformly distributed over the height of the detector, a pattern that will not easily be confused with nucleon decays.

The other background problem in nucleon decay experiments is due to cosmic-ray neutrino interactions either inside the detector or in the rock surrounding the detector. The integrated flux of neutrino-induced muons entering our detector from the surrounding rock is 4 to 7×10^{-8} $m^{-2}sec^{-1}$, about 10^{-3} times the flux of direct cosmic-ray muons, or about 400 per year. Only 4 of these will escape being vetoed by our wall counter shield. These remaining 4 particles will have single particle trajectories that originate at the detector surface and will have a broad range of energy deposit. These particles will easily be distinguishable from nucleon decay.

Because of the energy resolution of our detector, only those neutrino interactions that deposit 938 ± 50 MeV in the detector are potential background problems for nucleon decay. Except for neutral current interactions, the detector will see the total neutrino energy in the signal generated by the interaction secondaries. The flux of muon neutrinos at 938 MeV is 3.7×10^{-2} cm^{-2}sec^{-1}sr^{-1}GeV^{-1}, or 3.7×10^{-3} in a 100-MeV window. The electron neutrino flux in the same window is 1.7×10^{-3} cm^{-2}sec^{-1}sr^{-1}. Using the sum of these fluxes integrated over solid angle and the known neutrino interaction cross section of 0.5×10^{-38} cm^2, we find 6 neutrino interactions per year under the nucleon mass peak in a 1000-ton detector. There will be a similar number of neutrino interactions in the 100-MeV windows below and above the nucleon mass peak. We can use these two sets of background events to estimate the number that lie under the nucleon mass peak and thus carry out a subtraction. This process should result in a net background of about 2 events per year. In carrying out this neutrino background rejection we have not relied on distinguishing the track and energy deposit patterns produced by neutrino interactions from those that result from nucleon decay. While an additional background suppression factor might undoubtedly be obtained from pattern recognition, we feel that over-reliance on this approach is risky in exploring a phenomenon as new and rare as nucleon decay. By fully utilizing the energy resolution of the detector, however, we very effectively reject background and simultaneously make it possible to investigate the details of the decay modes by measuring the individual neutrino escape peaks.

An attractive feature of the anticoincidence system is that we will have available a very large number of vetoed background events due to either muons or neutrinos. This collection of labeled background events will make it easy for us to study the behavior of muon and neutrino interactions in our detector and thus convince ourselves that any unvetoed interactions have been completely rejected.

The tracking spectrometer will therefore be able to provide convincing evidence for essentially all conceivable nucleon decay modes for lifetimes up to 3×10^{32} years (see fig. 3). With the introduction of a momentum-balance requirement for the decay secondaries, we believe that, for favorable decay modes (such as $e\pi$), we can in fact attain the 10^{33} year lifetime range before being limited by background.

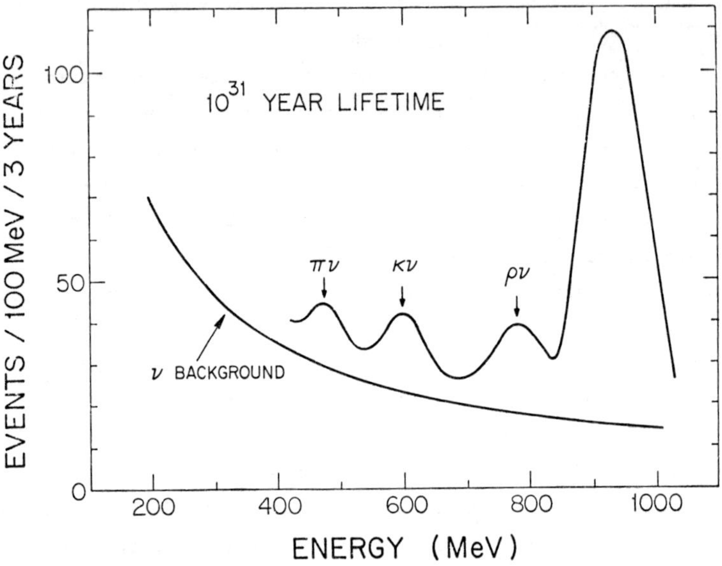

3. Expected distribution of nucleon decay events and neutrino background for the tracking spectrometer. We have assumed branching ratios given by typical SU(5) models, a nucleon lifetime of 10^{31} years, and an energy resolution of 10%. The neutrino background is known rather well near 1 GeV, but the shape of the neutrino distribution is uncertain at lower energies. We have estimated the neutrino background by scaling the known neutrino interaction rate at 1 GeV by E^{-1} at energies below 1 GeV. As explained in the text, further suppression of the neutrino background can be achieved by use of a momentum balance requirement for the decay secondaries.

A prototype of the tracking spectrometer is now under construction and will be installed in the Homestake mine in August of 1981. We plan to have a 250-ton section of the detector in operation by the summer of 1982 and the full 1000 tons in operation one year later.

It is a pleasure to thank the Homestake Mining Company and Mr. Al Gilles for their generous cooperation on these experiments. We also wish to acknowledge the essential contributions of Dr. R. Davis, Jr., Dr. W. Frati, and Mr. R. Punkinnen.

		Events without veto signal	Events with veto signal	Multiple muon decay events	Total events	Total muon decays
RUN 1	upward	14	11	7	32	42
	sideward	13	7	3	23	30
	downward	32	10	5	47	54
	total	59	28	15	102	126

RUN 2	upward	9	9	4	22	26
	sideward	6	13	7	26	38
	downward	12	21	5	38	45
	total	27	43	16	86	109

TABLE I. Characteristics of muon decay events observed during the two runs.

		NUMBER OF MODULES TRIGGERED						
		2	3	4	5	6	>6	Total
RUN 1	without veto	3	5	9	8	1	1	27
	with veto	3	5	4	2	1	3	18
RUN 2	without veto	2	8	2	3	0	0	15
	with veto	5	3	2	5	3	4	22

TABLE II. Module number distributions for upward and sideward muon decays.

2 MODULE EVENTS WITHOUT ANTICOINCIDENCE SIGNAL

	Event no.	Date (days)	Direction	Energy (MeV)	Delay (μs)
RUN 1	1	1559.455	up	980	2.3
	2	2171.070	side	40	5.2
	3	2195.484	up	580	5.5
RUN 2	4	1914.977	up	1010	2.5
	5	2244.850	up	400	0.8

2 MODULE EVENTS WITH ANTICOINCIDENCE SIGNAL

RUN 1	1	1708.751	side	690	4.7
	2	1758.907	up	890	3.7
	3	2076.364	side	2370	0.7
RUN 2	4	1834.866	up	1060	2.5
	5	1936.172	up	910	5.1
	6	2056.809	up	840	4.7
	7	2060.138	up	1210	3.0
	8	2248.827	side	440	1.2

TABLE III. List of two module upward and sideward muon decay events.

Decay mode	2-module partial detection efficiency	Branching ratio to muons	Internal survival fraction (ref.14)	2-module total detection efficiency	SU(5) branching ratio (ref.17)	Decay mode detection efficiency
$p \to e^+\rho^0$	.133	1	.5	.067	.024	.0016
$\to e^+\eta$	.333	.29	1	.097	.069	.0067
$\to e^+\omega$	.250	.91	1	.228	.214	.0488
$\to \nu\pi^+$	.053	1	.5	.026	.143	.0038
$\to \mu^+K^0$	.167	1	1	.167	.183	.0306
$\to \nu\rho^+$	.142	1	.5	.071	.010	.0007
$n \to \nu\rho^0$	.098	1	.5	.049	.006	.0003
$\to \nu\eta$	.181	.29	1	.052	.015	.0008
$\to \nu\omega$	.182	.91	1	.166	.047	.0078
$\to \nu K^0$	.029	.70	1	.020	.011	.0002
Overall detection probability for water detector (B_μ)						.055

TABLE IV. Calculation of B_μ, the overall nucleon decay detection probability. The total detection efficiencies are obtained by multiplying the partial detection efficiencies by the corresponding branching ratio to muons and internal survival fractions. The decay mode detection efficiencies are calculated by multiplying the total detection efficiencies by the SU(5) branching ratios.

FOOTNOTES AND REFERENCES

*Work supported in part by the U.S. Dept. of Energy under contract DE-AC02-81-ER-40012 and in part by the U.S. National Science Foundation under grant AST-79-08670.

1. J.C. Pati and A. Salam, Phys. Rev. Lett. 31, 661 (1973).

2. H. Georgi and S.L. Glashow, ibid. 32, 438 (1974).

3. For an excellent and thorough review of grand unification and nucleon decay, see P. Langacker, SLAC-PUB-2544, to be published in Physics Reports.

4. M. Goldhaber, P. Langacker, and R. Slansky, SCIENCE 210, 851 (1980).

5. F. Reines and J. Schultz, Surv. High Energy Phys. (USA) 1, 89 (1980).

6. R.I. Steinberg and J.C. Evans, Jr., Proc. Int. Conf. on Neutrino Physics and Neutrino Astrophysics, Baksan Valley (Acad. of Sciences of the USSR, Moscow, 1978), p. 321.

7. E. Fireman, ibid., p. 53.

8. The thickness of this detector (12.5 cm) was much less than the expected range of nucleon decay products (typically about 1 m). Because of this limitation, most of the nucleon decays which this instrument would have detected would have taken place in the rock surrounding the detector. Much of the nucleon decay information would therefore have been lost, in contrast to the situation with the present 120 cm-thick Homestake detector.

9. F. Reines and M.F. Crouch, Phys. Rev. Lett. 32, 493 (1974).

10. J. Learned, F. Reines, and A. Soni, ibid. 43, 907 (1979).

11. Preliminary results of this experiment have been reported previously by M. Deakyne et al., Proc. XVth Rencontre de Moriond, March 1980; Proc. First Workshop on Grand Unification, Durham, N.H., April 1980.

12. M. Deakyne, W. Frati, K. Lande, C.K. Lee, R.I. Steinberg, and E. Fenyves, Proc. Neutrino '78, West Lafayette, April 1978; Proc. Conf. on Nucleon Stability, Madison, Dec. 1978.

13. We thank Prof. T. Stanev for providing us with his electromagnetic cascade code. (T. Stanev and Ch. Vankov, Computer Phys. Comm. 16, 363 (1979); Bulg. J. Phys. 5, 433 (1978)).

14. D.A. Sparrow, Phys. Rev. Lett. 44, 625 (1980).

15. G.L. Cassiday, J.W. Keuffel, and J.A. Thompson, Phys. Rev. D7, 2022 (1973).

16. S. Miyake, Proc. 13th Int. Conf. on Cosmic Rays, Denver, 1973, p. 3638.

17. G. Kane and G. Karl, Phys. Rev. D22, 2808 (1980).

18. J.F. Donoghue, Phys. Lett. 92B, 99 (1980).

19. E. Golowich, Phys. Rev. D22, 1148 (1980).

EXPERIMENTS AT THE SOUDAN MINE: OPERATING AND PROPOSED

Michael A. Shupe
School of Physics
University of Minnesota
Minneapolis, Minnesota 55455

Abstract

We report on the initial operation of a 30 ton dense calorimeter and on the proposed construction of a 1000 ton calorimeter at the Soudan mine in northern Minnesota--both optimized for the detection of proton decay.

1. Progress at Soudan 1.

Data acquisition has begun in the 30 ton detector installed by the University of Minnesota (J. Bartelt, H. Courant, K. Heller, M. L. Marshak, E. A. Peterson, K. Ruddick, M. Shupe) and Argonne National Laboratory (D. S. Ayres, J. Dawson, T. H. Fields, E. N. May, L. E. Price) on the 23rd level of the Soudan mine. The first few weeks of full operation have yielded the expected large number of through-going cosmic ray muons, as well as two other classes of cosmic ray events: 1. Events where the incoming muon creates a knock-on electron which showers in the detector. 2. Events with multiple tracks, all through-going and parallel. These are assumed to be muons resulting from energetic air showers. The most spectacular event of this type contains 12 tracks.

Before discussing these events further, we summarize the relevant physical properties of the detector, some of which have been discussed in greater detail elsewhere.[1,2] The detector consists of 3456 proportional tubes of 1 inch diameter imbedded in a rectangular solid (3m. X 3m. X 2m.) composed of taconite concrete. The tubes are approximately 3 meters long, arranged in 48 horizontal layers of 72 tubes each, with alternate layers crossed. They are operated on a gas mixture of 91% argon and 9% carbon dioxide. Each is equipped with a 32 level logarithmic pulse height analyzer, which is sensitive for a total of 7 microseconds, and can detect and time muon decays within this range. A high voltage plateau curve from one of the tubes is shown in Figure 1. Rates from natural radioactivity are typically 3 Hz in a single tube, with tubes near the edge of the detector counting a factor of two higher because of radiation from the mine walls. The tubes are self triggering, with each layer of 72 "ored" together. A coincidence of at least three layers is required to generate an event trigger.

Taconite concrete was first fabricated by man in October, 1979, consisting of 22% Portland cement, 67% taconite concentrate, and 11% water. Taconite powder is 95% iron oxide which results as a byproduct of the refining of low grade ore strip mined in northern Minnesota. Taconite concrete is roughly 57% iron. It has a specific density of 2.49, a radiation length of 6.22 cm., and a nuclear absorption length of 25 cm. With 2.54 cm. tubes placed on 4.00 cm. centers, the net detector density is 1.85, the radiation length 8.42 cm., and the absorption length 34 cm.

With these physical parameters, the detector behaves as a dense calorimeter; where particle identification is performed on the basis of range, ionization, and track topology. A Monte Carlo calculation has been used to simulate the behavior of proton decay products in the detector. A 500 MeV muon will range out in 95 ± 10 cm., hitting 16 ± 4 tubes, with negligible deflection of the track. Its energy is determined by range to 10% (sigma). Negative muons undergo nuclear capture at the end of their range, and the decay electron is not observed. The decays of positive muons are observed with 35% efficiency (geometry plus absorption of the electron). A charged pion of 300 MeV will travel approximately half a meter, hitting about 8 tubes, and is likely to undergo one to several scatters before absorption-- which serve to aid in pion identification. Photons and electrons at 300 MeV shower electromagnetically, with most of the

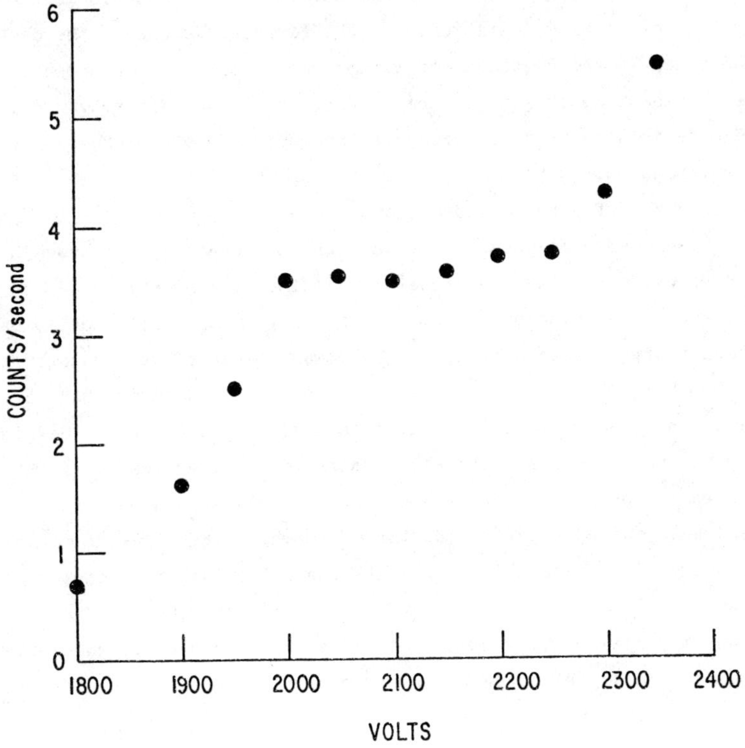

Figure 1. High voltage plateau curve of one of the 3456 proportional tubes in the 30 ton detector.

shower energy contained in a length 20 cm. About 7 tubes are hit in such a shower, leading to an energy resolution of $20\%/\sqrt{E}$ (sigma). Tube counting gives a better energy determination than ionization at this level of shower statistics. The fiducial volume of the detector is between 16 and 24 tons, and is somewhat decay mode dependent. The best present estimate of the 30 ton detector sensitivity, including all effects such as absorption of hadronic decay products by the parent nucleus, is to proton lifetimes up to 1.0 to 2.0 X 10^{30} years in 2.5 years of detector operation.

Most cosmic ray muons pass through the detector, as shown in the two views of a typical one-event display in Figure 2. But as indicated in discussion above, cosmic ray muons will occasionally interact with dramatic results. The mean energy of muons surviving to the depth 1800 hg/cm^2 is 250 GeV. A muon at this energy has a 4% chance of knocking on an electron of greater than 1 GeV in the detector, and the resulting electromagnetic shower can hit several hundred tubes. A typical event of this class is shown in Figure 3. The cosmic ray muon enters the top of the detector, induces a shower beginning in layer 20, and exits the bottom of the detector. Some showers are quite large, and may originate in the surrounding rock, as shown in Figure 4. Events of class 2 correspond to energetic air showers. Multiple- muon events are observed several times per hour, with the largest seen in preliminary scanning having 12 clearly discernible tracks. This event is shown in Figure 5.

Events of class 1 also constitute a potentially serious background for proton decay. Such an event, if not clearly contained, would mimic the decay $n \to \pi^- e^+$. Several measures will be taken to diminish this background. An active shield of scintillation counters to detect incoming tracks will be installed in July and August, 1981. The efficiency of all tubes will be known by late June. Finally, the fiducial volume cut will be made more conservative in the direction of highest background.

At present, the 30 ton detector is operating with a duty factor of 1/3, during the hours from midnight to 8:00 a.m., and being adjusted and calibrated during the daytime and evening hours. This factor should rise to 2/3 during June, where it will probably remain for much of the summer. We anticipate publishing cosmic ray results

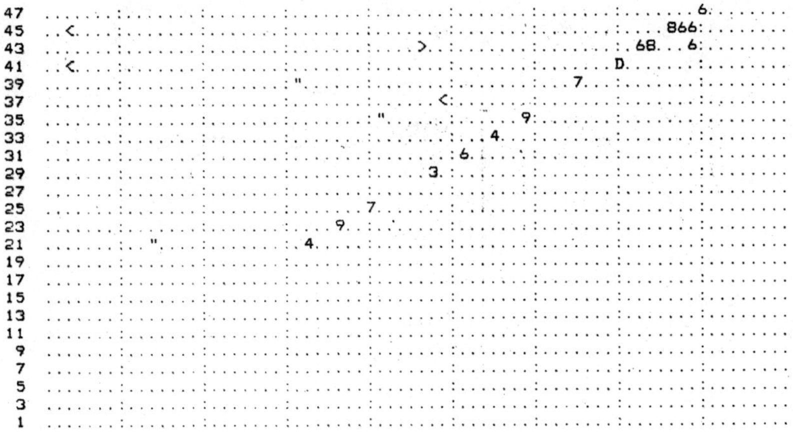

VIEW FROM NORTH

VIEW FROM WEST

Figure 2. Two views of a cosmic ray muon passing through the 30 ton detector. This track passes in the top and out through the south face, with a possible delta ray in layer 43.

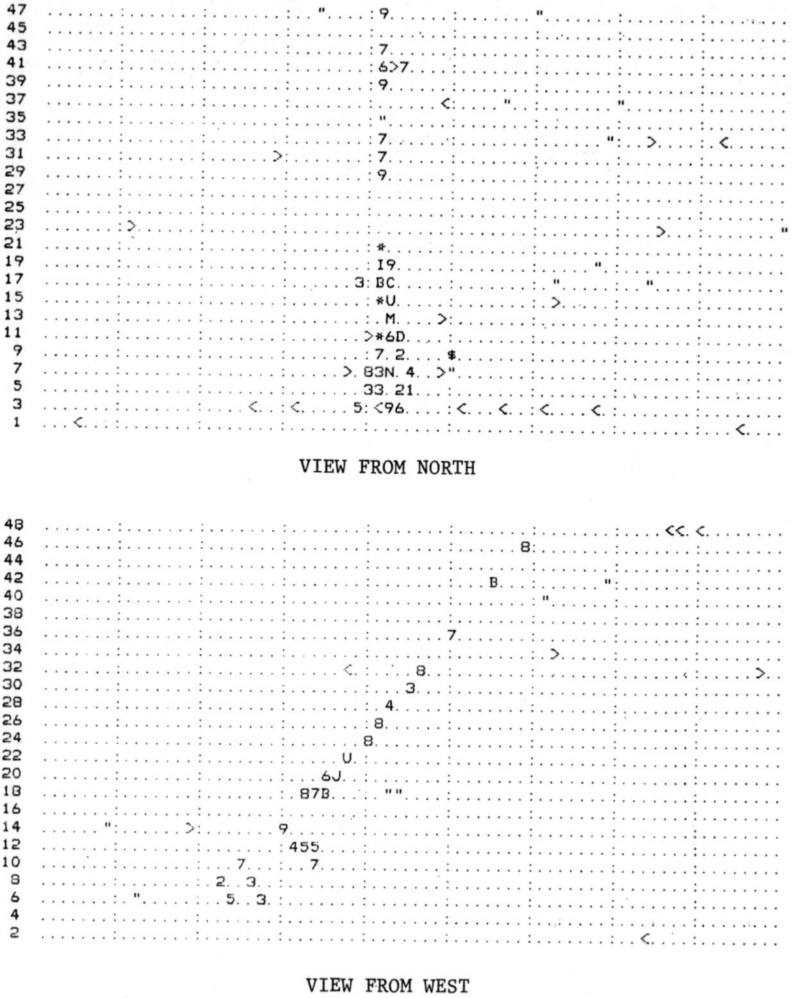

Figure 3. Two views of a single muon passing through the 30 ton detector. A knock-on electron is created in layer 20, resulting in an electromagnetic shower in the bottom of the array.

VIEW FROM NORTH

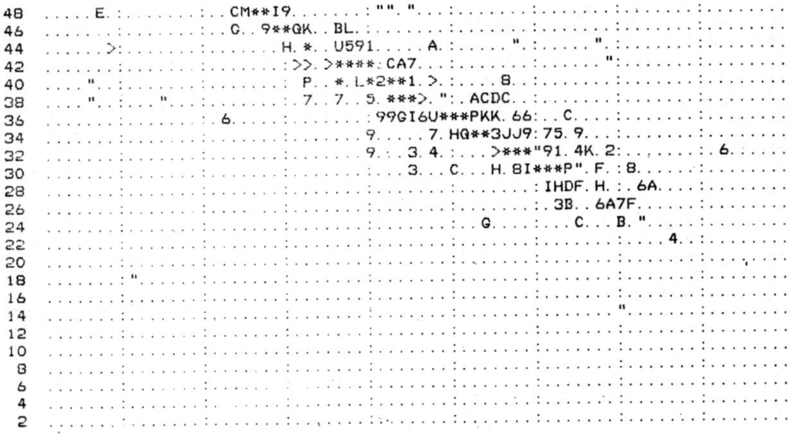

VIEW FROM WEST

Figure 4. A large electromagnetic shower in the 30 ton detector. This event is likely to have been caused by an incoming muon creating a knock-on electron in the material just above the detector.

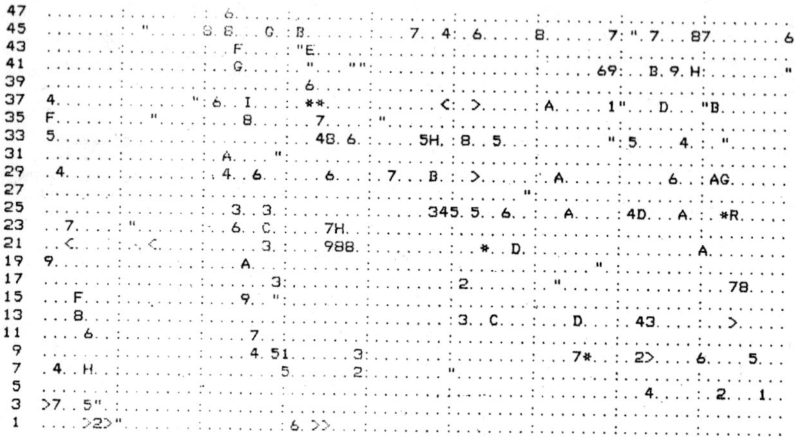

VIEW FROM NORTH

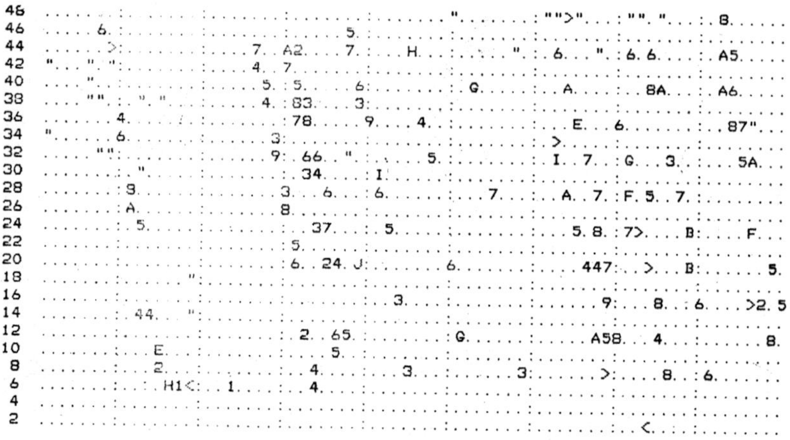

VIEW FROM WEST

Figure 5. Two views of the highest-multiplicity multi-muon event yet seen in the 30 ton detector. 12 (possibly 13) tracks are discernible in this event. We have also collected a large sample of multi-muon events of lower multiplicities.

in the fall, and are keeping an eye peeled for the first clear appearance of proton decay.

2. Development of Soudan 2.

A collaboration of the University of Minnesota, Argonne National Laboratory, and Oxford University (W. W. M. Allison, C. B. Brooks, D. H. Perkins, G. Saitta) has proposed that a one kiloton tracking calorimeter be installed on the 25th level of the Soudan mine at a depth of 1985 hg/cm^2. In the past year, extensive design development, Monte Carlo studies, and prototyping have been done. Several designs are under consideration. The one most fully developed to date is described in this section.[3]

Our original intention was to extend the proportional tube and taconite concrete technology employed in Soudan 1 to the one kiloton detector. However, studies of expected backgrounds showed that as the mass of the detector was increased, it would have to be more finely segmented. An acceptable match between mass and granularity implied that more than 300,000 proportional tubes would be necessary.

Experience with the 30 ton detector has shown us that problems which are small at the single tube level create significant delays when modifications are required for each detector element. With the aim of minimizing the number of sampling elements in the one kiloton detector, the group has demonstrated the feasibility of using large drift cells, employing drift distances of 50 cm. or more to the anode wire. In the presently proposed design, only 10,000 such elements would be needed for the one kiloton detector. Cell sharing would reduce the required number of channels of electronics to 5000.

The proposed detector is constructed of large planar drift chambers alternated with sheets of steel. The steel plate thickness is 0.5 cm. The drift chambers are constructed as modules with 1.0 cm. gas gaps, 50.0 cm. of width in the drift direction, and 5.0 m. in length. For convenience, each cell is suspended with the long axis vertical. The anode wire is run vertically near one edge of the cell, and the drift time to this wire is measured by the electronics. The vertical coordinates of anode wire avalanches are measured on 250 cathode pads, spaced 2 cm. apart.

A primary design requirement for this detector is to maintain ease of access and repair. The plates and drift chambers are grouped into twenty 50-ton modules. Each module contains 250 steel sheets and 500 drift chambers. To allow for some non-flatness of the steel, 2.0 cm. gaps are left between plates. The resulting dimensions of a module are then 1.0 m. in width, 5.0 m. in the direction in which plates and chambers are sandwiched (horizontal length), and 5.0 m. in height (the individual chamber length). The overall detector dimensions are thus 10.0 X 10.0 X 5.0 m. For access, the modules are suspended from a trolley system which runs on I-beams, with an extra spacing of one meter left at the ends. Figure 6. shows a design study of the mechanical support system. Also included is a table summarizing the physical properties of the detector.

In order to verify that drift cells with this large aspect ratio (1:50) operate without appreciable loss of pulse height or position resolution, a prototype cell was constructed at Argonne and tested with a beta source at various high voltages and with some attempts at field shaping. The results of measurements taken at 10 K.V. and 6 K.V. are shown in Figure 7. The attenuation over 50 cm. of drift is 30%, and the position resolution varies from about 1.9 mm. near the wire to 2.3 mm. at the longest drift. Such performance is more than adequate for the final detector.

Backgrounds in this detector have been studied extensively by Monte Carlo techniques. Mixed samples of signal and background events were generated at Argonne and hand-scanned at Minnesota. The design match between granularity and mass for the one kiloton detector was achieved by this process. Typical sample events are shown in Figure 8. Proton decay events in such a detector have the appearance of bubble chamber events. We feel that this ability to scan the raw data by hand constitutes an advantage of tracking calorimeters over water detectors, where considerable processing of Cerenkov light intensity and arrival time information must be done before events become "understandable."

There will be several cross checks on the accuracy of our Monte Carlo. The detector at Soudan 1 is taking cosmic ray data which we

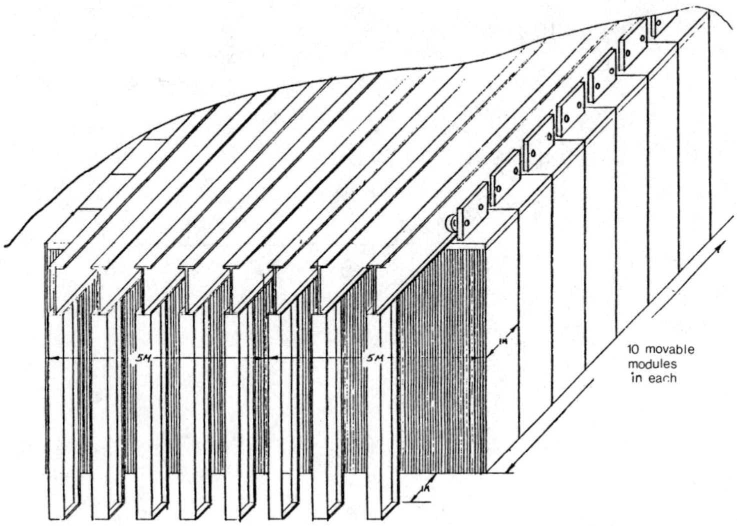

Parameters of the 1000-Ton Detector

Depth:	2150 ft = 1985 meters of water equivalent
Active Volume:	10 m x 10 m x 5 m = 500 m^3
Total Mass:	1000 metric tons = 6 x 10^{32} nucleons
Fiducial Mass:	648 metric tons = 3.9 x 10^{32} nucleons
Average Density:	2.0 g/cm^3
Radiation Length:	1.76 cm (iron), 7.04 cm (average)
Nuclear Absorption Length: (high energy)	17.1 cm (iron), 68.4cm (average)
Composition of Detector: (by weight)	88.6% Iron; 10.5% Carbon 0.9% Hydrogen
Drift Chamber Module:	5.0 m long x 50 cm wide x 1.0 cm gap
Maximum Drift Time:	50 μsec (for 10 kV/50 cm drift field)
Spatial Resolution:	±2 mm
Energy Resolution ($p \rightarrow e^+\pi^\circ$):	±25%
Chamber Arrangement:	Four $(5\ m)^3$ cubes, each with 250 crossed layers of 10 chambers each = 10,000 chambers total
Active Shield:	Drift chambers on all six sides

Figure 6. Design study of the 1000 ton tracking calorimeter. A tabulation of the detector properties is shown below.

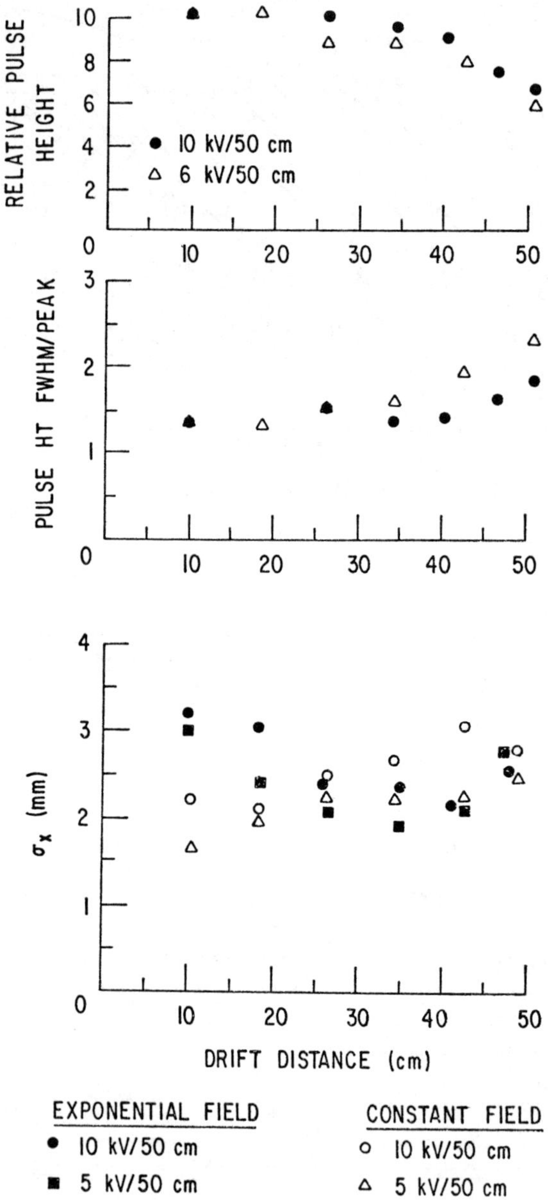

Figure 7. Results of tests of a large drift cell prototype for 1000 ton detector development.

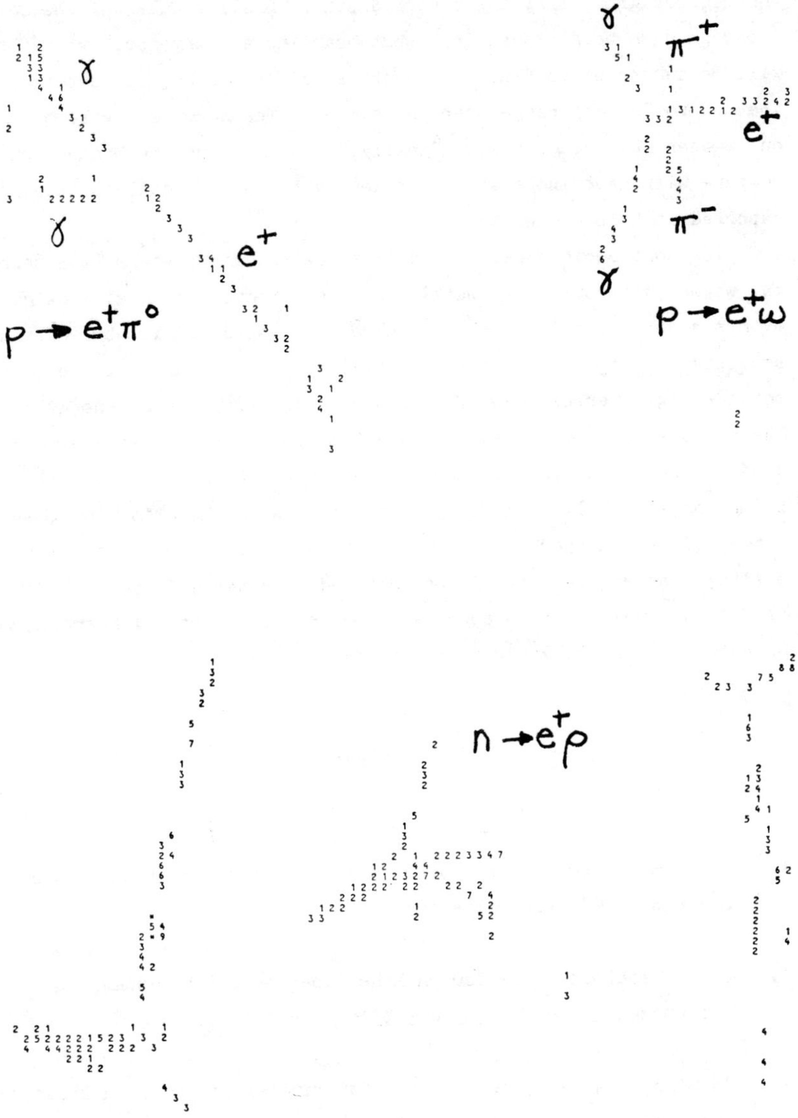

Figure 8. Proton decay and background events generated by Monte Carlo program as part of the 1000 ton detector studies.

will compare with our present understanding of the muon and neutrino induced backgrounds at these rock depths. Small modules of the Soudan 1 design have been placed in a test beamline at Argonne, where they will be subjected to fluxes of protons, pions, muons, and electrons in the 300 to 500 MeV range over the summer. The same test will be done on Soudan 2 prototypes. Finally, we will run the Monte Carlo to compare with neutrino beam data taken by the NUSEX collaboration and reported at this conference.

The most serious background in tracking calorimeters (and perhaps in water detectors as well) is from single-pion neutrino-induced events such as $\nu_e N \rightarrow e^- p \pi^o$. These processes limit the ultimate sensitivity of the proposed detector. Broad cuts will be made on total energy, vertex identification, coplanarity, and energy flow. These cuts are expected to pass 64% of all proton decay events and 1.5% of the single-pion background (all other background contamination being negligible). In a 5 year run, with the criterion that the number of proton (neutron) decay events equal the number of background events, the sensitivity of the proposed detector will be to lifetimes of 2×10^{32} years. With approval from the Department of Energy, we hope to have the detector operating by late 1983.

REFERENCES

1. "A Dense Detector for Proton Decay," University of Minnesota preprint, 1979 (unpublished).

2. E. A. Peterson, "The Soudan Mine Experiment," to appear in the procedings of Orbis Scientiae , 1981.

3. "Soudan 2: A 1000 Ton Tracking Calorimeter for Nucleon Decay," Minnesota-Argonne-Oxford collaboration, 1981 (proposal to the Department of Energy).

THE MONT BLANC FINE GRAIN EXPERIMENT ON NUCLEON STABILITY

G. Battistoni(1), E. Bellotti(2), G. Bologna(3), P. Campana(1), C. Castagnoli(3), V. Chiarella(1), D. Cundy(4), B. d'Ettorre(3), E. Fiorini(2), E. Iarocci(1), G. Mannocchi(3), G. Murtas(1), P. Negri(2), G. Nicoletti(1), L. Periale(3), P. Picchi(3), M. Price(4), A. Pullia(2), S. Ragazzi(2), M. Rollier(2), O. Saavedra(3), L. Trasatti(1) and L. Zanotti(2).

(1) Laboratori Nazionali dell'INFN - Frascati

(2) Istituto di Fisica dell'Università and INFN - Milano

(3) Laboratorio di Cosmogeofisica del CNR - Torino

(4) CERN, Geneve (Switzerland)

INTRODUCTION

We will briefly report on the experiment NUSEX (for Nucleon Stability Experiment) presently being prepared to be run in an underground laboratory located in the Mont Blanc Tunnel. The present status of the experiment and some preliminary results obtained at CERN by exposing a part of the apparatus to beams of pions, electrons and intermediate energy neutrinos will be discussed in some detail.

THE MONT BLANC LABORATORY

Our experiment will be installed during the second part of this year in a laboratory (Fig.1) located approximately in the middle (Fig.2) of the highway tunnel connecting France and Italy under the Mont Blanc (Fig.3 and 4). We have measured the background of atmospheric muons in this laboratory (see Table I). In this table we have conventionally taken for all the American Sites, apart from Homestake, the depth of 1800 metres of water equivalent (m.w.e) and the same muon background as reported by

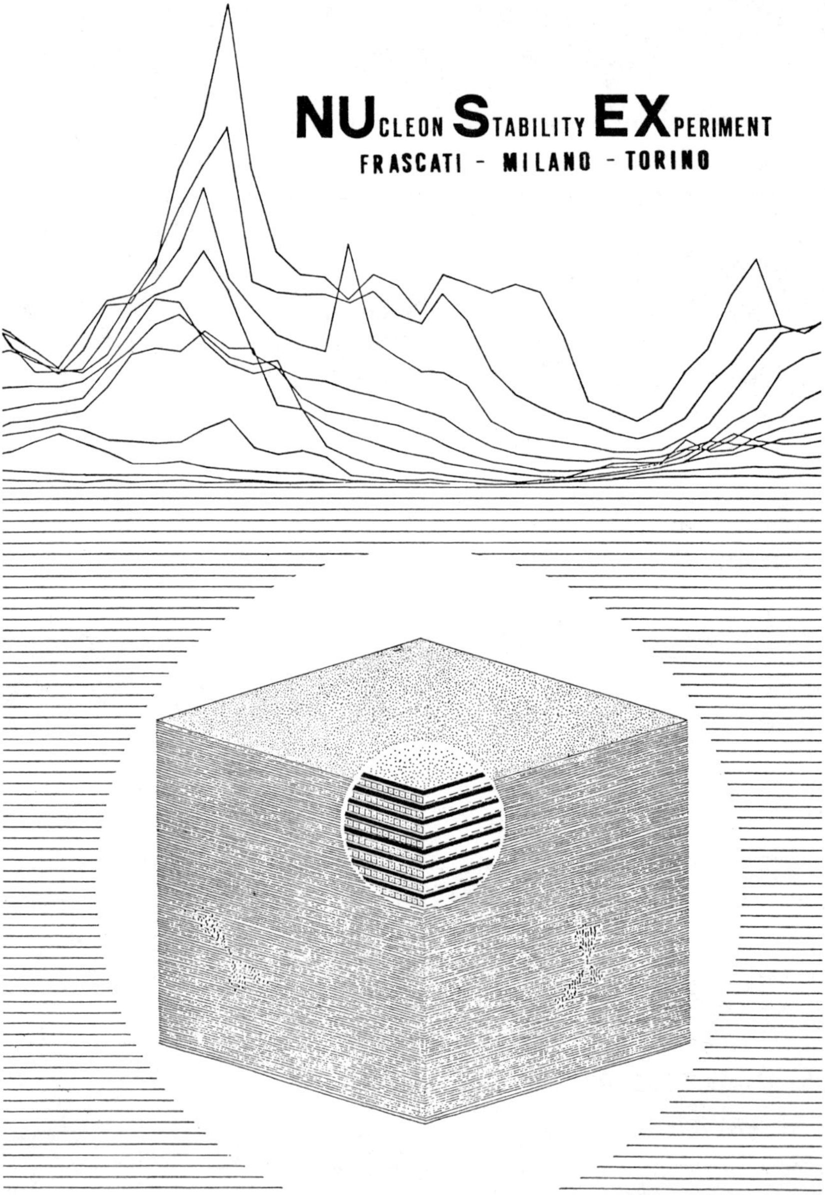

FIGURE 1 - The NUSEX Experiment.

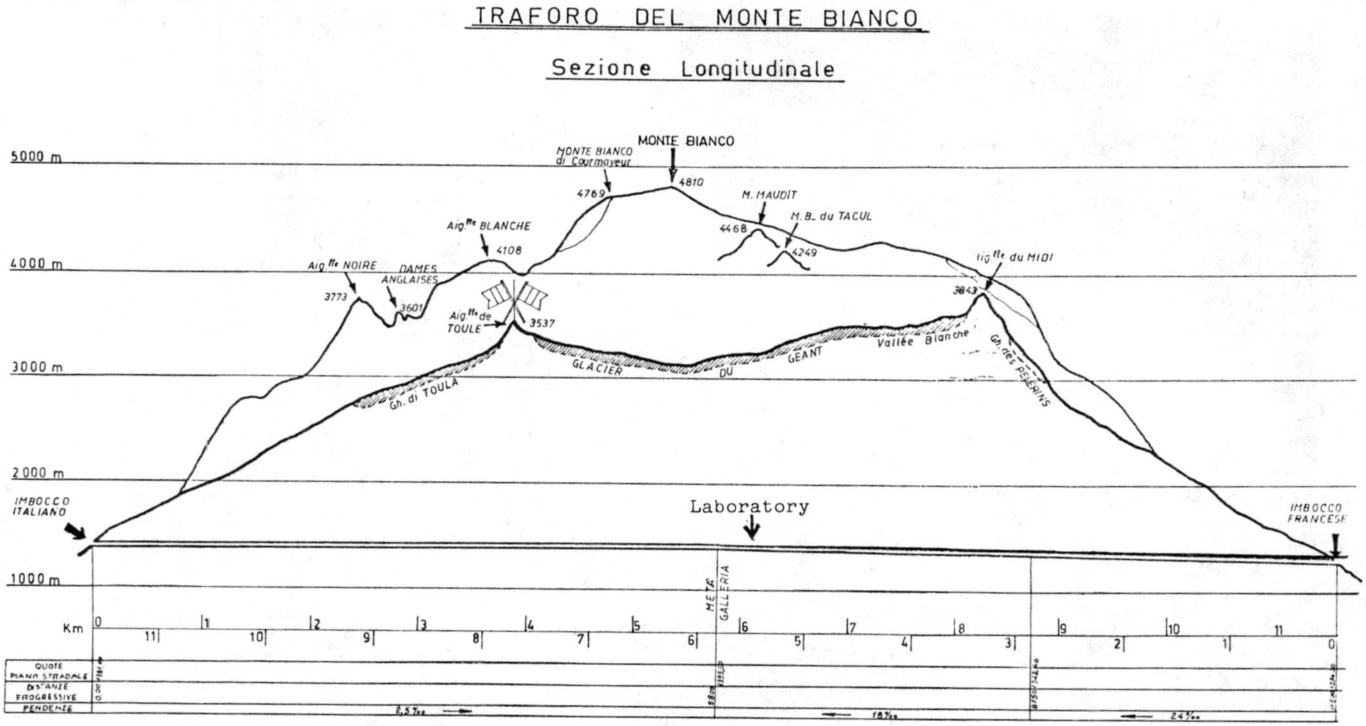

FIGURE 2 - Cross section of the Mont Blanc Tunnel.

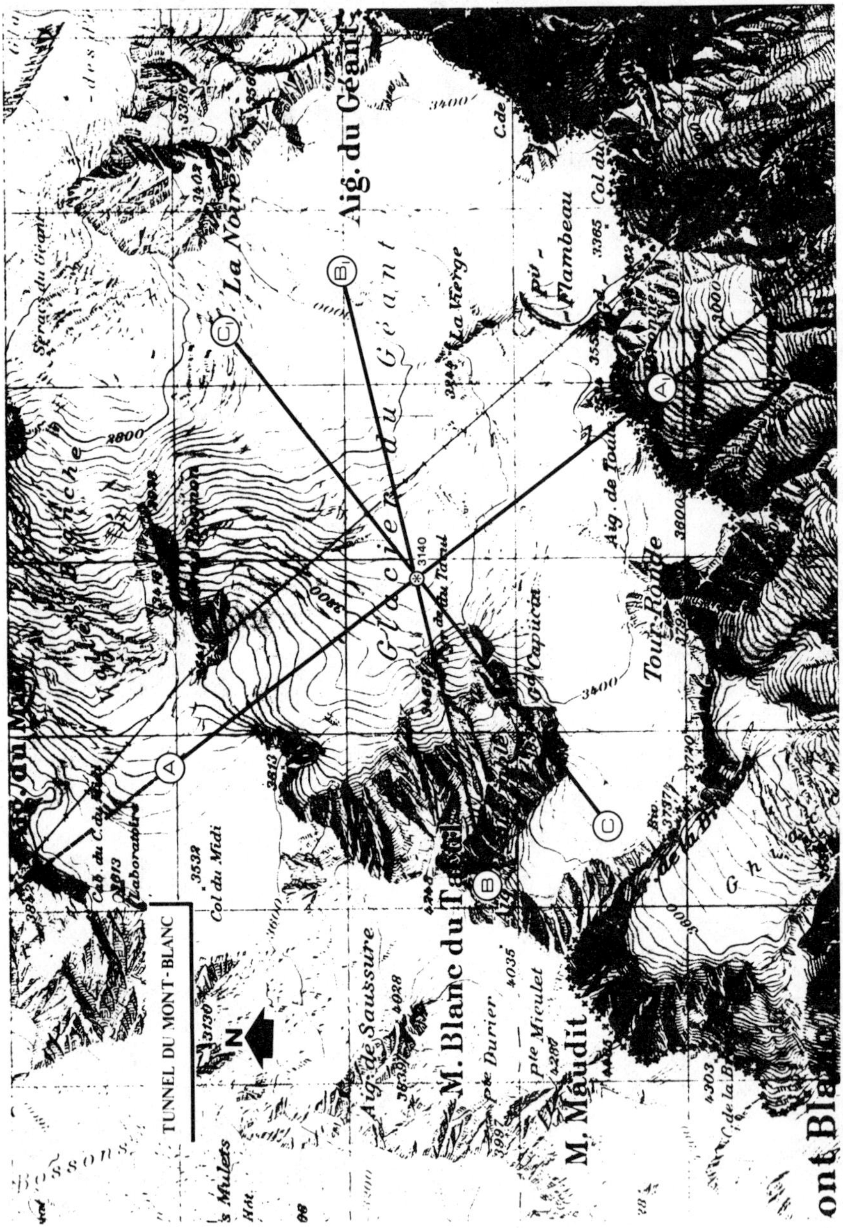

FIGURE 3 - Position of the Laboratory in the Mont Blanc.

FIGURE 4 - View of the entrance of the tunnel (Italian side).

TABLE I

Muon background in a cubic detector of 10 m side.

Location	Nature of the site	Vertical depth (in m.w.e.)	Muon background (per year)
Simplon (Europe)	Railway tunnel	5,600	50,000
Mont Blanc (Europe)	Highway tunnel	5,200	17,000
Frejus (Europe)	Highway tunnel	4,200	500,000
Gottard (Europe)	Highway tunnel	3,700	800,000
Gran Sasso (Europe)	Highway tunnel	4,000	500,000
Homestake (U.S.A.)	Gold mine	4,400	800,000
Kolar Gold Field (India)	Gold mine	7,600	3,000
Morton, King and Soudan (U.S.A.)	Mines	1,800	30,000,000

D. Cline [1]. The expected rate of muons crossing our set-up in Mont Blanc is about 5-6 per day, while the number of interactions in the detector by neutrons produced by atmospheric muons in the rock should be a few per year, i.e. totally negligible in a nucleon stability experiment. The rate and energy spectrum of gamma and neutron radioactivity in the laboratory has been measured [2]. The expected noise in the detector should not exceed 100,000 Hz, and should not cause any problem in the trigger for useful events.

The only background in our experiment should therefore come from interactions of atmospheric neutrinos, and will be discussed in detail later on.

THE DETECTOR

The detector (3.5 x 3.5 x 3.5 m^3) (Fig. 5) consists of a sandwich of 134 iron plates (1 cm^2 thick), interleaved with limited streamer tubes, with 1 cm^2 cross section and 3.5 m length. The tubes [3] are made of plastic, with 100 μm diameter wires and with resistive graphite coating as cathode. A system of orthogonal pick-up strips external to the tube modules, is used to localize tracks in two dimensions on each tube layer (Fig. 6). From calibration measurements made on electron beams between 0.15 and 1 GeV/c, the expected energy resolution is 20% on average in the $\pi^o e^+$ channel.

TESTS ON THE SYSTEM AND ON THE APPARATUS

In order to test the performance of our apparatus in view of an experiment on nucleon decay, we have constructed a test module with the same granularity and with tubes of the same length, but with a volume of 3.5 x 1 x 1 m^3 (Fig.7).

A first exposure has been carried out with negative beams of electrons and pions with momenta ranging from 150 MeV to 2.0 GeV/c. The orthogonal views of the tracks of these particles are shown in Fig.8.

In a second run, presently being carried out, the test model is being exposed to an unfocused neutrino beam produced by 10 GeV protons on a 60 cm long 1 cm diameter beryllium target. It has been in fact shown [4,5] that this beam has almost exactly the same spectrum as that of atmospheric neutrinos produced by the decays of cosmic ray muons, pions

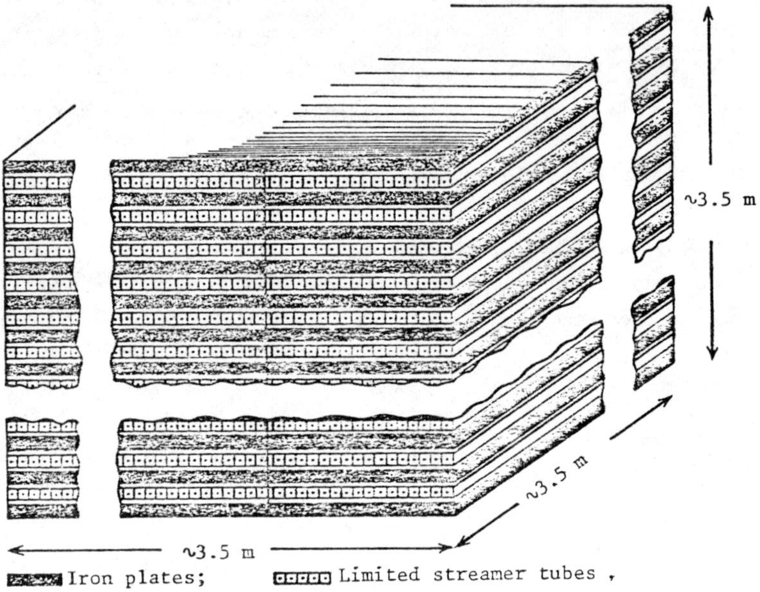

FIGURE 5 - Schematic view of the detector.

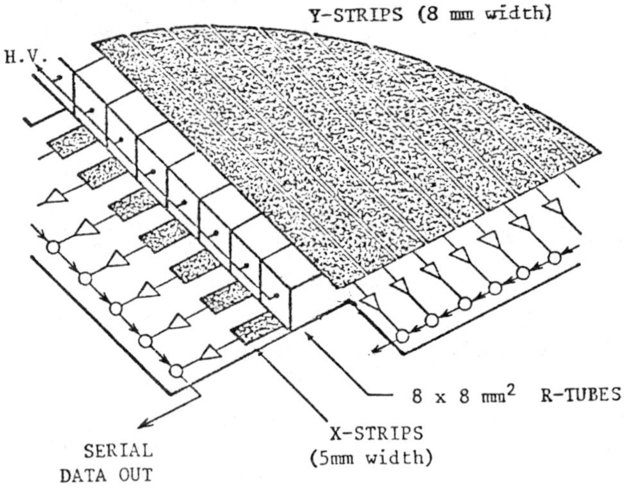

FIGURE 6 - The X-Y bidimensional read-out.

Figure 7a - View of the test model.

Figure 7b - View of the test model.

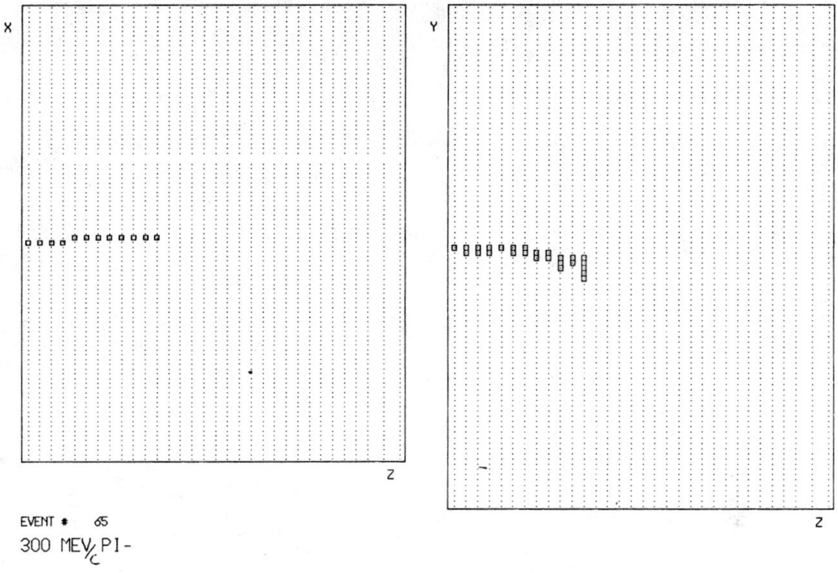

Figure 8. Events in the test model: a) pions

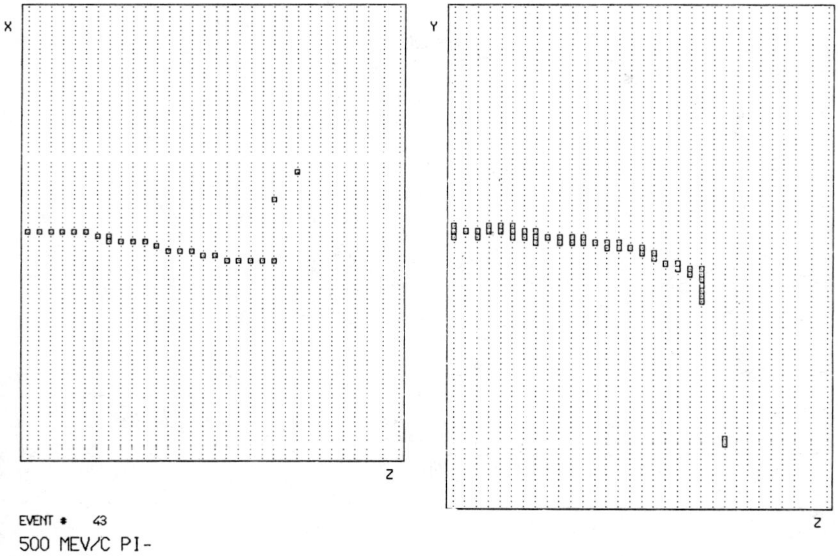

Figure 8. Events in the test model: b) pions

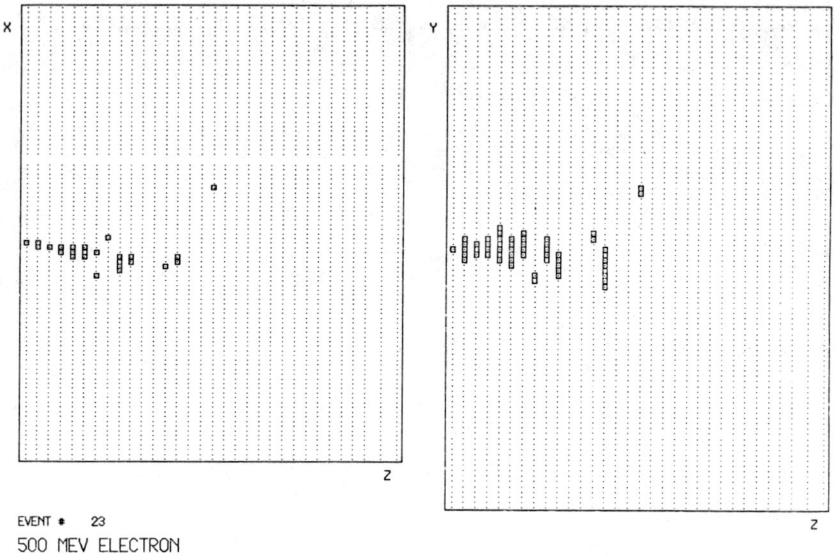

Figure 8. Events in the test model: c) electrons

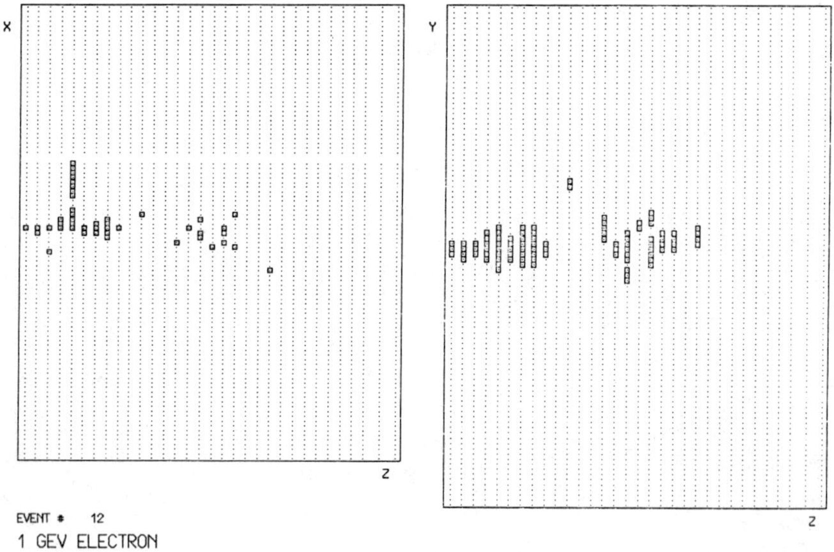

Figure 8. Events in the test model: d)electrons

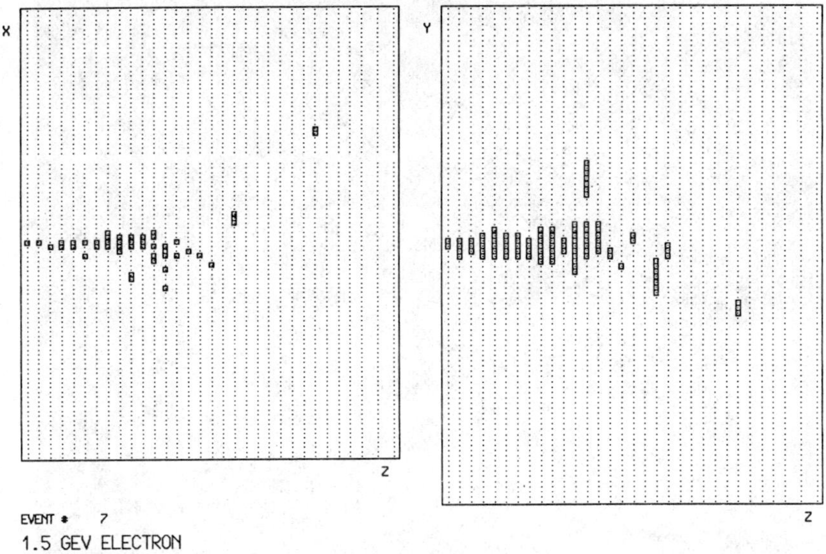

Figure 8. Events in the test model: e) electrons

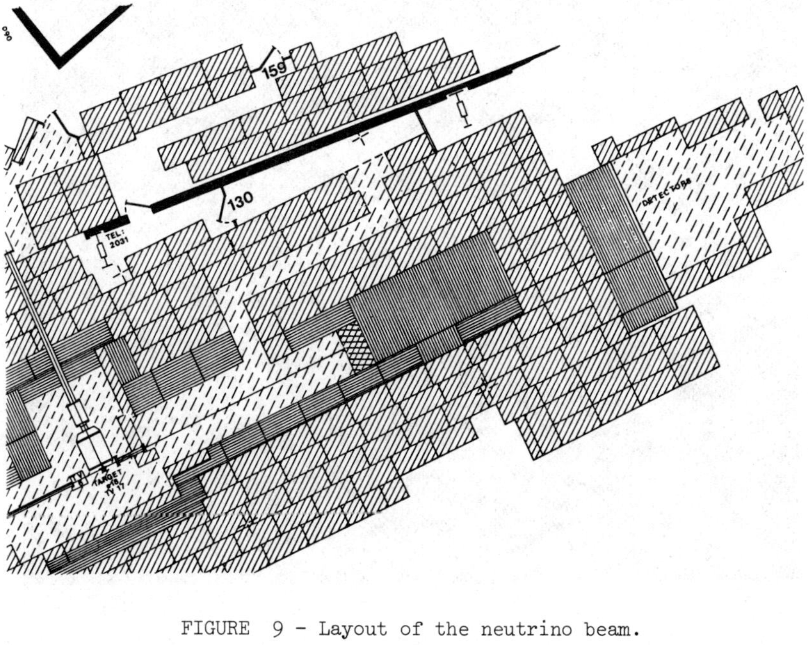

FIGURE 9 - Layout of the neutrino beam.

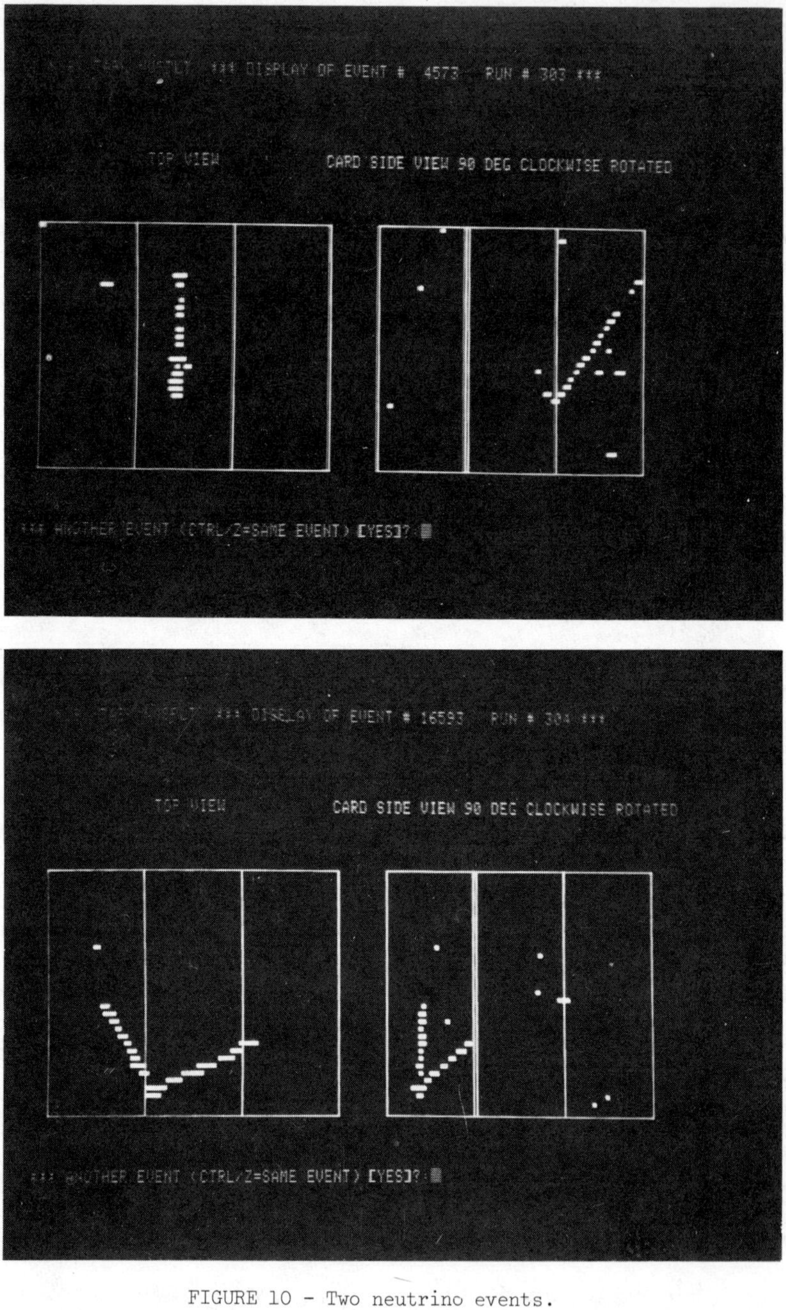

FIGURE 10 - Two neutrino events.

and kaons. The beam layout is shown in Fig.9. The proton beam is dumped on the beryllium target, followed by a shield of 6 m iron and 4 m concrete. The detector is heavily shielded in all directions by blocks of concrete in order to reduce the very large background produced by the beam. We expect to collect about 700 neutrino events, corresponding to an exposure of about 4×10^{18} protons on target. Two elastic events are shown in Fig.10.

PRESENT AND FUTURE OF THIS EXPERIMENT

The Mont Blanc Laboratory already exists and has been used for experiments of smaller sizes. At present it is being renovated in view of our experiment. Two special walls to reduce the intense traffic noise are being mounted and new electric and telephone lines, plus an air conditioning system, are being installed. This work is foreseen to be concluded by the end of July. Mounting of iron planes, together with transversal strips, will start at the beginning of September. The limited streamer tubes of our detector are being constructed at CERN and should be ready by that date. The experiment should therefore be operational by the end of this year.

Plans for a second generation experiment are being considered by part of the NUSEX Collaboration together with a group of the University of Rome. A multikiloton detector of a calorimetric type is being considered, and would be mounted in a laboratory in the Gran Sasso Tunnel, in central Italy.

REFERENCES

[1] D.B. Cline: Experiment to detect proton decay and neutrino oscillations (using cosmic ray neutrino events), talk given to the 1980 "Ettore Majorana" School, August 1980.

[2] G. Battistoni et al. (Frascati-Milano-Torino Collaboration) Proposal for an experiment of nucleon stability with a fine grain detector Dec. 1979.

[3] G. Battistoni et al.: Nucl. Instr. and Methods 164, 57 (1979) and references therein.

[4] D. Cundy: Background calibration for proton-lifetime detectors - Preprint (Oct. 1979).

[5] E. Fiorini: Experiments on lepton and baryon stability and oscillation phenomena - Invited paper at the Royal Society Meeting on Gauge theories of the Fundamental Interactions, London 29-30 April, 1981.

A FINE GRAIN DETECTOR IN THE FREJUS TUNNEL

R. Barloutaud
Département de Physique des Particules Elémentaires, CEN-Saclay

1. Introduction

This talk reports the status of a proposal to study the nucleon instability using a very fine grain detector placed in an underground laboratory in the Fréjus tunnel. This proposal has been submitted last October by the groups of Orsay, Ecole Polytechnique and Saclay [1] ; some other groups intend to join this collaboration in the near future.

2. General Description of the Detector

The interest to use a fine grain calorimeter detector to study nucleon decay is mainly to visualize directly the tracks coming from the decay products, to identify them as well as possible, to measure their momentum and to get some informations on their charge. It is clear that in order to get all this information at energies of the order of few hundred MeV one needs a very fine grain detector. The fiducial mass of the detector must be of the order of one kiloton (Kt) if one wishes to reach a lower limit of 10^{32} years or to achieve a serious study of the decay modes for a lifetime of 10^{30} - 10^{31} years in a few years.

The detector which will be quickly described in this talk, has been designed to meet as well as possible the above requirements at a reasonable cost. It is made of three different elements :

1. A spatially fine resolution detector formed by plastic flash chambers with tubes of $(0.5\ \text{cm})^2$ cross section.
2. A trigger and time detector realized by Geiger tubes planes. The resolution time (0.15 μsec.) is small enough to allow the detection of $\mu^+ \to e^+\nu\bar{\nu}$ decays.

3. A massive element, the main source of the nucleons, which is made of iron plates of 3 to 4 mm thickness. This small thickness is essential for good pattern recognition and for energy resolution. As an example, a 200 MeV/c electron or muon crosses perpendicularly about twenty 3 mm iron plates.

This detector is modular ; this allows one to increase its mass continuously when only a part has been put in operation. This is an essential property in an experiment on nucleon decay, in which only an upper limit on the expected signal is known.

A detector of about 1 Kt fiducial mass, which corresponds to about 1.5 Kt total mass, will have a sensitivity of the same order of magnitude as the largest Cerenkov device (IBM detector [2]) and is able to reach a lifetime of the order of 10^{32} years. Above this value the problem of background is such that a much heavier detector will be necessary to isolate clearly the nucleon decay events. I will then describe a 1.5 Kt apparatus, keeping in mind that, if the lifetime appears to be smaller than $\simeq 10^{31}$ years a part of it could be already very useful to get important results.

Table 2-1 summarizes the main characteristics of the detector ; most of them are not yet completely frozen and will be decided after the final tests.

Total mass	1.5 Kt
Fiducial mass	1 Kt
Section	$(\simeq 6 \times 6)\ m^2$
Density (g/cm^3)	2.2 to 2.5
Length	15 to 20 m
Flash tube internal section	$(0.5 \times 0.5)\ cm^2$
Number of flash chamber planes (or iron plates 4 or 3 mm thick)	1100 to 1500
Number of flash tubes	1.1 to 1.5 10^6
Number of Geiger planes (2.4 cm iron interval between trigger planes)	$\simeq$ 200
Number of Geiger tubes ($\simeq 1.5 \times 1.5\ cm^2$ section)	$\simeq 8\ 10^4$

Table 2-1

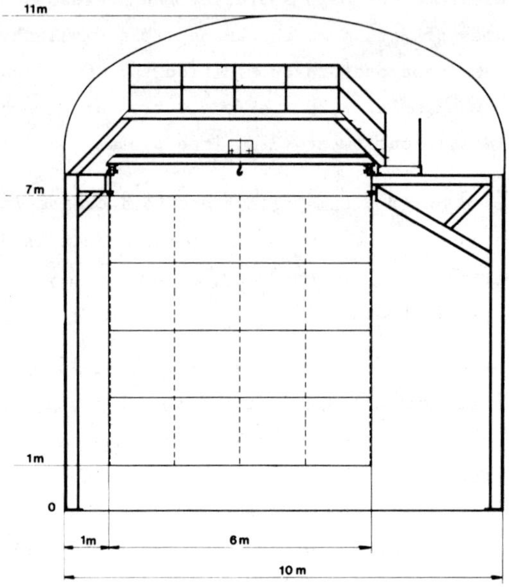

Fig. 2-1

Suspension of the chamber plane in the laboratory

The flash chamber elements and the trigger planes will be suspended and assembled as sketched in Fig. 2-1. The successive planes are oriented so as to have the tubes of consecutive elements oriented alternatively horizontally and vertically both for flash chambers and for Geiger planes.

3. Flash Chambers

The elements of flash chambers of 6 × 1.5 m^2 are made of two polypropylene flash planes with parallel tubes separated by one iron plate and surrounded by two half-thickness iron plates as shown in Fig. 3-1.

Although the flash chamber technique is now familiar [3-4], its extrapolation to a length of about 6 m has been tested. An electric field of 5 to 10 kV/cm has been applied to a stack of flash planes filled with a mixture of 70 % neon and 30 % helium. The electric pulse obtained with a distributed zig-zag delay line has a rise time of about 150 nsec. The efficiency of the flash tube reaches an 80% plateau above ≃ 8 kV/cm

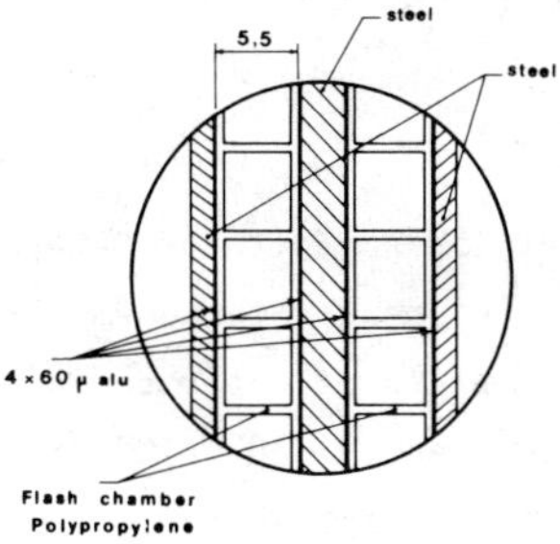

Fig. 3-1

Structure of a flash chamber element.

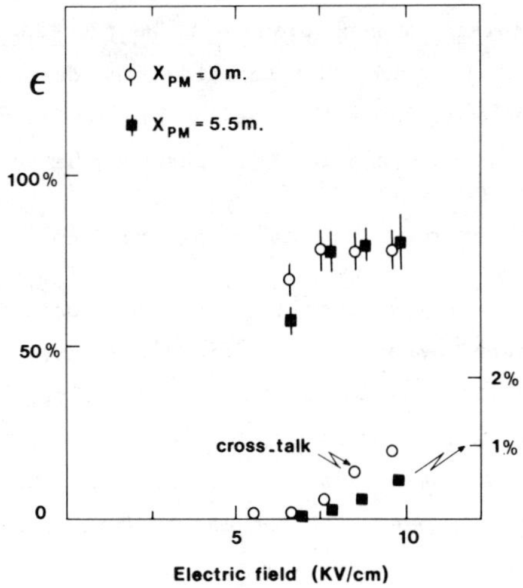

Fig. 3-2

Efficiency and proportion of cross-talk versus the electric field for different positions (X_{PM}) of the crossing particle.

while the proportion of background partly due to cross-talk phenomenon is less than 5 %. The efficiency does not depend on the position of the track along the tube (Fig. 3-2). These results are quite similar to those obtained in the large flash chamber calorimeter build at FNAL [4].

Although these tests were performed with an optical read out of the flash tubes, the possibility of an electrical read out was also tried. The methods developed at Wuppertal with inside electrodes collecting a charge inside the plasma [5] as well as the one used at FNAL [4] and Irvine [6] with outside electrodes have equally been successfully tested. The tube position read out by shift registers seems to be the most reliable for a reasonable cost [5, 6].

4. Geiger Tubes

Extruded planes of aluminium Geiger tubes of about (1.5 × 1.5) cm^2 section, filled with argon and a small percentage (2 %) of non explosive organic vapour (methylal) are under study. The position of a group of 8 tubes will be recorded by a shift register while a TDC will record the crossing time of a particle through a Geiger plane. A signal to trigger the flash chambers will be obtained by a coincidence of three consecutive planes with a resolution time of about 0.15 μsec. The time of the pulse for each plane will be recorded up to about 6 μsec after each trigger in order to detect the delayed e^+ decay from a stopping μ^+ and eventually the passage of slow cosmic particles (such as grand unification monopoles) [7] through the detector ($0.005 < \beta < 0.1$).

5. Underground Lab in the Fréjus Tunnel

The Fréjus tunnel in the Alps, located about 150 km south of Geneva, has been open to car traffic since last summer. The mean thickness of rock above its center is 1680 m corresponding to 4500 meters of water equivalent. In the beginning of 1980, a 44 meter gallery and a $(10 \times 10 \times 8)\ m^3$ room have been excavated allowing the excavation to continue for a larger laboratory even with the car traffic on (Fig. 5). The working conditions inside the present room are excellent (temperature 22°C, very small humidity, very easy access). A series of tests have been made recently with large scintillators and streamer counters to measure the radioactivity level and the muon flux.

The walls of the present room are made of limestone rock which is weakly radioactive. The radioactivity which has been measured is probably due to a small amount of radon or thoron gas always present in underground cavities : its intensity is such that the expected trigger rate in the proposed detector will be much smaller than the atmospheric muon trigger rate, and the background in the flash chamber negligible.

The atmospheric muon flux is at this depth about 10^{-6} times lower than on the earth's surface. Taking into account the mountain profile it has been evaluated to be 3000 μ/m^2.year. Using four large scintillators in coincidence the rate has been measured, it amounts to (3500 ± 1300) μ/m^2.year. This corresponds for the 1.5 Kt proposed detector to about one muon crossing the apparatus every minute. The efficiency loss due to the deadtime of the flash chamber is then negligible. All the atmospheric muons can be triggered which will allow us to test on line the detector efficiency and to keep all interesting cosmic ray events.

A laboratory of 30 m × 10 m × 10 m, including the present room (Fig. 5-1), which could contain a modular detector up to about 2 Kt, is expected to be built and outfitted in less than one year's time.

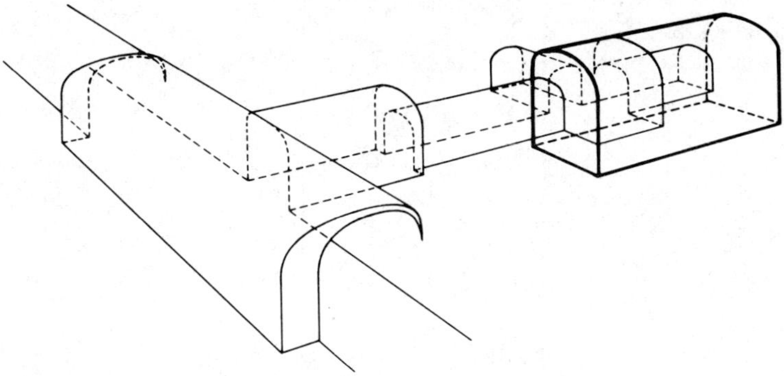

Fig. 5-1

Sketch of the cavity in the Fréjus tunnel. The light lines represent the present cavity, the heavy lines the proposed cavity.

6. Expected Performances of the Detector

The performance of a fine grain detector for particles under $\simeq$ 0.5 GeV/c momentum may be evaluated by Monte-Carlo simulations. However it is safe to check these evaluations with experimental results. This has been realized with a test calorimeter made of planes of streamer tubes of (2 $\times$ 2) cm^2 sections separated by 4 mm iron plates (0.22 Xo). From the experimental results obtained with 200 to 500 MeV/c electrons and pions beams hitting this detector at various angles, and a Monte-Carlo extrapolation to the proposed (0.5 $\times$ 0.5) cm^2 grain, the following performances have been evaluated.

6.1. Separation Electron-Pion

This separation which is essential in a nucleon decay experiment is based on the difference between showering and non showering particles as seen in Fig. 6-1. The average number of hits per plane which is almost energy independent and quite different for the two types of particles leaves only 20 % ambiguous tracks on average between 200 and 500 MeV/c with the (2 $\times$ 2) cm^2 streamer tubes ; this proportion is expected to be smaller than 10 % with the (0.5 $\times$ 0.5) cm^2 flash tubes.

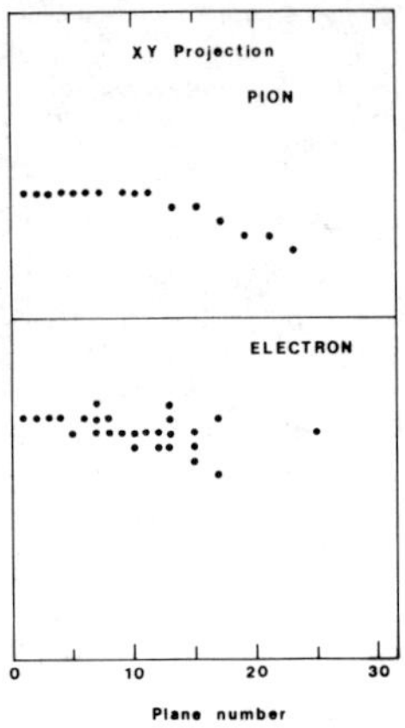

Fig. 6-1

Typical appearance of a 500 MeV/c electron and pion in the test calorimeter (4 mm iron thickness, (2×2) cm^2 section streamer tubes) (experimental data).

6.2. Energy Resolution

With thin iron plates the number of hits is proportional to the energy for the electrons. About 80 tubes are hit by an electromagnetic shower of 500 MeV - perpendicular to the plates.

After averaging over all directions, this corresponds to an energy resolution $\Delta E/E \simeq 15\ \%/\sqrt{E}$ GeV.

For the charged pions the estimate of the energy requires the study of the pattern and the evaluation of the losses by nuclear interaction. That must be achieved event by event, and will give rise to a reliable evaluation of the pion energy.

6.3. Space and Angular Resolution

The space resolution of 0.5 cm leads to an angular resolution for a pion, a muon or an electron which varies with energy but should not exceed a few degrees at 200 MeV/c.

6.4. Pattern Recognition

One of the projection of an $e^+\pi^\circ$ event simulated by Monte-Carlo is shown in Fig. 6-2. The small grain and the small iron sampling (3 mm iron)

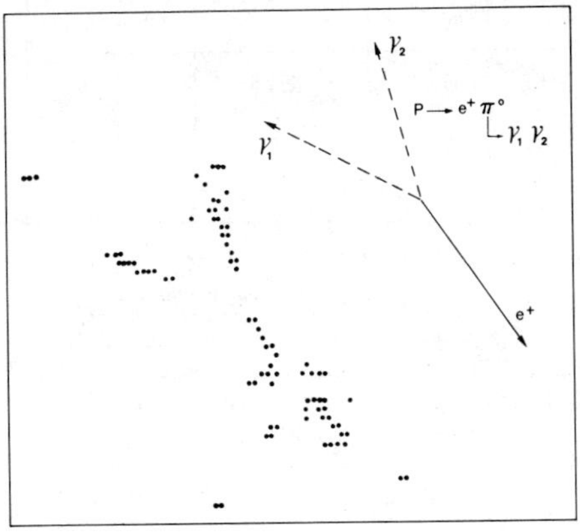

Fig. 6-2

Appearance of an event of the type $p \rightarrow e^+\pi^\circ$ in the detector (Monte-Carlo).

leads to a clear identification of the event where about eighty tubes are hit.

6.5. Sensitivity

The total extension of a nucleon decay event in the detector will be of the order of 70 cm. Taking a 50 cm thickness non-fiducial region all around the detector is enough to contain almost all events and to protect the fiducial region against interactions of neutrons not associated with atmospheric muons. This leads to a fiducial volume of about 2/3 of the total volume.

The number of expected well identified $e^+\pi$ and e^+ + multipions events is evaluated in Table 6-1, where the trigger and the identification efficiency have been taken into account as well as the probability of non reinteraction of the decay pion in the iron nucleus ($\simeq$ 60 %). In order to separate efficiently the ν background simulating nucleon decay, a cut in total energy and colinearity is also needed.

For a 10^{31} year nucleon lifetime, 12 well identified events and less than one simulating ν event are expected. That means that a lifetime determination of 10^{31} years can be obtained in about one year, while a lower limit of about 10^{32} years may be achieved in a few years.

Decay/year.Kt fid. for $\tau = 10^{31}$ years		$e^+\pi$ *	$e^+ + n\pi$ * n>1	Unabsorbed pion	After** acceptance		Simulating ν background
60	30 p	10	13	12.5	5	12	< 1
	30 n	18	9	15.5	7		

* Based on calculated branching fractions

** The acceptance for a fully identified e^+ + pions event is the product of three comparable fractions : trigger acceptance (80 %), good identification ($\sim$ 80 %) and total energy cut ($\sim$ 80 %).

Table 6-1

7. Conclusion

In conclusion it is interesting to compare the expected qualities of water Cerenkov and ionisation fine grain detectors.

- The sensitivity of a 1.5 Kt fine grain detector is comparable to that of a 4 Kt Cerenkov detector, both having a fiducial mass of approximately 1 Kt.
- The modularity of the fine grain detector allows a continuous extension and permits a start with a fraction of the detector.
- The confirmation of an effect with a different technique seems highly desirable.
- The detection of low energy pions or muons is only possible with ionisation detectors.
- While a dense detector is more expensive per ton, it can be installed in the smaller laboratories typically found at greater depth, thereby reducing the trivial muon background.

It is also worth noting that whatever is the nucleon lifetime, above about 10^{32} years an accurate study of the neutrino background will be necessary before much larger detectors are designed.

References

[1] P. Bareyre, R. Barloutaud, L. Behr, G. Chardin, B. Degrande, B. Dudelzak, E. Engels, J.Y. Grandpeix, B. Grelaud, M.A. Jabiol, S. Jullian, U. Nguyen-Khae, C. Paulot, A. Rousset, Ph. Roy, G. Szklarz, B. Tallini. "Proposal to Study the Instability of the Nucleon Using a Calorimetric Detector", Orsay-Palaiseau-Saclay Collaboration (Oct. 1980).

[2] IMB Detector. See D. Sinclair's Talk in these Proceedings.

[3] M. Conversi et al., Nuclear Instr. Met. 151 (1978) 93.
L. Federici et al., Nuclear Instr. Met. 151 (1978) 103.

[4] F.E. Taylor et al., Internal Report Fermilab Conf. 79/86.

[5] R. Grassov ; J. Thierjung - Thesis from Gesamthochschule Wuppertal (March-April 1980).

[6] R.C. Allen, G.A. Brooks and H.H. Chen, University of California Preprint Irvine (1981).

[7] M.J. Longo, University of Michigan Preprint HE 80.39 (1980).
G. Lazarides, CERN TH 3008 (1980).

STATUS REPORT ON HARVARD, PURDUE, WISCONSIN PROTON DECAY EXPERIMENT*

Chuck Wilson
Department of Physics
Purdue University
West Lafayette, Indiana 47907

In the past year, since the First Grand Unification Theory Workshop, considerable progress has been made toward the installation of the Harvard, Purdue, Wisconsin (H.P.W.) detector. There have also been significant modifications in its design. In this report I will restrict myself to the progress and modifications made since the previous workshop. For a discussion of the motivation behind our particular design approach see the report by D. Winn at the previous workshop.

Since we wished to put several layers of veto counters and shielding around our water volume we needed to build a free standing tank to hold the water and be able to support our detector array. We found that the simplest and cheapest such container was a cylindrical tank, such as oil storage tanks. Further, since the tank had to be assembled in place a wooden tank was found to be the simplest to prefabricate and assemble underground.

The tank will be 38 ft. in diameter, 24 ft. high and will contain 780 tons of water or 4.25×10^{32} nucleons. The tank pieces will arrive at the Park City mine site the last week of June and assembly, which takes about three weeks, will begin immediately. When the tank is completed, a hypalon liner will be placed inside the tank to maintain water purity and the walls will then be completely lined with mirrors. The top and bottom mirrors will be flat while the side mirrors will be curved.

The photomultiplier tubes will be attached to circular hoops, 4 PMT to a hoop, which in turn will be suspended from the ceiling (Fig. 1). The mirrors, hoops and support structures are currently being fabricated and will be shipped to the mine site around the middle

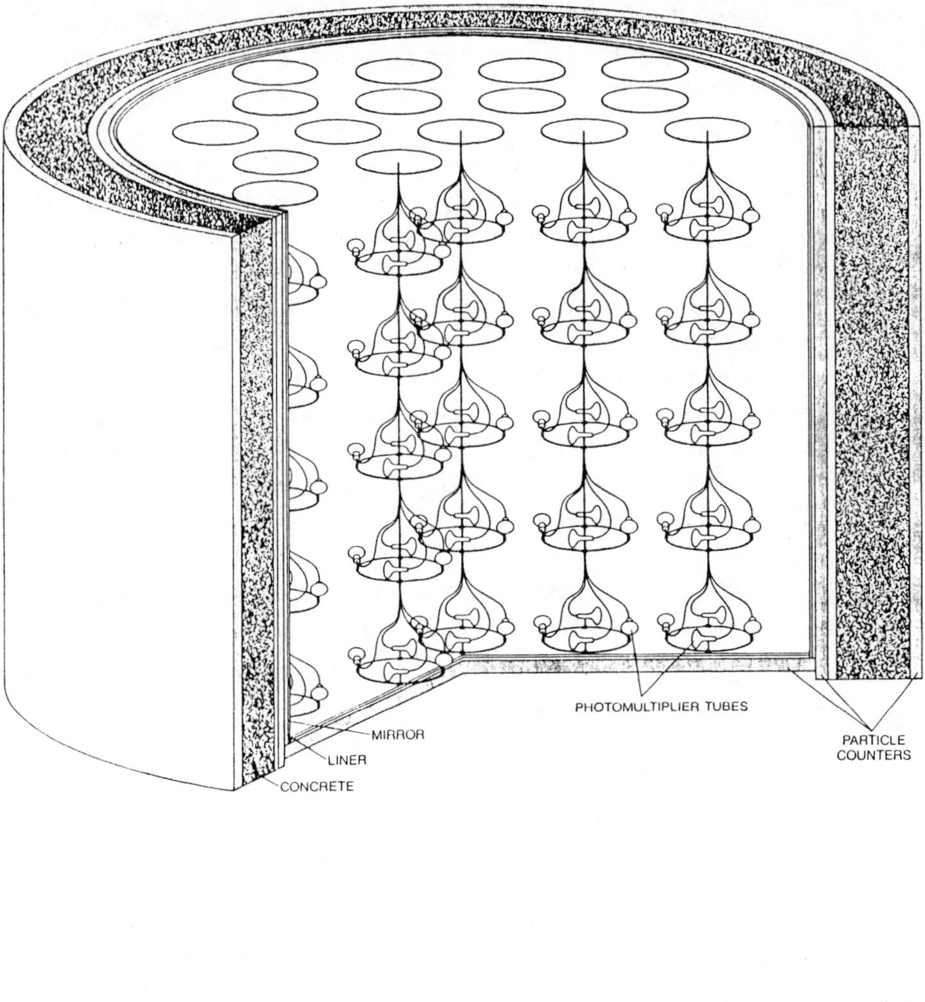
PHOTOMULTIPLIER TUBES
MIRROR
LINER
CONCRETE
PARTICLE
COUNTERS

Figure 1

of July. Installation will then begin and take about 1-2 months to complete.

In parallel with the assembly of the structure going on inside the tank, the electronics and water purification system will be assembled outside the tank. There is an abundant supply of clean ground water near our experimental chamber which we will use for our experiment. This water will be filtered and deionized before being placed in the tank. The water will then be recirculated through filters and D.I. tanks to maintain the clarity of the water needed for this experiment.

The electronics have been designed and are under construction. The heart of the system is a Multi-Time Digititer (MTD) developed at CERN. This device records, in very fast memory, the arrival time of discriminator signals from each PMT at a 200 Mhz rate. This gives an error of ±2.5 ns in the timing signal. Interpolating techniques are currently being studied which can reduce that error by a factor of 2 or more.

The exterior of the tank will be surrounded by proportional tubes run in the geiger mode and using charge division for position determination. These tubes will detect cosmic rays entering the tank. By knowing the entrance and exit points of the cosmic rays we not only can reject these events as not being proton decay, we can use this to help calibrate the system. Knowing the precise cosmic ray trajectory we can determine the PMT systems ability to reconstruct tracks.

With the large number of PMT required for this experiment and the need for individual control of high voltage, discriminator threshold and timing signal widths, the high voltage controllers and discriminators have been designed with complete computer control of all their relevant functions. This will allow us to rapidly set all PMT gains and thresholds equal and to monitor and correct long term shifts in tube characteristics.

The data acquisition is divided into two steps. An LSI 11/23, resident in the mine chamber, will directly control the hardware and read data via Camac. It will then perform some simple and fast tests of the data to determine if the event is a proton decay candidate. These tests should reject about 90% of the events as obvious cosmic rays. The data will then be transmitted via a 1 Mbaud coaxial data transmission line to a PDP 11/45 just outside the mine entrance. Additional, more sophisticated, filtering will then be done and the resultant events will be recorded for later offline analysis.

The system as described should be installed and operational by October of this year. Our design has built into it the ability to add additional features. In particular, space has been set aside to place concrete shielding and additional proportional tubes around the tank to reduce neutral hadrons produced in the rock walls around our chamber. This addition can be put in place within 6 months of completing the basic detector. Figure 1 shows a cutaway drawing of our proposed detector.

*List of Experimenters:

Harvard: J. Blandino, W.A. Huffman, A.M. Lutz, C. Rubbia, D.R. Winn

Purdue: E.C. Fowler, J.A. Gaidos, G. Kullerud, R. McHenry, J. Negret, T.R. Palfrey, R.B. Willmann, C.L. Wilson

Wisconsin: U. Camerini, D. Cline, W.F. Fry, R.J. Loveless, B. March, J. Matthews, R. Morse, T. Meyer, M. Procario, D.D. Reeder, A.H. Szentgyorgyi

IRVINE-MICHIGAN-BROOKHAVEN (I.M.B.) NUCLEON DECAY SEARCH STATUS REPORT

C. B. Bratton[1], W. Gajewski, W. R. Kropp, J. Learned[2],
F. Reines, J. Schultz, H. Sobel, D. W. Smith, C. Wuest,
The University of California at Irvine

R. M. Bionta, B. Cortez[3], W. Foster[3], T.W. Jones[4],
J. M. LoSecco, E. Shumard, D. Sinclair,
J. L. Stone, L. R. Sulak, J. C. van der Velde
The University of Michigan

M. Goldhaber
Brookhaven National Laboratory

[1]Also at Cleveland State University
[2]Also at the University of Hawaii
[3]Also at Harvard University
[4]Also at University College, London

Introduction

The Irvine-Michigan-Brookhaven collaboration is currently constructing a 8700 tonne water Cerenkov detector to measure, or establish a lower limit to, the nucleon lifetime. The detector which should be complete by the middle of October, is located at a depth of 1670 m.w.e. in the Morton Salt Mine near Cleveland, Ohio.

Activities during the past year have been dominated by the construction of the laboratory and of the phototube, electronic and software systems. In house tests of the phototubes and data acquisition chain have demonstrated their reliability, the feasibility of "on-line" rejection of cosmic ray muons, and have yielded information relevant to "off-line" methods of event reconstruction.

This report will concentrate on these activities and point out where actual experience or measurement has confirmed or modified features of the proposed detector as described previously in several articles.[1,2,3]

Main Features of the I.M.B. Detector

Any design of a nucleon lifetime experiment must be influenced by the following factors:

(i) Nucleon lifetimes are predicted to be in the range $10^{31\pm2}$ years in most standard grand unified theories.[4]

(ii) Most models[5] predict sizeable branching ratios into two body final states such as $e^+ \pi^\circ$ and $e^+ \pi^-$.

(iii) Dense detectors gain sensitivity per unit volume considerably slower than the density since the distortion of the products from the decay of bound nucleons is more severe the heavier the nucleus.

(iv) The detector must reject the unavoidable background from atmospheric neutrino interactions to a level ideally small, or at least comparable to the ultimate sensitivity for nucleon decays.

(v) Cosmic ray muons and their associated backgrounds, which clearly depend on depth, must be rejected again to a level small compared to the detector's sensitivity.

The I.M.B. detector (Figure 1) will have 2048 5" phototubes (EMI 9870) viewing a 22.5m x 17m x 18m volume, i.e., 6885 tonnes, of water. This surface array of tubes both serves to detect Cerenkov light from events within the water volume and as an active shield against cosmic ray muons.

Monte Carlo studies reported in the proposal[2] have shown that a 2m thick veto region in front of the tube array is adequate to reject incoming particles. In water the radiation and interaction lengths are 0.36m and 0.8m respectively, and such a veto region is comfortably larger than the spacial extent of nucleon decay events ($\sim$ 1m) and the expected spacial resolution ($\sim$ 1m). However, should a 3m or 4m veto region be required, the experiment would still be sensitive to nucleon lifetimes at the upper end of the range given above. The fiducial volumes within 2m, 3m and 4m veto regions contain respectively 2, 1.3, and 0.8×10^{33} nucleons.

Such a detector is optimally sensitive to two body decay modes, since the back to back Cerenkov cones provide an excellent event signature. This signature also enables the neutrino background to be suppressed to the same order as the detector's ultimate sensitivity ($\sim 10^{33}$ yrs). Of course, three body decays will be looked for and

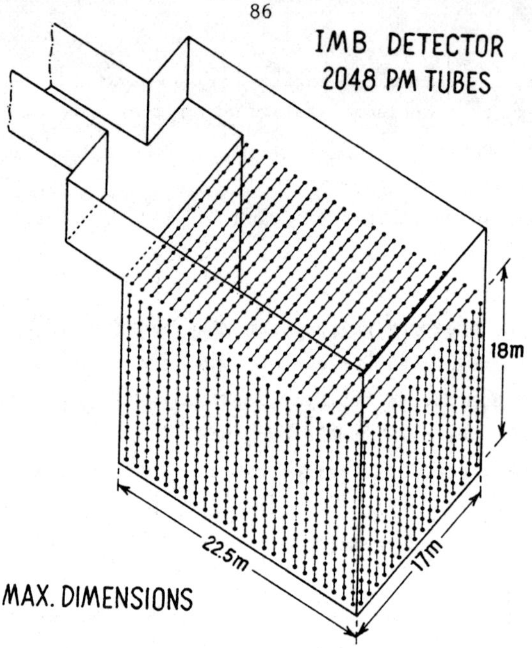

Figure 1. The I.M.B. Dectector

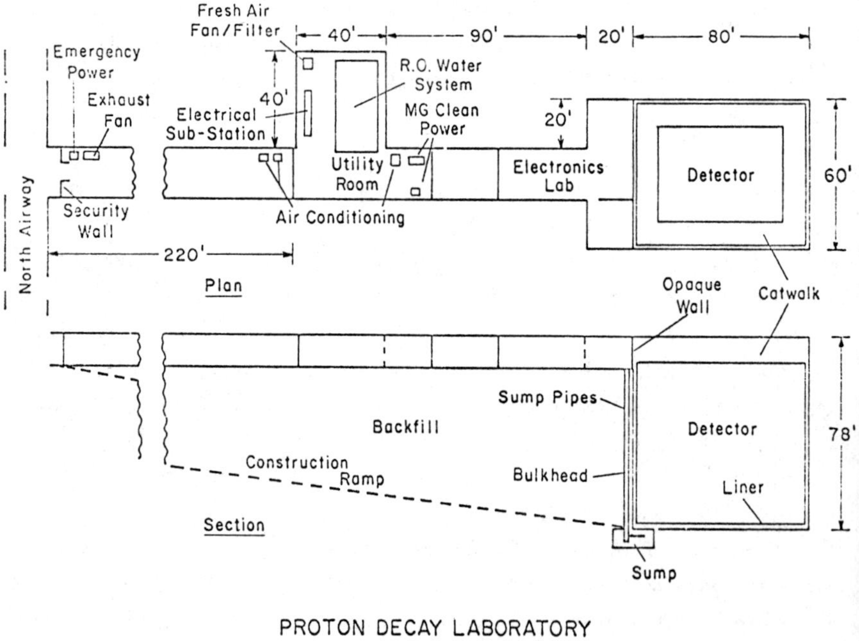

Figure 2 and 3. Disposition of the Laboratory

attempts made to reconstruct them and to establish the neutrino background for such modes.

A water detector has the advantage that a total detector mass of the order of 10 kilotons can be constructed at a reasonable cost. Furthermore, water contains 10% free protons and the various estimates[6,7,8] of the nuclear losses in oxygen are not so large as they are for heavier nuclei such as iron.

Cavity Excavation and Laboratory Construction

Excavation of the 24.0m x 21.0m x 20m detector cavity was completed in October 1980. Rock bolting and chain meshing of the walls, and the excavation of the 350' drift shaft required for egress of the Dosco continuous miner, were done as the excavation proceeded downwards. At an early stage in this work the utility room and an apron area for the electronics laboratory were excavated. These tasks took considerably longer than the mine operator's first estimate and excavation was finally halted 5' short of the design depth because hard shale was encountered. The size[9] of this unique cavern may only be fully appreciated by those who worked from October to January surveying the walls, covering 2000 or so rock bolts with teflon, inserting cushioning material behind the chain mesh where necessary, and laying geotextile (for drainage) and conveyor belting on the floor.

At the same time drainage sumps and a 65' high x 18' wide bulkhead were installed at the end of the drift shaft. The delayed completion of the excavation meant that the backfill of the drift shaft had to be done during the busy salt season at the Morton Mine with the consequent shortage of plant and manpower.

The general disposition of the laboratory and detector is shown in Figures 2 and 3.

Construction of the Detector (Pool) Area

In January 1981, a "unistrut" catwalk (Figure 4), including cable trays and ventilation ducts, was installed on top of the beams which span the pool area. This was followed by the installation of a double liner made out of 2.5mm dense polyethylene sheet and separated by a drainage geotextile. Each liner is a completely sealed tub with its own drainage system. The liner material itself will extend 14% and 700% before yield and break respectively, its reflectivity has been measured to be less than 1% and it causes negligible water contamination.

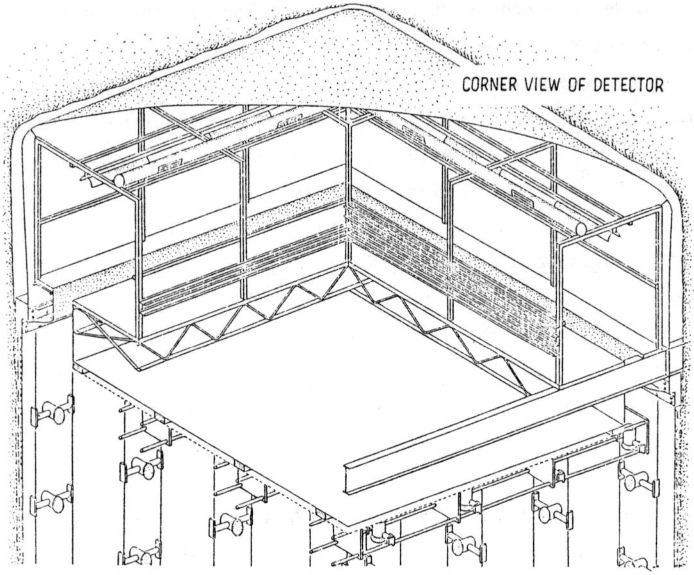

Figure 4. The Catwalk

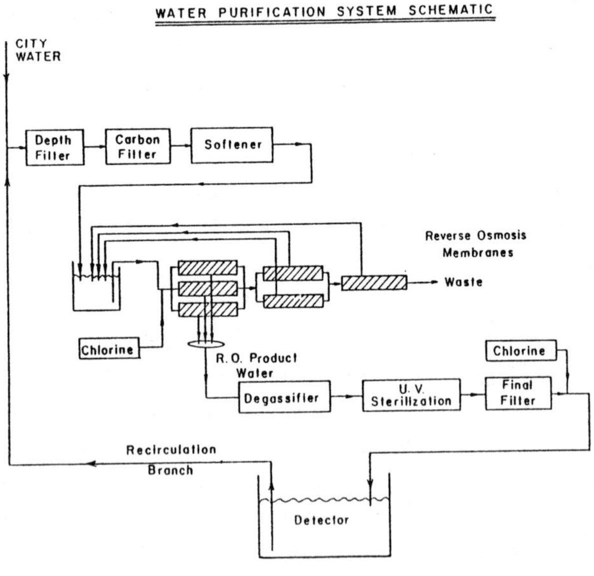

Figure 5. The Water Purification System

The main construction in the pool area was completed by a vinyl "air dome" and an opaque wall which provide an air seal for cleanliness, and isolate this area from the electronics laboratory and the utilities in the tunnel.

Laboratory Construction in the Tunnel, Electronics and Utilities Areas

The tunnel, utility room and apron area near the detector had to be transformed from dusty, irregularly shaped cavities into a working laboratory. Final topping of the backfill, rock bolting and meshing of the walls were completed in April 1981, and all concreting has now been completed.

Fairly demanding standards are required of the services for the detector.

Firstly, the power required to operate the detector is approximately 0.3 Mw mostly used by the Reverse Osmosis system (see the next section). A substation at the entrance of the detector is needed to accommodate this load. Furthermore, a motor generator will be used to isolate a clean power supply for the electronics and computer since there are large voltage excursions and noise levels on the Morton power lines.

Secondly, the air quality is vital for the experiment. Air circulated in the pool area is cleansed of salt particles, diesel fumes and other foreign materials with a system of high efficiency filters so as to maintain water clarity. In the utility room, air quality must be maintained to avoid corrosion of the equipment, while the computer and electronic systems obviously need an air-conditioned environment.

Thirdly, the safety of University and Morton personnel is an overriding consideration. The mine environment together with the need to operate the computer and other equipment for many hours unattended, mandate a fire control system. A halon system has been designed, which together with other measures, should prevent serious fires in the electronics laboratory and utility room.

The equipment and plant necessary to meet the above requirements are already stored underground at the mine. It is anticipated that all the above services will be installed by the middle of August.

The Water Purification System

A Cerenkov detector, which is instrumented on its surface, is limited in size by the absorption length of its characteristic radiation. Transmission measurements were made on different water samples and it was established that the reverse osmosis technique provides a readily available source of pure water whose attenuation length is 50m at the point of minimum attenuation around 450 nm. This technique, which is commercially available in large throughput systems, produces a large quantity of waste water. Since this is undesirable underground a unique system was designed to process and concentrate its own waste water without reducing the quality of the products (see Figure 5).

The addition of a deionizer at no extra cost would increase the attenuation length at 450 nm to 60m or greater. However, the deionizer will not be added until it is shown that Rayleigh scattered photons, which degrade event reconstruction, are less than the current assumption in the Monte Carlo program developed at Michigan.

All materials to be used in the pool have been tested so as to determine their contamination rates. During operation the water will be continuously recirculated at 50 gal. per min. at which rate it is estimated that contaminants would be removed with a time constant of about a month.

Components of the water system are all down the mine and the anticipated completion date is again the middle of August. Since it will take about six weeks to fill the pool, the detector should be full of water by the middle of September, barring unforeseen setbacks.

Phototube Testing and Mounting

The EMI tubes are delivered to Irvine where they are tested for gain and noise. They are then sent to Michigan to be further tested, mounted in their water-tight housing and attached to their cables. The complete assemblies are baked and then subjected to a leak test at the bottom of a redundant elevator shaft which is full of water.

Gain matching enables sixteen tubes to be supplied from one H.V. terminal and noise rate selection enables the average rate per tube to be 2Khz.

Already $\sim$1000 tubes have been delivered to the mine, and at the present production rates, the full complement of 2048 tubes should be on the catwalk by the middle of July.

Data Acquisition Electronics, Production and Testing

The data acquisition system is shown in Figure 6. The pulse height (Q) and firing time (T_1) of the discriminator on any tube, as well as the time of any delayed pulse (T_2), are stored on three

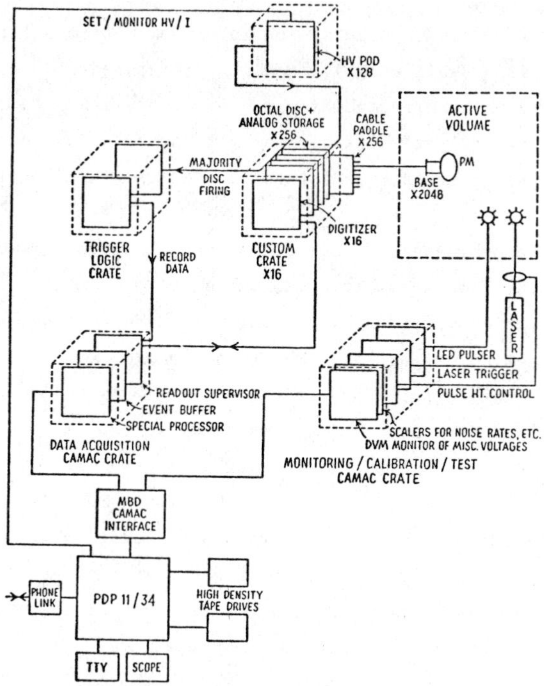

Figure 6. Data Acquisition System

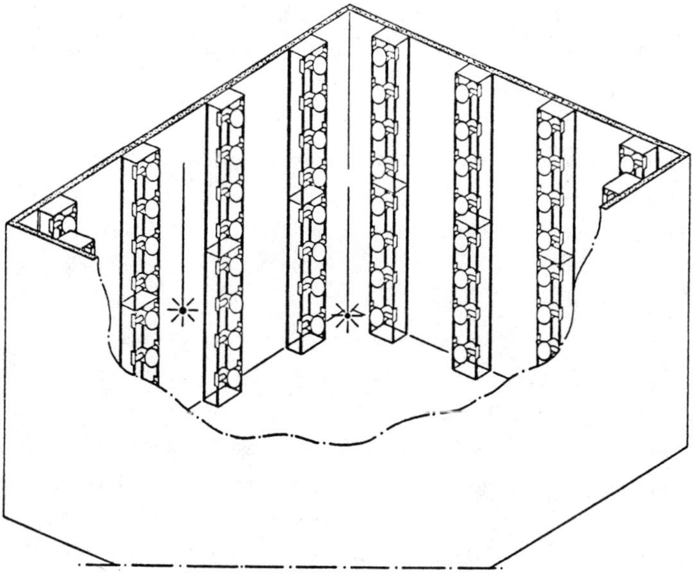

Figure 7. The 128 Tube Room

capacitors. If a trigger occurs, the information from all the tubes is digitized (6144 9-bit words), read out and stored in one of five buffers. Rapid pattern recognition can be accomplished in a special processor unit to filter events, or all of the data can be dumped through the (MBD) interface to a fast tape drive via a PDP 11/34 computer.

All of the overhead hardware (computer, NIM logic, power supplies, supervisor, etc.) has been borrowed, purchased or constructed and most have been operational for over six months. In house construction of the custom electronics for each tube (base, analog storage, digitizer, etc.) is now complete for >1500 units. Assembly is followed by rigorous testing and bake-in at high temperature.

The proving ground for this work is a mock-up black room filled with 128 phototubes (see Figure 7). Light sources in this room consist of flashes produced by cosmic rays traversing scintillator blocks and computer triggered nanosecond pulses from both a nitrogen laser and a green LED array. The upper and lower half-points of the wavelength response of the detector are bracketed by the wavelengths of the two test light sources which will also be used in the mine.

Currently 512 channels of the electronics are under long-term surveillance to uncover any possible drifts, and to evaluate long term reliability. By June 1 all electronics for 2048 phototubes will have been added to this system for final test. Following a further month's test, the system will be ready for delivery to the mine.

Tests done on the 128 tubes in the dark room demonstrate that the time resolution of the tubes are worse than that assumed in the proposal, which was ± 3 ns, and that it is feasible to achieve an average noise rate of 2 Khz per tube (see Figure 8).

Signal Processing

Triggering the Detector

Triggering schemes will obviously evolve under real life conditions as the detector becomes operational. One simple possibility is to require a 3 tube coincidence from the summed discriminator outputs from each "quad" of 64 tubes. This output is made with a coincidence time of ∿ 50 ns (the time light takes to travel across a quad). Any two such quads out of a total of 32, firing within 150 ns (the time light takes to traverse the whole detector) would then initiate the trigger. The threshold for such a trigger is about 20 Mev for single showering

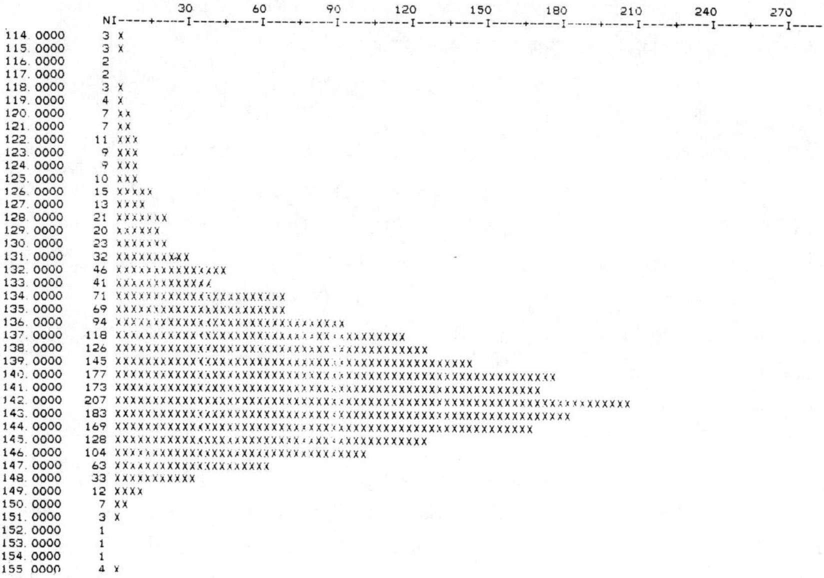

(a)

(b)

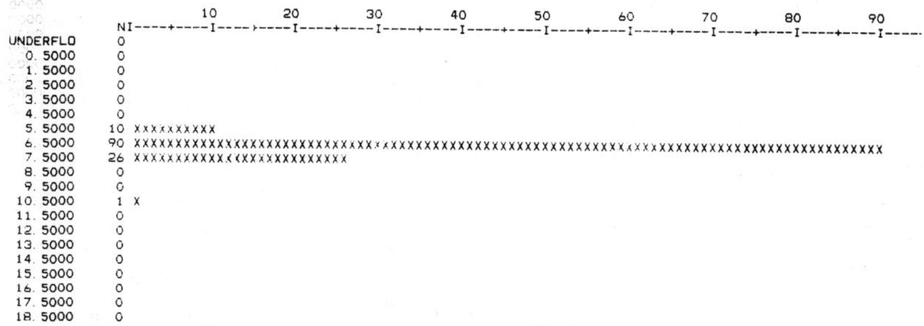

(c)

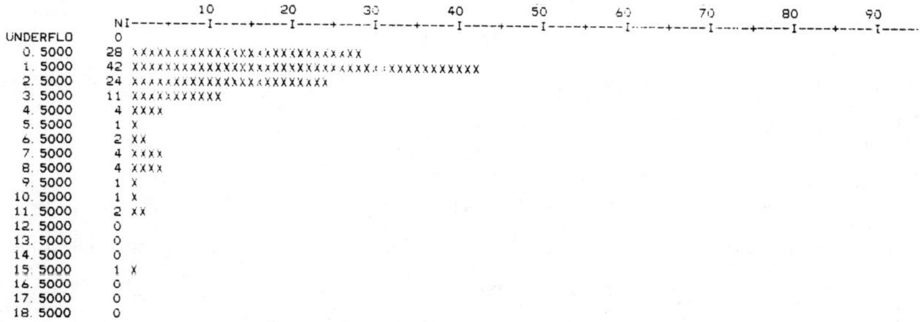

Figure 8. (a) Timing Response of Tube (nsec bins)
(b) Distribution of r.m.s. Timing for 128 Tubes (nsec bins)
(c) Distribution of Noise Rates for 128 Tubes (Khz bins)

tracks, which is comfortably below that of a decaying nucleon. The motivation for keeping the trigger threshold low is the desire to study possible decay modes with little visible light, as well as low energy neutrino interactions.

Once a trigger is initiated the T_1, T_2 and Q data can be recorded for each phototube. The T_1 and Q data together with the coordinates of the struck tube are basic pieces of data for the reconstruction, and the T_2 data would be used to look for decaying muons. The tubes that do not fire are used in the cosmic ray rejection algorithm to be described in the next section and can be used in off-time reconstruction as well.

On-Line Software and Cosmic Muon Rejection

A very powerful algorithm[10] for recognizing cosmic ray muons which traverse the detector approximately 3 times per second has been developed. The algorithm makes use of various combinations of nearest-neighbor tubes firing. Then it uses only binary (on-off) data from the bit patterns which permits rapid on-line computation in about 30 ms. If desired the algorithm can be used to reject traversing muons up to a factor of 500 which would result in a tape write every 3 minutes. On the other hand, the buffering capability in the data acquisition system would enable essentially all of the cosmic ray events to be written on tape if so desired. When used in its most stringent form the "nearest neighbor" algorithm saves > 95% of two body nucleon decays.

Programming for this has been accomplished and tested in the special processor unit which operates upon data in the buffer stack at the CAMAC crate level (see Figure 6). Most of the on-line software has been written and debugged using the 128 tube setup at Michigan. Obviously, the software will evolve in the mine but an initial operating system is ready.

Off-Line Reconstruction and Monte Carlo Studies

Improvements continue to be made in the reconstruction of events in the detector. Although the timing resolution of the tubes is now known to be worse than that assumed in the proposal, which degrades the vertex finding for two body nucleon decays from 30 cm to 75 cm, new techniques have improved the reconstruction of track angles. Showering and non-showering tracks can be reconstructed in angle to 12° and 5° respectively, and the total event energy is known to ± 10%. A cube plot of an $e^+\pi^0$ event, and projections used in reconstruction are shown in Figure 9.

Single muons of energies around 1 Gev and originating from the fiducial volume can be reconstructed to 5° in angle and 70 cms in vertex position. This work is being extended for those muons which survive the nearest neighbor algorithm and to electrons.

The next step will be to attempt reconstruction of two forward going tracks such as in neutrino events and then three body nucleon decays. Many of the techniques found to be successful for two body nucleon decays and for single tracks will be useful in these tasks.

In this manner, the expected resolutions and capabilities of the detector, based on measured tube response are being discovered, and it will then become possible to refine the estimates of neutrino and other background.

The work described above is based on a Monte Carlo program developed at Michigan. The energy resolution given by this program for showering tracks is somewhat better than that given by a program developed independently between 1978 and 1980 at U.C.L. In this program the showering development is given by the E.G.S. program [11] and formed part of an independent design of a surface array detector for nucleon decay. It is hoped to resolve this difference, and perhaps to establish how much Rayleigh scattering should be included in the near future. An independent program has been written at Irvine and it is intended that there should be two completely separate programs for event generation and analysis to insure reliability.

<u>Other Physics and Conclusion</u>

Other interesting physics investigations appear to be possible with this unique instrument

(i) Neutron-anti neutron oscillations;
(ii) Neutrino oscillations;
(iii) Atmospheric neutrino interactions;
(iv) Neutrinos from other sources, such as bursts from stellar collapse;
(v) Anomalous events as seen in the Kolar experiment;
(vi) Muon distributions and anisotropies.

The feasability of these ideas of course depend on the outcome of the studies described in the previous section.

Many of the above are directly relevant to current ideas of grand unification. However, in spite of the elegance and attractiveness of these recent speculations there is a lack of really convincing experi-

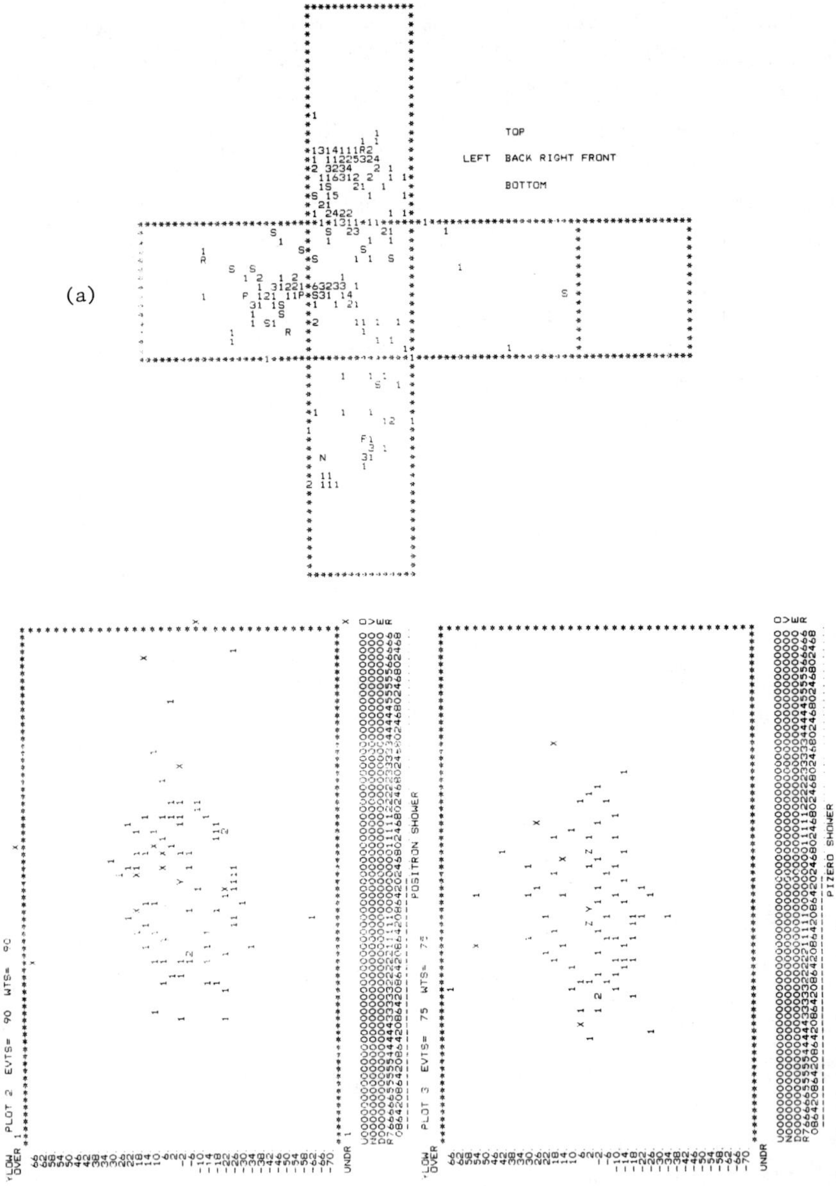

Figure 9. In 9(b) X indicates an indirect photon
Y indicates the direction
of the electron or pizero
X indicates the direction
of the two gamma rays

mental evidence. Certainly, any experiment that would claim that matter is unstable must present evidence whose strength is commensurate with such a remarkable discovery.

As the completion of the detector approaches members of the I.B.M. collaboration are guardedly optimistic that in the near future it will be possible to confront recent theoretical predictions.

REFERENCES

[1] M. Goldhaber, et al., Proceedings of the International Conference on Neutrino Physics, Bergen, June 1979, C. Jarlskog (ed.), p. 121.

[2] Irvine-Michigan-Brookhaven Collaboration, D.O.E. Proposal 1979.

[3] L. Sulak, The Irvine-Michigan-Brookhaven Nuclear Decay Facility. Status Report, Proceedings of the First Workshop on Grand Unification, University of New Hampshire, April 1980, P. H. Frampton, S. L. Glashow and A. Yildiz (eds.).

[4] See article by P. Langacker at this conference for a review.

[5] See G. Kane and G. Karl, Phys Rev D22, 2808(1980) for a recent review.

[6] D. A. Sparrow, Phys Rev Lett, 44, p. 625 (1980).

[7] R. C. W. Henderson and T. W. Jones, reported in T. W. Jones, Nucleon Lifetime Experiment--Collaboration with the IMB Experiment, Rutherford Laboratory Proposal 1980 (unpublished).

[8] C. Longuemare, private communication.

[9] See photograph in the article by S. Weinberg, Scientific American, June 1981, p. 65.

[10] The IMB Nucleon Decay Detector, Data Acquisition and Triggering System. B. Cortez, G. W. Foster, J. M. LoSecco, E. L. Shumard, J. C. van der Velde. Proceedings of CERN Microprocessor Conference, May 1981.

[11] R. L. Ford and W. R. Nelson, The EGS Code System: Computer Programs for the Monte Carlo Simulation of Electromagnetic Cascade Showers (version 3), SLAC Report No. 210, June 1978.

REVIEW OF NEUTRON-ANTINEUTRON OSCILLATION EXPERIMENTS

K. Green
Rutherford and Appleton Laboratories
Chilton, Didcot
OXON, UK

1. Introduction

In a conference devoted to Grand Unification and to nucleon decay experiments in particular, it is perhaps not inappropriate at the end of such a meeting to discuss a phenomenon which, if observed, would owe its existence not so much to Grand as to "Petit" Unification. Such sacreligeous thoughts come from the spirit of enquiry which is now prepared to question the concept of baryon number conservation. In their simplest forms, Grand Unification symmetries of the type SU(5) naturally predict nucleon decay as second order in a $\Delta B=1$ amplitude which transmutes quarks into leptons through the exchange of massive bosons (X) of around 10^{15} GeV; they do not allow for $\Delta B=2$ processes of which neutron (N) antineutron ($\bar{N}$) oscillation is the classic example. Unified theories with partial unification at mass scales of 10^{5} GeV or greater can accommodate $N\bar{N}$ transitions at a level which is amenable to experimental test and which if observed would imply a "flowering of the desert" between the electro-weak unification mass of 10^{2} GeV and the so called grand unification mass of 10^{15} GeV.

Several experiments are now either underway or being planned to search for $N\bar{N}$ oscillations and in this paper I shall review the background to these experimental searches and the present status of the various proposals.

2. Phenomenology

The existence of a massive boson (Y) which transmutes quarks into antiquarks allows a 6-quark coupling to take place and convert a neutron into an antineutron.

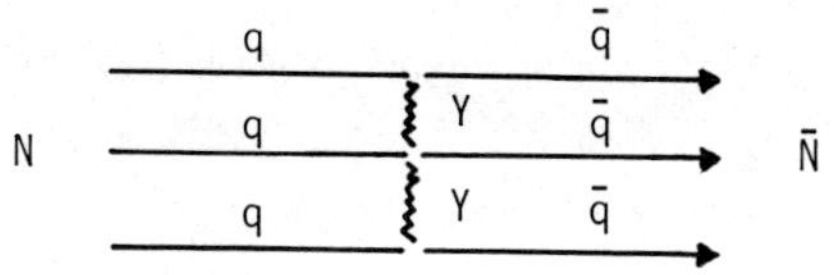

A process of this type is compatible with existing particle physics data if $M_Y \gtrsim 10^5$GeV and would have a coupling constant $g(\Delta B=2) \propto M_Y^5$. The consequences of this coupling are bizarre. For example the physical mass and the interaction eigenstates of the neutron will no longer be the same but a linear combination of N and $\bar{N}$; thus the physical neutron (N_1) and antineutron (N_2) states will be

$$N_1 = N \cos\Theta + \bar{N} \sin\Theta \text{ with energy } E_1=E_0+\tfrac{1}{2}\{\varepsilon^2 + \Delta E^2\}^{\frac{1}{2}}$$

$$N_2 = -N \sin\Theta + \bar{N} \cos\Theta \text{ with energy } E_2=E_0-\tfrac{1}{2}\{\varepsilon^2 + \Delta E^2\}^{\frac{1}{2}}$$

where ε is the transition energy between the N and $\bar{N}$ states, ΔE any externally induced change in the N and $\bar{N}$ energies and $\Theta = \tan^{-1}\{\varepsilon/\Delta E\}$ is a mixing angle. The two states will evolve with time as e^{-iE_1t} and e^{-iE_2t} respectively.

An initial neutron (N) will therefore develop in time a component of antineutron ($\bar{N}$) leading to the phenomenon called $N\bar{N}$ oscillation. This oscillation behaviour, governed by a characteristic time $\tau_{N\bar{N}}=h/\varepsilon$, is very similar to that of $K^0\bar{K}^0$ oscillations [1], muonium-antimuonium [2] and hydrogen-antihydrogen [3] conversion; all of which have similar formalism to describe the amplitude oscillations and is well described in the literature.

The probability $P(N\bar{N})$ of an antineutron developing at time t in an initial pure neutron beam is given by:

$$P(N\bar{N}) = \frac{\varepsilon^2}{\varepsilon^2+\Delta E^2} \sin^2 \{\frac{(\varepsilon^2+\Delta E^2)^{\frac{1}{2}}}{\hbar} t\}$$

where ε is the transition energy between N and $\bar{N}$ states (equals for example 6.6×10^{-21}eV if the oscillation time $\tau_{N\bar{N}} = 10^5$ secs) and ΔE is an externally induced energy perturbation between the initial N and final $\bar{N}$ states. This perturbation can be either of nuclear origin reflecting the different strong interactions of neutrons and antineutrons, or magnetic as a consequence of N and $\bar{N}$ having the same spin angular momentum but magnetic moments of opposite sign. The presence of a finite ΔE strongly suppresses the maximum amplitude of the $\bar{N}$ growth whilst rapidly increasing the oscillation frequency.

There are two extreme situations of environment to consider:-

1. "Free" neutrons $\Delta E = 0$ when $P(N\bar{N}) = (\frac{t}{\tau_{N\bar{N}}})^2$ at small t

and 2. $\Delta E \neq 0$ e.g., inside nuclei when $P(N\bar{N})_{max} = (\frac{\varepsilon}{\Delta E})^2$

and the $\bar{N}$ amplitude oscillates with frequency $\omega = (\frac{\Delta E}{\hbar})$

In practice the condition $\Delta E = 0$ need only be satisfied within the constraints of the uncertainty principle. Thus provided $\Delta E \cdot t < \hbar$

$$P(N\bar{N}) = (t/\tau_{N\bar{N}})^2.$$

The $\bar{N}$ amplitude growth can however be suppressed by external perturbations as $(\Delta E)^{-2}$ or $(\Delta B)^{-2}$. These relations are the basis for observing the phenomenon and verifying the validity of any observed signal.

It is interesting to consider numerically some realistic situations:

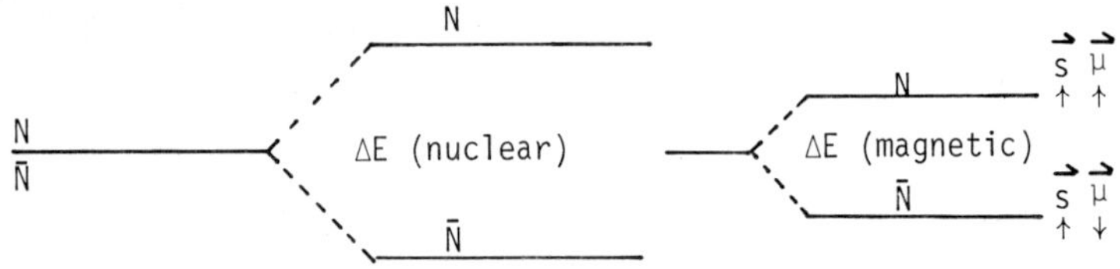

Neutrons "Free"	In Nuclei	In Magnetic Fields (B)
$\Delta E = 0$	$\Delta E \sim 10$ MeV	$\Delta E \simeq 6\times10^{-12}$ eV gauss^{-1}
$P(N\bar{N}) = (t/\tau)$ at small t	$P(N\bar{N})_{max} = (\frac{10^{-21}}{10^{7}})^2 = 10^{-56}$	$P(N\bar{N})_{max} \simeq 10^{-18}$
$\omega = \frac{\varepsilon}{\hbar} = \frac{1}{\tau}$	$\omega = 10^{23}$ sec^{-1}	$\omega = 10^{4}$ sec^{-1} gauss^{-1}

3. Experimental "Limits" from Nuclear Stability

Nuclear matter is remarkably stable. Neutron-Antineutron transitions inside nuclei would contribute to any instability and at this conference there have been very full discussions on the various proposals and experiments to extend the sensitivity in nucleon decay to 10^{32} years or more per nucleon. For the moment however it is conventional to take $\tau_p \gtrsim 3 \times 10^{30}$ years [4] but this limit is decay mode dependent and in particular relies on the observation of stopping muons as secondary products from nucleon decay occurring within nuclei. Clearly the generation of an antineutron within a nucleus will lead to annihilation on a spectator nucleon, the release of 2 GeV of energy and the subsequent break up of the nucleus as the 4 or 5 pions from the annihilation are produced. The question arises as to what is the $\bar{N}$ annihilation probability inside nuclei and what is the likelihood of π^+ mesons escaping from within a nucleus. A Monte-Carlo calculation of this latter effect has been made by Jones [5] based on Gargamelle neutrino data which indicates for single π production from nucleon decay a survival probability of $\simeq 40\%$. Pions from $\bar{N}$ annihilation are all very close to the Δ resonance region and their survival is expected to be significantly less; perhaps a factor of 10 should be allowed for in deducing $N\bar{N}$ oscillation limits from nucleon stability experiments. A conservative approach to the question of the relevant nuclear stability limit for $N\bar{N}$ limits is to take the decay mode independent limit of Bennett [6] who in observations of tracks from spallation reactions in cleaved mica crystals set $\tau_p \gtrsim 10^{27}$ years. The question of the overlap integral (a) to take in calculations with the annihilating $\bar{N}$ and the spectator nucleon is also debatable. Feinberg [2] has suggested $a = 10^{-2}$ in discussions of the process $2p \rightarrow 2e^+$ inside nuclei and in the following table limits are shown for a variety of parameters.

In a simple model $N\bar{N}$ oscillations will be occurring inside nuclei with a frequency of oscillation $\omega = (\Delta E/\hbar)$ and a maximum amplitude of $\bar{N}$ generation $P(N\bar{N}) = \{\varepsilon/\Delta E\}^2$ where ΔE is the difference in nuclear potential seen by the N and $\bar{N}$. These antineutrons will annihilate to give an effective nuclear stability lifetime of "τ_p"

i.e. $\frac{1}{"\tau_p"} = P(N\bar{N}) \times$ Probability of Annihilation

$$= \left(\frac{\varepsilon}{\Delta E}\right)^2 \cdot a \cdot \left(\frac{\Delta E}{\hbar}\right) \text{ with } \varepsilon = \hbar/\tau_{N\bar{N}}$$

$$\therefore \quad \tau_{N\bar{N}} = \left\{\frac{\hbar}{\Delta E} \cdot "\tau_p" \cdot a\right\}^{\frac{1}{2}}$$

"τ_p"	ΔE	a	$\tau_{N\bar{N}}$
10^{27} yrs	1000 MeV	10^{-2}	2×10^4 secs
10^{27} yrs	10 MeV	10^{-2}	2×10^5 secs
10^{30} yrs	10 MeV	1	5×10^7 secs

As can be seen from this table neutron oscillation times as short as 10^5 secs cannot be excluded from nuclear stability data alone.

More detailed and sophisticated calculations using optical models have been made by Sandars [7], Chetyrkin et al [8] and Kazarnovskii et al [9]. These relate the width of decay (Γ) to the $\bar{N}$ potential taken from antiproton X-ray data, $V_{\bar{N}} = U_{\bar{N}} - i\, W_{\bar{N}}$, in a nucleus (A,Z) by:

$$\Gamma = \frac{2\,(\hbar/\tau_{N\bar{N}})^2\, W}{(U_N - U_{\bar{N}})^2 + W_{\bar{N}}^{\,2}}$$

and give limits of 3×10^5 secs to $\tau_{N\bar{N}}$ if "τ_p" $\gtrsim 10^{27}$ years
and 1×10^7 secs to $\tau_{N\bar{N}}$ if "τ_p" $\gtrsim 10^{30}$ years.

The $N\bar{N}$ potential from antiproton X-ray data averages over all nucleons in the nucleus and will not reflect any core repulsion existing within nuclei.

4. Theoretical "Possibilities"

There are as yet no strong theoretical predictions for expecting $N\bar{N}$ oscillations to occur at any particular value. Moreover there are some, as with the well loved SU(5) symmetrical theories which specifically prohibit $\Delta B=2$ amplitudes to first order. There are though a wide range of theoretical ideas which, allowing for intermediate mass scales between 10^2 and 10^{15}GeV, give the possibility of $\Delta B=2$ processes and $N\bar{N}$ oscillations in particular. Pati and

Salam [10] have classified a range of these theories but it was Glashow [11] who initiated much of the recent theoretical activity by his observation that present nuclear stability limits only set $N\bar{N}$ oscillation times at $\gtrsim 10^5$ secs. Glashow also argues on purely dimensional grounds that if $N\bar{N}$ processes were first order in an effective interaction which gave rise to nucleon decay then

$$\tau_{N\bar{N}} \sim \{M/\tau_p\}^{\frac{1}{2}}$$

which equals 10^6 secs if M is taken as a typical hadronic mass (10GeV) and τ_p is 10^{30} years.

More detailed predictions and possibilities have been discussed by Kuo and Love [12], Chang and Chang [13] and in a series of papers by Marshak and Mohopatra [14, 15]. In this latter work a left-right symmetric theory $SU(2)_L \times SU(2)_R \times U(1)$ is constructed which restores parity conservation at an intermediate mass scale and with three Higgs mesons of $\sim 10^4$GeV generates a $\Delta B=2$ process which leads to $N\bar{N}$ oscillations at a level of 10^6 secs.

A full review of the theoretical background to $N\bar{N}$ oscillations has been given at the First Workshop on Grand Unification by Mohapatra [16] and since then by Chang [17].

In conclusion it would seem that experimental limits on nuclear stability only restrict $N\bar{N}$ oscillation times to $\gtrsim 10^5$ secs and that partial unification theories with intermediate mass scales of $\sim 10^5$GeV could accommodate $N\bar{N}$ oscillations at a level of 10^6 to 10^8 secs. The rest of this paper will be devoted to a study of how experiments in practice can investigate the phenomenon to this level of sensitivity.

5. Conceptual Experiments

In section 2 it was shown that an $\bar{N}$ component will develop in an initially pure neutron beam at small t with probability

$$P(N\bar{N}) \simeq (t/\tau_{N\bar{N}})^2$$

where t is the observation time on any neutron and that the neutrons are "free" from external perturbations which would suppress the growth of the $\bar{N}$ amplitude. That is provided $\Delta E \cdot t < \hbar$ the neutron external interaction energy is zero within the constraints of the uncertainty principle and the initial N and final $\bar{N}$ state wave functions are degenerate. The requirements therefore to carry out such an experiment on "free" neutrons are:

1. A source of "free" neutrons
2. A long observation time (t) on each neutron
3. Minimal external energy perturbations (ΔE)
4. A means of detecting a $\bar{N}$ component
5. Shielding against and rejection of background events.

5.1 Neutron Sources

Free neutrons are copiously produced in fission reactors and from high energy spallation reactions. These neutrons are produced with energies of a few MeV but are then slowed down and moderated to form a low density neutron gas. It is conventional to call those neutrons with velocities 2200 metre sec^{-1} (thermal); those with several hundred metre sec^{-1} (cold) and those with velocities less than 7 metre sec^{-1} and which can be totally contained within material bottles (Ultra Cold Neutrons). Typical moderated fluxes (ϕ) of thermal neutrons are:

$\phi \sim 10^{15}$ $n.cm^{-2}$ sec^{-1} from fission reactors
$\phi \sim 10^{13}$ $n.cm^{-2}$ sec^{-1} from spallation reactors.

Reactors are in general steady state devices producing neutrons continuously whilst spallation sources are pulsed with duty cycles 10^{-5} to 10^{-2}. These features can be exploited in the various experiments.

5.2 Long Observation times (t)

With a $\bar{N}$ component developing as t^2 it is clearly advantageous to make t as long as possible. Typical observation times will be

$t \sim 10^{-3}$ sec/metre of flight path - thermal neutrons
$t \sim 10^{-2}$ sec/metre of flight path - cold neutrons
$t \sim 10^{2}$-10^{3} secs for stored ultra cold neutrons (UCN)

with the maximum governed by the β-decay lifetime of free neutrons.

5.3 Minimize External Energy Perturbations (ΔE)

a) Magnetic : $\Delta E = 2\mu\cdot B = 1.2 \times 10^{-11}$ eV $gauss^{-1}$

therefore to prevent suppression of the $\bar{N}$ component we need

$$B \lesssim \frac{\hbar}{\Delta E} = \frac{5 \times 10^{-5}}{t\ (sec)} \text{ gauss}$$

Multilayer mu-metal shields can be constructed [18] to give shielding factors against the residual earths field ($\sim$1 gauss) up to $\sim 10^6$. Thus in principle it is possible to allow $\bar{N}$ components to develop over times up to $\sim$100 secs. However magnetic shields become very bulky and costly at this level, aside from the general problem of external perturbations induced if UCN storage methods are used to enhance the observation time. This latter point will be discussed in section 4.3 b).

Multilayer cylindrical magnetic shields give a shielding factor

$$S_T = S_T^{1} + S_T^{2} + S_T^{3} + S_T^{1} * S_T^{2} * S_T^{3} * \{1-(\frac{R_2}{R_1})^2\}\{1-(\frac{R_3}{R_1})^2\}$$

where S_T^{i} are the individual shielding factors for cylinders of radius R_i and a 3 layer example has been given

$S_T^{i} = \mu t/2R_i$ where t is the thickness of mu-metal

$\simeq 20$ for practical shields 1 metre in diameter and 1mm thick.

Magnetic shielding is not then difficult for neutron observation times of 10^{-2} to 10^{-3} sec where residual fields $\lesssim 10^{-3}$ gauss are needed.

b) Nuclear: Within nuclei energy perturbations to neutrons of around 10MeV are to be found. As this limits observation times (t) to $\lesssim 10^{-22}$ secs they have to be avoided at all costs. This means consideration must be given to vacuum conditions and to the penetration of neutrons into nuclear matter, as for example during neutron reflection from walls. For UCN this penetration in the walls of the containment bottle is $\sim 1000A^{0}$ and a potential of $\sim 10^{-7}$eV is seen; such an energy will strongly suppress the growth of the $\bar{N}$ component unless the real parts of the N and $\bar{N}$ scattering potential are equal. In general this will not be so and the growth of $\bar{N}$ will be limited to the time between bounces (typically 10^{-1} sec) rather than the many hundred seconds for total storage. In addition the imaginary part of the $\bar{N}$ scattering potential will rapidly absorb the $\bar{N}$ component and limit the number of available bounces. For these reasons the use of UCN do not appear to be suitable for studying $N\bar{N}$ oscillations in spite of their obvious attractive feature of long observation times.

In a similar way the use of polished guides to contain and transport neutrons by reflection again inhibits the growth of an $\bar{N}$ component.

5.4 Detection of $\bar{N}$ Component

Antineutrons annihilate rapidly on other nucleons with a cross-section (σ) dependent inversely on their velocity (β) and which can be parameterised as

$$\sigma \simeq \frac{33}{\beta} \text{ mb.}$$

This is a good approximation to existing data at energies $\gtrsim$ 1 MeV and predicts $\sigma = 10^5$ barns for a velocity of 100 metre sec^{-1}.

In the annihilation process nearly 2GeV of energy is released in the form of pions with energies around 300MeV. Typically [19] on nuclei

$$\bar{N}A \rightarrow (2.8 \pm 0.1)\pi^{\pm}$$
$$\rightarrow (2.1 \pm 0.1) \text{ protons}$$
$$\rightarrow (1.8 \pm 0.1)\pi^{0}$$

A detector of antineutrons therefore needs

a) to collect as much of the 2 GeV annihilation energy as possible

b) to identify multiplicity as a characteristic signature

c) to identify nuclear fragments as a characteristic signature

d) to identify hadrons as a characteristic signature.

An important point to stress again is that a unique identifier of a coherent build up of an $\bar{N}$ component will be its suppression in the presence of an externally applied magnetic field. Hence running the detection system with a finite magnetic field allows an unambiguous measurement of the background signal to be made; a feature unique to this type of experiment.

5.5 Shielding against Background

Background events in the $\bar{N}$ detection system will come from cosmic rays or from the neutron source. These events can be measured by "switching off" the $\bar{N}$ coherent build up through a magnetic field perturbation but it is never-the-less of great importance to suppress the background as much as possible. In the limit of no background the experimental sensitivity to $\tau_{N\bar{N}}$ goes as $\{\text{No events}\}^{-\frac{1}{2}}$ but when statistical considerations of background levels have to be considered it goes as $\{\text{No events}\}^{-\frac{1}{4}}$.

The shielding will need to be a combination of passive and active scintillator layers to reject cosmic ray events and a means to ensure minimum radiation levels from the intense flux of neutrons which pass close to the $\bar{N}$ detector. In addition any active cosmic ray veto shield will itself need to be shielded from self vetoing of true events from the annihilation products.

5.6 Experiment Types

Three experiments may be considered: in core; stored UCN and beam line experiments. They may be characterised by neutron density (ρ_N) and sensitivity $\tau_{N\bar{N}}$ in a time T_{expt}.

$$\rho_N = \frac{\phi_N(\text{flux})}{V_N\ (\text{velocity})} = \frac{\phi_c\ (\text{capture flux})}{2200\ \text{m.sec}^{-1}}$$

$$\tau_{N\bar{N}} = t\ \{\phi_N\ T_{expt}\}^{\frac{1}{2}}\ \frac{1}{\sqrt{N_{N\bar{N}}}}$$

Type 1:	In Core:	densities high	$\rho_N \sim 10^9$-10^{10} N.cm^{-3}
		observation time low	$t \sim 10^{-5}$ sec
		sensitivity	$t\sqrt{N}$ day^{-1} $\sim 10^7$ sec

Major difficulties with this type of experiment are that $\bar{N}$ detection probabilities are small and that $\bar{N}$ coherent build up is suppressed unless voids are created within the moderating core.

Type 2:	UCN:	densities low	$\rho_N \sim .1$ - 50 N.cm^{-3}
		observation time long	$t \sim 100$ - 1000 secs
		sensitivity	$t\sqrt{N}$ day^{-1} $\sim 10^5$ to 10^7 secs

The $\bar{N}$ coherent build up is restricted to single bounces (section 4.4) which drastically reduces the sensitivity unless clever [20] but probably impracticable schemes are adopted to compensate for the differing phases between N and $\bar{N}$ scattering

Type 3:	Beam Line:	densities	$\rho_N \sim 10^3$ - 10^5 N.cm^{-3}
		observation time	$t \sim 10^{-3}$ - 10^{-2} sec
		sensitivity	$t\sqrt{N}$.day^{-1} $\sim 10^5$ to 10^7 secs

The neutron beam lines can either be derived from pulsed sources (Spallation Neutron Sources) in which cosmic ray background rejection is good whilst source related backgrounds can be high or uniform from fission reactor. From reactors cold neutrons with short flight path or thermal beams with longer flight paths can both yield similar sensitivities.

All the experiments now planned come into this latter category and will now be considered in more detail.

6. "Real" Experiments

6.1 CERN-ILL-Padua-Rutherford-Sussex Collaboration [21]

This experiment is now running at the Institut-Laue-Langevin (ILL) in Grenoble, France. The Institute, a joint British-French-German venture, has as its major facility a 53MW Research Reactor. The neutron flux in the core $\phi \sim 1.5 \times 10^{15}$ n.cm^{-2}.sec^{-1} feeds many curved neutron beam guides. One of these (H18) from the liquid deuterium cold source has a flux $\phi_N = 10^9$ N.sec^{-1} over 20 cm x 3 cm with a mean velocity of $\simeq$ 200 metre sec^{-1} and is used to feed this experiment.

The experimental apparatus is shown in figures 1 and 2. Neutrons from the H18 guide are transported under high vacuum through a "field free" region generated by a triple layer mu-metal shield covering 6 metres in length, of which 4 metres are useful for the coherent build of an $\bar{N}$ component. The residual magnetic field is 10^{-4} - 10^{-3} gauss. At the end of the flight path the neutrons are dumped in a Lithium Fluoride (Li^6F) and Boron Carbide (B_4C) beam stop. The high neutron absorbtion cross-section ($n+Li^6 \rightarrow t+\alpha$) ensures a rapid neutron attenuation whilst any $\bar{N}$ component which might exist will annihilate into pions plus nuclear fragments.

The $\bar{N}$ annihilation detector is essentially a calorimeter and measures the total energy deposited in a 790 Kgm lead-scintillator sandwich. This calorimeter is 10% active and subtends a solid angle to $\bar{N}$ annihilation products of 25%. Preliminary calibration of the detector has been made in an antiproton beam at CERN and it is intended to calibrate the full system at the end of data taking in 1982. The calorimeter is triggered by two trigger planes (P_B and P_S) between it and the $\bar{N}$ annihilation region. The whole assembly is surrounded by a cosmic ray veto shield (A) of plastic scintillator, shielded from self vetoing of annihilation products by 20 cm of iron. The energy deposited into the calorimeter (CENT) is recorded every time a coincidence $CENT.P_B.P_S.\bar{A}$ is recorded. The data acquisition rate is $(2.40 \pm .02)10^{-2}$ sec^{-1} and a typical energy spectrum is shown in figure 3. About 8% of these events (150 day^{-1}) have energies $\gtrsim$25MeV and require closer inspection off-line. This off-line analysis of event topology is now proceeding and to date (April 1980) some 30 days of data have been taken; an equal mixture of magnetic field 'on' and field 'off' data. In addition reactor off data is taken to study the cosmic ray background alone. Running with neutrons will continue until the end of 1981 when more than 100 days of data will have been accumulated.

The sensitivity of the experiment, in the absence of background, $t\sqrt{N.day^{-1}}$ is 3×10^5 secs.

For the future this experiment could improve its identification criteria for $\bar{N}$ events through the provision of a multiplicity detector and greater positional information before the calorimeter. In addition it might be possible to improve the cosmic ray veto inefficiency from the present level of 5×10^{-4} to 10^{-5} or 10^{-6} by the addition of extra

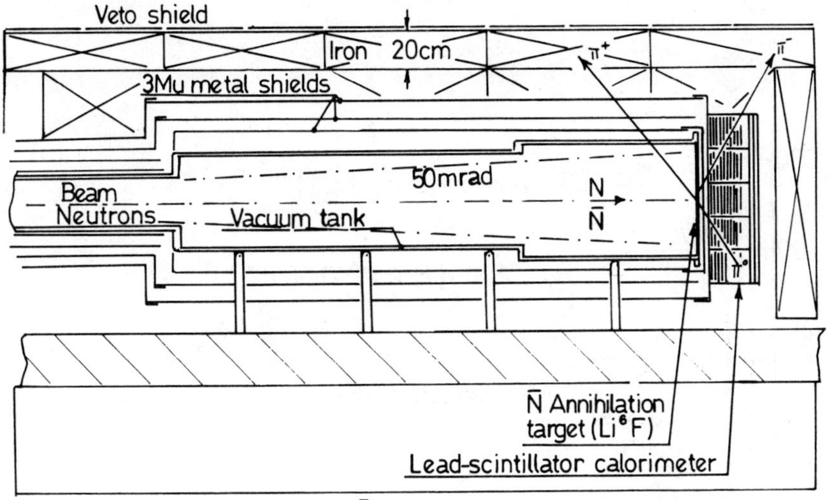

Figure 1. The layout of the $N\bar{N}$ experiment by the CERN-ILL-Padua-Rutherford-Sussex at the ILL, Grenoble.

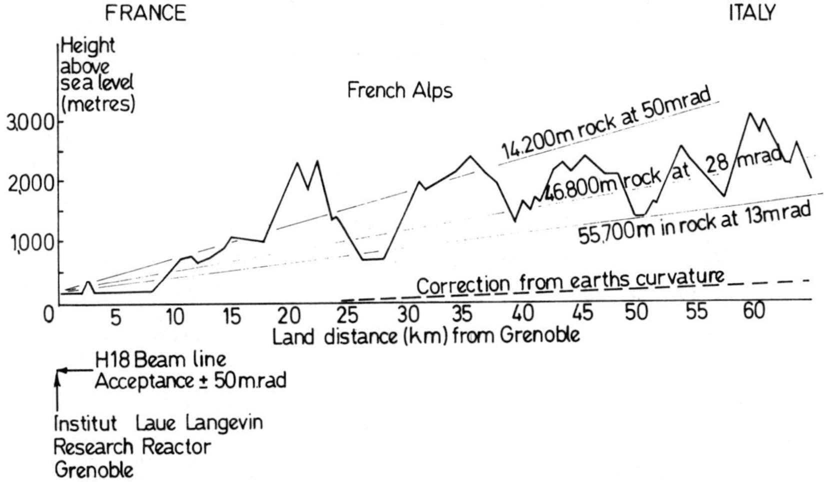

Figure 2. The ground profile in the direction of the H18 neutron beam line at ILL. Any cosmic ray muons passing along the neutron guide have to pass through at least 14,000 metres of rock before reaching the detector. Atmospherically produced muons therefore have an energy cut off in excess of 6000GeV and a flux less than 10^{-9} sec^{-1} in this direction.

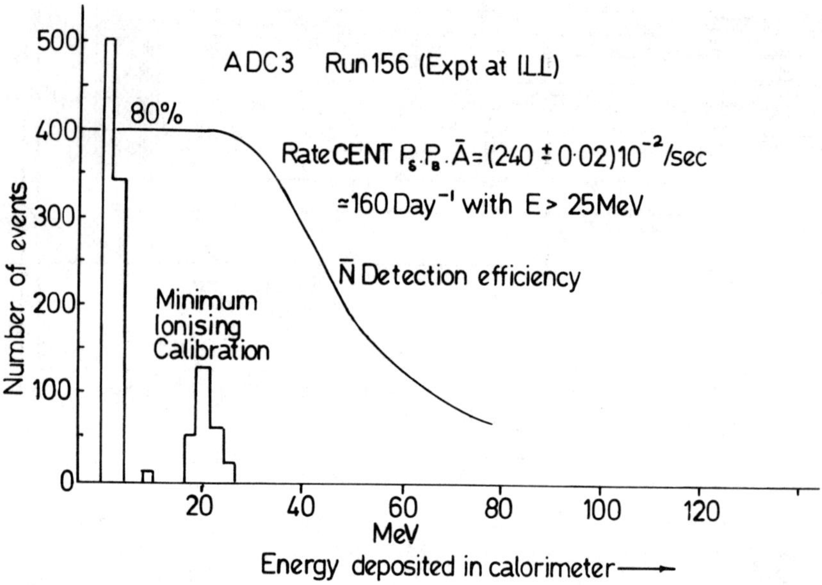

Fig. 3 A sample of data taken at ILL with the apparatus shown in figure 1. About 150 events day^{-1} have an energy deposition $\gtrsim$ 25MeV in the calorimeter and if no further analysis is done this sets a limit of $\sim 10^5$ secs to the N$\bar{N}$ oscillation time.

passive and active veto shields. Neither of these changes would improve the intrinsic sensitivity but could materially help in distinguishing true $\bar{N}$ events from cosmic ray induced backgrounds. Increased sensitivity will only come from an increase in flux and this will require moving closer to the source, either the ILL reactor or a pulsed spallation source and accepting faster neutrons by looking directly at the source. Direct line of sight to the source will dramatically worsen the source associated background and for the moment we choose to remain in the very clean environment of the curved, cold H18 neutron guide at ILL.

Some or all of these suggestions are incorporated in the various proposals now being made and I will now discuss in detail a proposal for an experiment at Oak Ridge which incorporates many of these improvements.

6.2 Harvard-Oak Ridge-Tennessee Proposal [22]

This proposal is for an experiment to be carried out at the 30MW reactor at Oak Ridge. The so called "Goat Hole Facility" is proposed with a 42" diameter vacuum tube looking directly to the reactor core over a 42cm hole where a core flux $\phi \sim 1.5 \times 10^{14}$ n.cm^{-2} sec^{-1} exists. Only 1 in 10^4 of the incident neutrons will survive the free flight of 20 metres when they will be dumped in a stop of Li^6D. Three metres before this dump a 25μ CD_2 foil will intercept 4×10^{13} neutrons per second in the beam and annihilate any $\bar{N}$ component which may have coherently developed over the 20 metres of magnetically shielded flight. A cross-section of the proposed apparatus is shown in figure 4 and the detection apparatus for identifying $\bar{N}$ annihilations in figure 5. The detector covering 60% of the annihilation solid angle is 60% active and weighs 3.022 tonne; segmented lead glass and lead scintillator sandwiches are proposed to measure the multiplicity and record the energy deposited from each event.

An elaborate cosmic ray veto shield is proposed comprising a 2 layer active plastic scintillator shield and a passive shield of 1.5 metre concrete plus 1 metre of iron. The dominant background in this experiment is seen as wide angle bremstrahlung from cosmic ray muons and this shield is necessary to reach the planned sensitivity t√N day^{-1} of 1×10^7 secs.

The proposal is complete and the experiment awaits funding.

6.3 Pulsed Source Experiments

In any experiment where the steady state cosmic rays constitute a

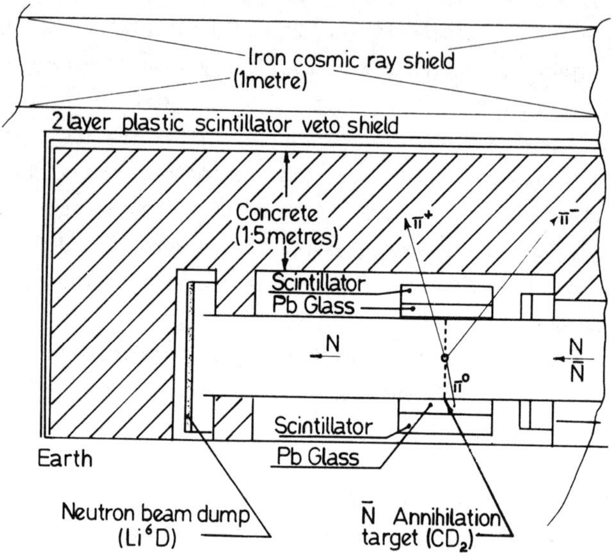

Fig. 4 A second generation experimental proposal from the Harvard-Oak Ridge-Tennessee collaboration. In this layout the $N\bar{N}$ annihilation target of a CD_2 foil is well separated from the Li^6D neutron beam dump; the detector which surrounds the annihilation region is segmented and shielded from cosmic rays by a double layer of plastic scintillation counter and extensive passive shields of concrete and iron.

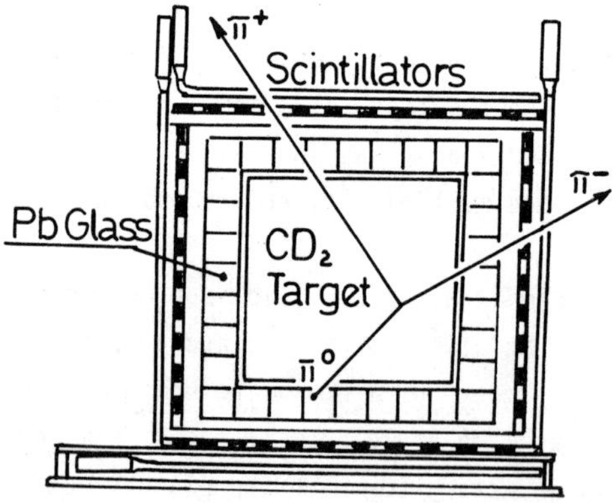

Fig. 5 The segmented detector from the layout of figure 4. The neutrons and any antineutrons enter perpendicular to the paper. Multiplicity as well as energy deposition are used as criteria for $\bar{N}$ identification.

background problem it is clearly advantageous to pulse the source and reap the advantage as the ratio of duty cycles. Pulsed neutron sources have this advantage and can be either of the pulsed reactor type or spallation type (SNS) where short intense pulses of medium energy protons impinging on heavy nuclei "boil off" neutrons at energies around an MeV. (For example, the SNS at Rutherford will have 200μA of proton current in bunches of 200nsec at 50Hz, a duty cycle of 10^{-5}). The full duty cycle of the proton machine cannot be exploited as in the process of moderating the neutrons to thermal energies or below the pulse width broadens typically as $10\lambda(A^{o})$ μsec. For this reason cold neutrons ($\lambda \gtrsim 5A^{o}$) essentially lose the pulse structure of the machine but thermal neutrons with a wavelength acceptance of 20% will allow a 5% duty cycle to be obtained with fluxes $\phi \sim 10^{8}$n.cm^{-2}.sec^{-1}. This 20x reduction in effective cosmic ray background is clearly of advantage and a proposal along these lines has been made at LAMPF [23]. This proposal from the LAMPF beam dump claims a sensitivity to $\tau_{N\bar{N}}$ of 1 x 10^{5} secs. day^{-1}.

A proposal to pulse a research reactor in Pavia, Italy [24] has also been made and a sensitivity of 3 x 10^{5} secs day^{-1} is expected with a 20 metre flight path. The reactor, at 200KW power, can be pulsed 10 times day^{-1} with a 10^{3} gain in power and a 30 msec pulse length.

6.4 Other Experiments

A detailed proposal has been made by Anderson [25] to use the Omega West Reactor in Los Alamos. This experiment, as with the Oak Ridge proposal, looks at the reactor core and transports $\sim 10^{13}$ neutrons sec^{-1} over a 10 metre magnetically shielded flight path. The sensitivity t√N day^{-1} will be 2 x 10^{6} secs.

Proposals have also been made to use the KEK pulsed neutron facility in Japan. One [26] suggests the use of guided cold neutrons as a means of limiting lateral dimensions to the experimental apparatus and another [20] proposed an UCN experiment. Both however suffer, as discussed earlier, from problems of the loss of $\bar{N}$ coherence as neutron reflection occurs from a nuclear medium.

The Pavia group in collaboration with the National Bureau of Standards propose also an extension to their pulsed reactor proposal in which using the NBS reactor in Washington and a 100 metre flight path an ultimate sensitivity of 10^{8} secs could be reached.

7. Backgrounds

In all the experiments discussed the cosmic ray background is the primary problem. Although the dominant component in cosmic rays at sea level is the muon component (Fig. 6) it is the hadronic component (Fig. 7) which can cause the greater problems in experiments of this type. The integral muon flux is $\sim 10^{-2}$ cm^{-2} sec^{-1} st^{-1} whilst the integral nucleon component above 2 GeV, the threshold for pion production, is 1.6×10^{-5} cm^{-2} sec^{-1} st^{-1}. The secondary neutrino flux in cosmic rays is such that 0.12 events $year^{-1}$ occur in a detector of 1 tonne mass with visible energy greater > 1GeV. Such neutrino interactions could simulate $\bar{N}$ interactions but are sufficiently rare to be neglected. The background signals in the experiment at ILL and the proposal at Oak Ridge are summarized below:

	Experiment at ILL	Proposal to Oak Ridge
Veto Counter Inefficiency	5×10^{-4}	10^{-5} to 10^{-6}
Cosmic Ray Neutrons	10^{-4} sec^{-1}	2×10^{-8} sec^{-1}
Cosmic Ray Neutrinos	0.1 yr^{-1}	0.4 yr^{-1}
Reactor Neutrinos	100 yr^{-1} but E < 10MeV	
Cosmic Ray Muons		
Flux (cm^{-2} sec^{-1} st^{-1})	10^{-2}	5×10^{-3}
Energy cut-off	0.6GeV	1.8GeV
Penetration to detector	19 Sec^{-1}	1.8 sec^{-1}

In circumstances when the hadronic cosmic component is removed to acceptable levels by attenuation through passive shielding then the inherent inefficiency of charged particle active shields becomes important and rare processes like wide angle bremstrahlung from high energy muons assumes a dominant role. Detailed design of shielding is then very important.

8. Summary

Neutron-antineutron oscillations would be "weird physics" as described by S Weinberg at this conference. Never-the-less the

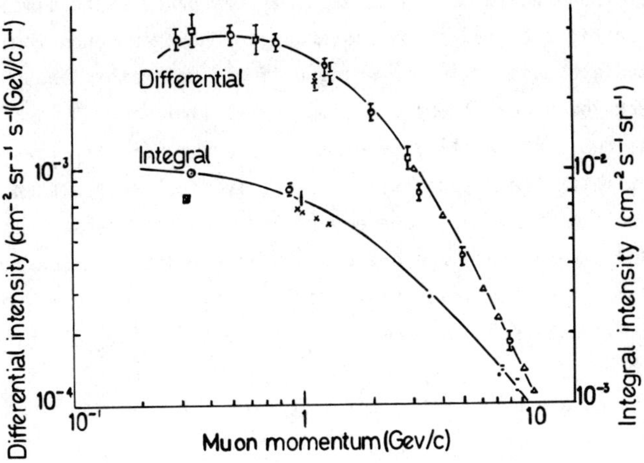

Fig. 6 The differential and integral cosmic ray muon spectrum at sea level [27].

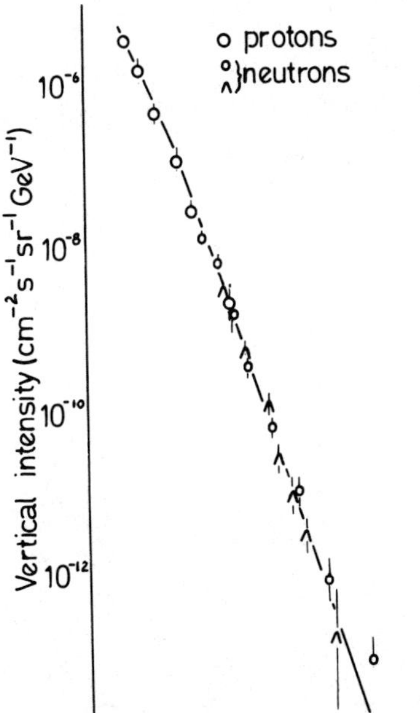

Fig. 7 The differential cosmic ray nucleon spectrum at sea level which can be parameterised as $N(E)dE = a\ e^{-b}\ dE$ with $a = 1.6 \times 10^{-4}\ cm^{-2}\ sec^{-1}\ GeV^{-1}\ st^{-1}$ and the exponent $b = (2.95 \pm 0.10)$, [27].

rewards if found would be high even if in the process the simpler ideas of Grand Unification had to be discarded. The phenomenon would imply an intermediate mass scale; opening up for accelerators above 100GeV the prospect of a desert which blooms. The limits on $N\bar{N}$ oscillation times from nuclear stability experiments are sufficiently low that experiments with free neutrons with sensitivities greater than 10^5 secs are well worthwhile. One experiment with this sensitivity is now running and several more are being proposed. During the next few years it should be possible to extend this sensitivity to 10^7 or 10^8 secs; we all await this outcome with interest.

In the accompanying table the status of the experiments and proposals is reviewed as of April 1981.

Table 1
Status of $N\bar{N}$ Oscillation Experiments
(April 1981)

	Facility	Status	Neutrons	t (sec)	ϕ_N n sec^{-1}	Duty Cycle	$\tau_{N\bar{N}}$ Sensitivity secs day^{-1}
CERN/ILL/Padua /Rutherford/Sussex	ILL 53MW Reactor	Data Taking	Cold Beam	3×10^{-2} (4.5 metres)	1×10^{9}	1.0	3×10^{5}
Harvard/Oak Ridge /Tennessee	ORR 30MW Reactor	Proposal	Thermal	8×10^{-3} (20 metres)	4×10^{13}	1.0	1×10^{7}
Los Alamos 1. Anderson	LASL 8MW Reactor	Proposal	Thermal	4×10^{-3} (10 metres)	1×10^{13}	1.0	2×10^{6}
2. Proposal No 647	LAMPF beam dump	Proposal	Pulsed	-	-	10^{-1}	1×10^{5}
Pavia/NBS 1.	LENA Pulsed Reactor	Proposal	Thermal	1×10^{-2} (20 metres)	3×10^{5}	3.5×10^{-6}	3×10^{5}
2.	NBS Reactor	Proposal	Thermal	1×10^{-2} (20 metres)	-	1.0	10^{7}

Table 2
Prospects for Future

Nuclear Stability Experiments	"τ_p" years	$\tau_{N\bar{N}}$/day	Free Neutron Experiments
Decay Mode Independent Limit		10^5 secs	← LAMPF (? 1982 +)
			← ILL (1981)
	10^{28}		← LENA (? 1982 +)
		10^6 secs	← LASL Ω West (? 1983 +)
Irvine Expt.	10^{30}		
Penn. Expt.			
Kolar Gold Field			← Pavia/NBS (? 1984 +)
		10^7 secs	← ORR (? 1982 +)
	10^{32}		
IMB, HPW, Soudan, Frejus) Proposals		10^8 secs	← Possible limit for $N\bar{N}$ Experiments

References

[1] P K Kabir, The C P Puzzle. (Academic Press. New York, 1967)

[2] G Feinberg & S Weinberg. Phys. Rev. 123 (1961) 1439

[3] G Feinberg, M Goldhaber & G Steigman - Phys. Rev. D18 (1978) 1602

[4] J Learned et al. Phys. Rev. Lett. 43 (1979) 907

[5] T Jones. Irvine-Michigan-Brookhaven Experiment - private communication

[6] C Bennet. This conference

[7] P G H Sandars. J. Phys. G: 6, L161 (1980)

[8] K G Chetyrkin et al. Phys. Lett. 99B (1981) 358

[9] M V Kazarnovskii et al. JETP Lett. 32, 82 (1980)

[10] J C Pati, A Salam & J Strathdee. Probings through Proton decay and $N\bar{N}$ oscillations IC/80/183 & J C Pati - this conference

[11] S L Glashow. Review talks on particle physics HUTP79/A029; HUTP79/A040; HUTP79/A059

[12] T K Kuo and S T Love. Phys. Rev. Lett. 45 (1980) 93

[13] L N Chang & N P Chang - Phys. Lett. 92B (1980) 103, errata 94B (1980) 551

[14] R N Marshak & R E Mohapatra. Phys. Rev. Lett. 441 (1980) 1316

[15] R N Marshak & R E Mohapatra. Phys. Lett. 94B (1980) 193

[16] R N Marshak & R E Mohapatra. Proc. First Workshop on Grand Unification, Durham, New Hampshire (1980) CCNY HEP-80/8

[17] N P Chang - VPI Workshop on "Weak Interactions as Probes of Unification", Blacksburg, VA 1980)

[18] edm experiment at ILL (Harvard-ILL-Rutherford-Sussex Collaboration)

[19] H J Besch et al. Z. Phys. A292 197 (1979)

[20] H Yoshiki. KEK Proposal 80-10 (1980)

[21] M Baldo-Ceolin, G Fidecaro, K Green. ILL Research Proposal (3-05-027); (CERN-ILL-Padua-Rutherford-Sussex Collaboration); (1980)

[22] R Wilson. ORR Research Proposal (Harvard-Oak Ridge-Tennersee Collaboration); (1981)

[23] R Ellis, J Moses, A Taylor. LAMPF Proposal No 647 (1980)

[24] T Pinelli, S P Ratti and P Trower. INFN Reactor Proposal at Pavia (1980)

[25] H Anderson. LASL Proposal (1981)

[26] T Kamae. KEK Proposal (1980)

[27] Cosmic Rays at ground level, Institute of Physics (1973) ed. by A W Wolfendale.

LIMITATIONS ON PROTON DECAY MODES FROM A PASSIVE DETECTION SCHEME

C. L. Bennett

Department of Physics, Princeton University
Princeton, New Jersey 08540

Passive proton decay experiments are based on the observation that a kiloton of material after one year will have experienced as many proton (or bound neutron) decays as a single gram of a billion year old piece of matter. The observations I will discuss are based on the following. First, following a proton decay event within a nucleus, such as ^{16}O or ^{56}Fe, a spallation of that nucleus is fairly probable. The size of the probability depends somewhat on the nucleon decay products, e.g. a pion will scatter or react approximately half the time in ^{16}O. However, even if nucleon decays were into all neutrinos, the "hole" left in the daughter nucleus would leave it with an excitation energy of up to about 50 MeV. This highly excited nucleus would then emit protons, neutrons and alphas until only a stable particle fragment was left. Second, it is known that the heavy recoiling fragments from such spallation reactions are highly ionizing enough to cause damage in solids such as mica, and indeed short (1 to 2 μ) spallation recoil tracks have been observed.[1]

To determine the total length of the tracks, samples of mica, mostly muscovite, were etched in 48% HF at 19°C with a cleavage plane propped open with a flake of mica. A relatively long etching time (2 to 3 hours) was used to produce tracks with a diameter of about 5μ in order to clearly reveal short tracks. Figure 1 shows some typical tracks as seen with a scanning electron microscope and with an optical microscope. A spontaneous fission track is shown in Fig. 1a, the rest are short tracks. The diamond shape of the tracks cross section is the result of preferential etching for certain directions in the crystal. Using interference contrast microscopy, with the optic axes of the mica turned 45° from extinction, the shallow pits of short tracks show up very readily and display a distinctive "blinking" from bright to dark as the focal plane passes through them. This is illustrated in Figure 1c (a pion induced spallation track) and 1d (a short fossil track).

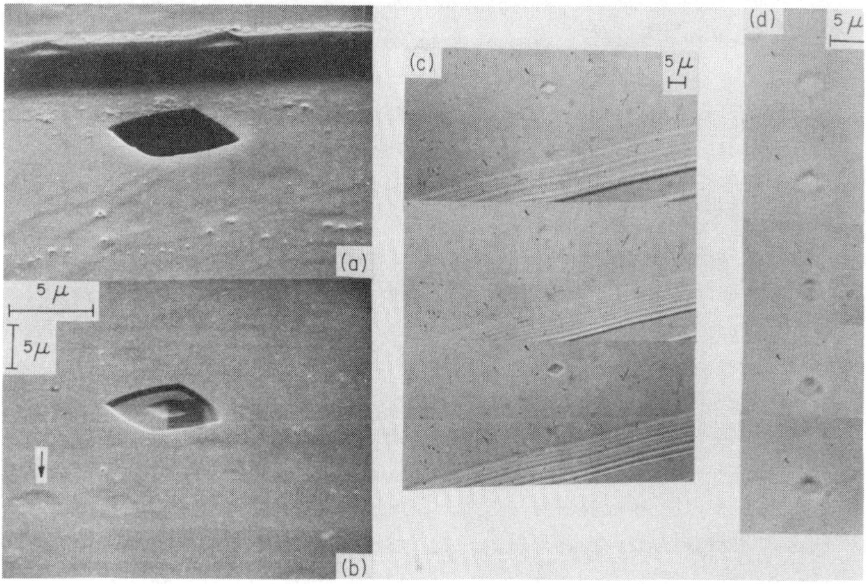

FIGURE 1

Several sources of background exist which also produce short tracks. Among these are the familiar events due to cosmic ray muons and neutrinos common to all proton decay experiments, but these may be attenuated by obtaining samples from sufficiently deep sources. For samples originating from about 2 miles depth or deeper essentially only the neutrino background remains significant. The expected track density resulting from these sources has been plotted as a function of depth in Figure 2. (Neutrino flux and cross sections are from ref. 2 and 3.)

For solid state tracks a far more serious background exists. This background is the result of partial annealling of spontaneous fission tracks from the natural ^{238}U in the minerals. Spontaneous fission tracks are normally 20μ long, however, thay can anneal at several intermediate locations along their length, creating unetchable blockage. In this way, a given surface density of fission tracks after partial annealling produces essentially the same surface density of short tracks but yields a volume density of spurious short tracks much larger than the original volume density of fission events. Since some annealling is inevitable, if for no other reason than that the crystals were certainly hot immediately after their formation, the density of short tracks can only be used to place a limit on the rate for possible proton decay modes.

To minimize these backgrounds, samples with the lowest possible uranium concentration, and coolest thermal history are needed. Since the uranium concentration is normally determined in the process of fission track dating, and partial annealling has the effect of shortening the average fission track length, the extent of the background to spallation recoil tracks can be at least qualitatively assessed.

Some results from two particular samples of mica are listed in the table. Considering that these uranium concentrations are considerably lower than the average of about 3 ppm and that deep samples will experience even more heating than surface samples in general, it will be hard to get a much better limit on the proton lifetime from nuclear tracks than that obtained from these data. From these data the short track producing decay rate of nucleons can be said to be less than about 5×10^{-28} yr^{-1}. It may be that the most useful feature of this result is to place a constraint on the decay rate into all neutrinos, but it is certainly far from competing with the "live" experiments that have already been done.

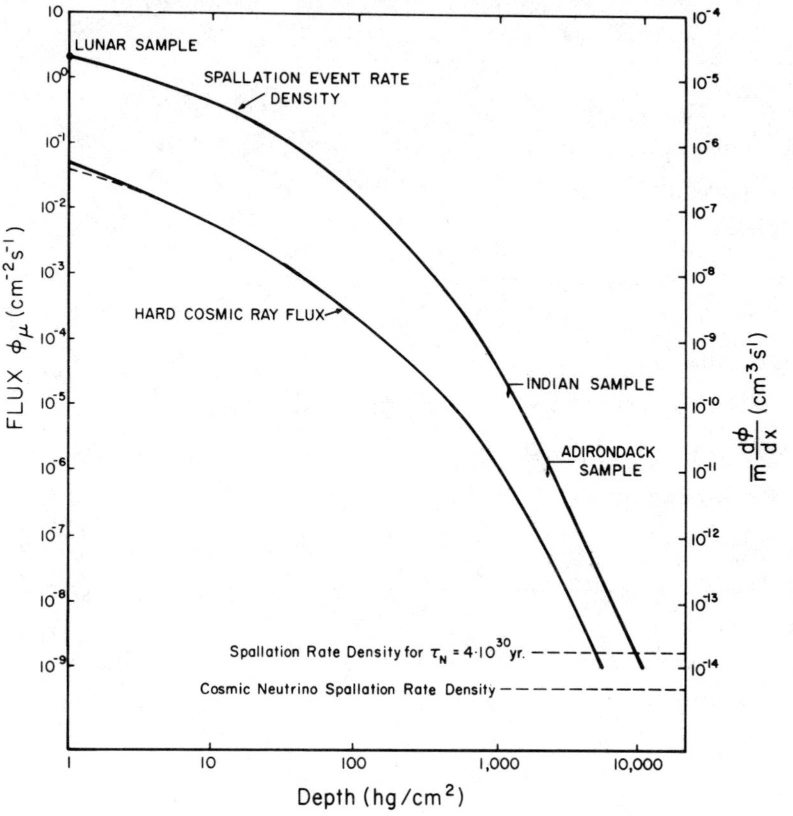
LUNAR SAMPLE
SPALLATION EVENT RATE DENSITY
HARD COSMIC RAY FLUX
INDIAN SAMPLE
ADIRONDACK SAMPLE
Spallation Rate Density for $\tau_N = 4\cdot10^{30}$ yr.
Cosmic Neutrino Spallation Rate Density
FLUX ϕ_μ (cm^{-2}s^{-1})
$\frac{d\phi}{m\,dx}$ (cm^{-3}s^{-1})
Depth (hg/cm^2)
1
10
100
1,000
10,000

Figure 2

TABLE I - TRACK DENSITIES IN VARIOUS SAMPLES

	Adirondack Biotite[b]	Indian Muscovite[c]
Induced Fission Track[a] Surface Density (cm^{-2})	$2.3 \pm 0.1 \cdot 10^5$	$1.15 \pm 0.05 \cdot 10^4$
Induced Fission Track Length (μ)	20	20
Spontaneous Fission Track Density (cm^{-2})	$3.5 \pm 0.5 \cdot 10^4$	900 ± 70
Spontaneous Fission Track Length (μ)	12.5 ± 3.7	15 ± 3.5
Spallation Track Density (cm^{-2})	150 ± 50	1100 ± 160
Uranium Concentration 10^{-9}g/g	51	2.6
Deduced Sample Age 10^9 yr	1.7 ± 0.6	0.8 ± 0.2

a) Total thermal neutron dose: $1.6 \cdot 10^{17}$ cm^{-2}

b) Sample collected by Prof. R. Hargraves at the site of the Benson open pit iron mine, Star Lake, NY.

c) Samples obtained from mica mines of the Nellore district, India, present mine depths <100 m.

References

1. R.L. Fleischer, P.B. Price and R.M. Walker, Nuclear Tracks in Solids, U. of C. Press, Berkeley, Calif. (1975).

2. J.L. Osborne, S.S. Said and A.W. Wolfendale, Proc. Phys. Soc. (London) 86, 93 (1965).

3. Particle Data Group, Phys. Lett. 84B, 208 (1979).

ULTIMATE LIMITS TO PROTON DECAY SEARCHES

John G. Learned
Department of Physics and Astronomy, University of Hawaii at Manoa
Honolulu, Hawaii

ABSTRACT

Proton decay searches become increasingly difficult above lifetimes of 10^{33} years. It is concluded that in the forseeable future no experiment will surpass $\sim 10^{34}$ years. If proton decay is not observed by the current round of experiments, then the approach suggested herein, of a deep ocean experiment, will permit achieving a lifetime limit near 10^{34} years. Utilizing a relatively dense DUMAND-like deep ocean detector of $(100\ m)^3$ size appears to be the only economically feasible approach for detectors containing $\gtrsim 10^{35}$ nucleons.

1. Introduction

Currently there are at least five dedicated proton decay searches in operation or nearing readiness throughout the world [1-6]. Several more are proposed or in the design stages [7]. The largest of these [2] is a water Cerenkov detector and will have a fiducial volume of $\sim 2.3 \times 10^{33}$ nucleons. Most of the other detectors [1,4,5,7] are fine-grained tracking detectors of density substantially greater than water, and less mass (6×10^{30} to 4×10^{32} nucleons). As indicated in the report at this workshop by the Japan-India collaboration working in the Kolar Gold Fields (KGF) [1], there are limits which could mean that the proton lifetime is in the range of 5×10^{30} years. If these limits become solid observations (and they will be verified or not within the next year), then fine-grained detectors will take the field to study details of branching ratios, muon polarization, etc. There is an implicit faith expressed by the various experimental groups that the theoretically predicted lifetime of $10^{31\pm1}$ years will be borne out, and that we can get on to the business of resolving details of grand unification. We wish to consider herein the plausible scenario whereby proton decay is not observed in the first round of experiments and that the lifetime limit is pushed during the next few years to about 10^{33} years. What then?

2. Limitations Upon Detector Size

There is no physical limit to how big we might make future detectors; it is, as usual, ultimately a matter of economics. We will not take up space here discussing alternatives in detail but simply assert that:

a) Passive techniques such as geological and radio-chemical searches are limited to lifetimes below present limits ($\sim 10^{30}$ years [8]).

b) Building a detector in space to avoid the terrestrial neutrinos would gain by $\sim 10^3$ in background suppression, if one could shield against the primary cosmic rays. This forces the detector to have shielding such as could be achieved by burying the detector under 10 m or more of material on the moon. Even so one must have an enormous detector of at least 10^5 tons to make progress. The cost of such an endeavor would be, for the next few decades, inconceivably large (billions of dollars), requiring a major lunar laboratory.

c) Tracking detectors are inherently inefficient because they require many channels that are infrequently utilized in a proton decay observation (as opposed to water Cerenkov detectors where individual photodetectors view a large fraction of the event volume, and perhaps 10% of the detector elements are involved in recording a decay). Moreover, the scaling laws for the size versus cost of tracking chambers don't offer much, if any, economy of scale beyond $\sim$1000 tons. Cleverness can help somewhat, but one is up against a tyranny of numbers of channels (which in current proposals is approaching 10^6)! Another way to look at this limitation is that manufactured goods (e.g., steel) cost of the order of \$2/lb., which is equivalent to \$4M/1000T. It will require great trickery to reduce this by a factor of 10 in a tracking chamber. Moreover, the possible large number manufacturing economies, which set in when savings due to automated assembly offset the tremendous tooling costs, typically only take place for manufacturing runs in $\gtrsim 10^6$ elements, and with startup costs in the many millions of dollars. As indicated in figure 1, a crude estimate of cost scaling would have, at best, a 10^5 ton tracking detector approaching the total U.S. high-energy physics budget.

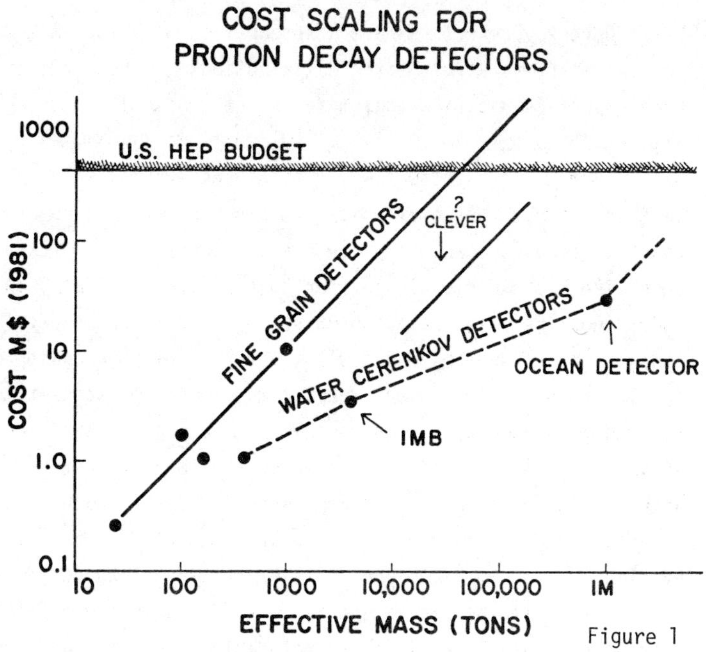

Figure 1

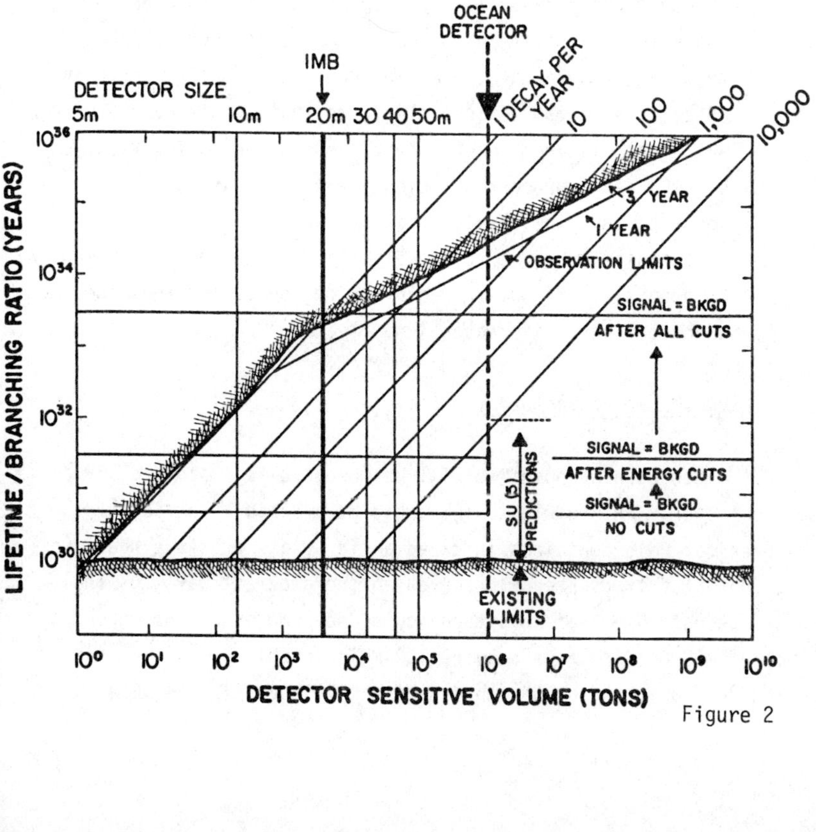

Figure 2

d) Water Cerenkov detectors have a somewhat complex scaling. For large arrays a volumetric distribution of tubes (rather than a surface distribution) makes sense. Assuming a detector that is large compared to the event volume(an equivalent volume within which all detectors would "see" the event), we generally want the largest detector that will give the minimum number of space-time points necessary for event reconstruction. Experience with Monte Carlo studies for this problem indicates that something like a minimum of $\sim$100 sensor hits is necessary. Since event volume increases with sensor size, the density of sensors required to achieve the minimum number of space-time records goes down with sensor size. At some point the event volume is no longer well contained by the total detector volume. There is thus an optimum sensor size for each total detector volume. In reality there are discrete sensor sizes available, and a maximum photomultiplier size of 20" diameter now exists. If the experiment is carried out in the ocean, this size is nearly as large as tolerable from the standpoint of random count rate due to ^{40}K decays (which produce a net random light flux of >150 photons/cm^2 sec). Tube noise increases with size in any case, so one may not envisage much larger detectors, even in pure water. Also, since the detector size is sufficiently large that events can be viewed well into distances where photon absorption is important, the gain in detector cost efficiency with increasing size is only logarithmic. The result of all these factors is that the cost versus size curve is relatively flat (compared to tracking detectors) until somewhere in the region of 10^6 tons, beyond which the curve increases linearly.

3. A Large Ocean Proton Decay Detector

For discussion purposes we will be more explicit about a large ocean detector. An example design would be a cubical lattice of 20" hemispherical PMTs contained in 27"-diameter glass pressure housings. Using the water transparency observed in the clearest parts of the oceans, we estimate that this tube would "see" an event throughout a volume of $\sim$8000 m^3. If we require 100 PMTs in this volume, we find a spacing of 4.3 m, and a 10^6 ton detector will have 24^3 = 13,824 such modules. This is a large but manageable number (it would take the factory several years to produce this many large tubes!). Based upon experience with the manufacturers and with the DUMAND cost studies [9], I estimate the total project cost at less than $40 m (the largest part of this being the phototubes themselves at $17 m).

The outline of a detector design presented above is not meant to be a serious proposal, at this stage. It must be refined by Monte Carlo studies and careful cost estimates, yet it cannot be far wrong in terms of total cost. The fact I want to emphasize here is that a 10^6 ton proton decay detector is plausible while a detector much larger is not.

4. Lifetime Limits Versus Detector Size

The lifetime limit achievable by a detector is proportional to the number of nucleons in the detector until one gets to the level of irremovable background (e.g., from neutrino interactions) (as shown in figure 2). One may still seek to observe the decays, but in the detector size regime dominated by background, one must look for a statistical increase in the neighborhood (in the multidimensional space of total energy, collinearity angle, energy balance, and other fitting parameters) where one expects proton decay. Having mapped the background over a large region, one may extrapolate smoothly into the proton decay space, leaving little systematic uncertainty about the background and permitting a statistical search for proton decay. For example, the 10^6 ton detector, containing 6×10^{35} nucleons, might be expected to produce 250 proton decay-like events due to neutrino interactions in one year. A signal could then be interpreted as an observation of anything more than, say, 330 events (five standard deviations), and we would be sensitive to a lifetime/branching ratio of 7.5×10^{33}. In several years of observation this would exceed 10^{34} years, as a limit on the lifetime. A catch in this whole business is, of course, the branching ratio. The above discussion applies only to the simple two-body modes ($e^+\pi^0$, $e^+\pi^-$, etc.). Should the decay be dominated by other mode (in the worst case $n \to 3\nu$), the lifetime limit will be, obviously, less.

5. Summary

The conclusion of this paper is that with presently available technology and resources, a proton decay detector as large as 10^6 tons could be realized. We now have a proton lifetime limit of a few times 10^{30} years, which by next year will be pushed to 10^{31} years, and within the next several years to 10^{32} years. If proton decay is not observed then after several years of careful work, the present generation of instruments will reach a limit of about 10^{33} years. Only then and with heroic efforts building a massive array in the ocean could we go further, to 10^{34} years or so. If the proton decay lifetime is longer than 10^{34} years, we will probably know that fact within the decade. There is little prospect of testing any theory with lifetime predictions beyond 10^{34} years. Thus if we do observe proton decay, which as of now seems to be the only possible laboratory observation of a consequence of grand unification theories, we will have been very lucky indeed.

REFERENCES

[1] The Kolar Gold Field detector by an Indian-Japanese collaboration.

[2] The Morton, Painesville, Ohio detector of the IMB collaboration.

[3] The Park City, Utah detector of the HPW collaboration.

[4] The Soudan, Minnesota detector of the MAO collaboration.

[5] The Mont Blanc detector of the Milano group.

[6] The Homestake, South Dakota detector of the Pennsylvania group.

[7] Other detectors are proposed or in design for Japan and several tunnels in Europe. Also several of the groups above have proposed second-generation detectors.

[8] J. Learned, F. Reines, and A. Soni, Phys. Rev. Lett. 43, 907 (1979).

[9] See the proceedings of the various DUMAND workshops, particularly the recent "1980 Deployment Workshop," A. Roberts, ed., Hawaii DUMAND Center, Univ. of Hawaii, 2505 Correa Road, Honolulu, Hawaii 96822 (published 3/81).

THEORETICAL EXPECTATIONS FOR PROTON DECAY

Paul Langacker
Department of Physics, University of Pennsylvania
Philadelphia, Pennsylvania 19104

ABSTRACT

In this talk I survey the theoretical expectations for baryon number violation within the context of grand unified theories. I first give a classification of many of the possible baryon-number violating effective operators, including a discussion of typical processes that they induce, their selection rules, their relevant mass scales, and typical underlying mechanisms. I then specialize to the dimension-six four-fermion operators that describe the exchange of lepto-quark di-quark bosons in many popular models such as SU_5, SO_{10}, E_6, etc. After describing the general structure of these operators I discuss the estimate of the unification mass, the predictions for the nucleon lifetime, predictions for the branching ratios, and nuclear effects. Special emphasis is given to the related questions of which results are common to a wide class of theories and which types of measurements could discriminate between models.

I. General Classification of Baryon-Number-Violating Interactions[1-6]

Most of this talk will concern the predictions for proton decay in SU_5 and similar models. However, it is useful to put the subject in perspective by first considering a more general class of baryon-number-violating theories, which have different selection rules, mass scales, and underlying mechanisms (e.g. in most of the non-standard models, baryon number violation is mediated by the exchange of Higgs bosons). It should be emphasized, however, that unlike the SU_5-type models there is generally no independent way to estimate the relevant mass scales in the non-standard theories. There is therefore no particular reason to expect them to yield nucleon or nuclear lifetimes in an experimentally accessible range.

A general baryon-number violating interaction can be described by an effective Lagrangian[1-3]

$$\mathcal{L} = C\, M^{4-d}\, O \quad , \tag{1}$$

where O is an effective operator of dimension d involving the light fields of the theory (e.g. quarks and leptons). M, which is assumed to be large compared to laboratory momenta, is the mass scale relevant to the interaction. M is typically the mass of the gauge boson, Higgs boson, or fermion that mediates the interaction, but it is not necessary to assume that the process proceeds via a tree level exchange. C is a dimensionless constant, typically of order e^{n-2}, where n is the number of external fields in O. (1) is analogous to the effective four-fermion Lagrangian

$$\mathcal{L}_W = \frac{G_F}{2\sqrt{2}} (JJ^\dagger + J^\dagger J) \tag{2}$$

of the weak interactions, for which d=6, n=4, and $G_F/\sqrt{2} = g^2/8\, M_W^2$.

An enormous simplification occurs for $M \gg M_W$, because in that limit $SU_2 \times U_1$ violating effects will be suppressed by powers of M_W/M. Hence, it is only necessary to consider $G_s \equiv SU_3 \times SU_2 \times U_1$ invariant operators O. $SU_2 \times U_1$ violation can be treated perturbatively by including operators involving a Higgs doublet ϕ, which can then be replaced by its vacuum expectation value.

$\Delta B \neq 0$ color singlet operators must involve at least three quark fields. Lorentz invariance then implies that the $\Delta B \neq 0$ operators must include ≥ 4 fermion fields and have $d \geq 6$. Therefore, the dominant operators for a given M are the d=6 four-fermion operators. It turns out that there are only six such operators[1,2] (up to family indices), all of the form $qqq\ell$, where q and ℓ represent quark and lepton fields, respectively. (The Lorentz, G_s, and family indices are displayed in the original papers). These operators can be generated by the exchange of gauge or Higgs bosons as in Figure 1a, and they can mediate such processes as $qq \to \ell^c q^c$, where the superscript indicates an anti-particle. All six satisfy the selection rule $\Delta B = \Delta L$ (e.g. $p \to e^+\pi^0$ but $n \not\to e^-\pi^+$). They also satisfy $\Delta S = -\Delta B$ for strangeness changing decays (e.g., $p \to \nu^c K^+$ but $n \not\to e^+K^-$) and a $\Delta I = 1/2$ rule for $\Delta S = 0$ decays. Operators in this class would lead to nucleon lifetimes in the experimentally

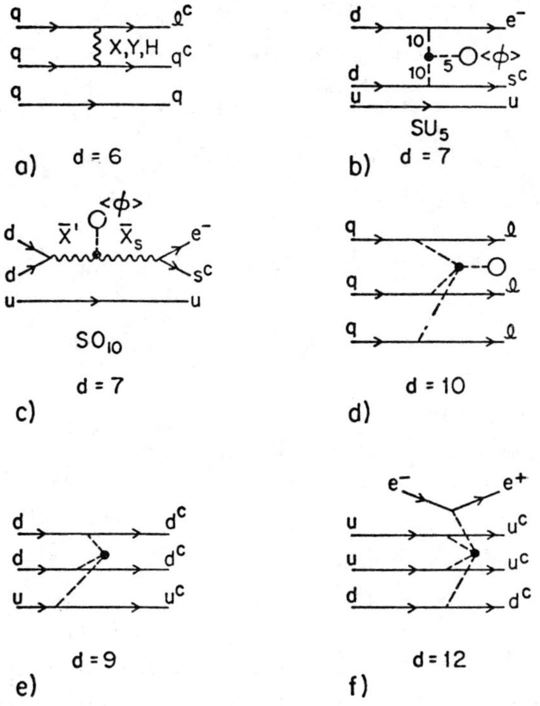

Figure 1. Typical diagrams leading to effective operators of various dimensions.

interesting range 10^{30}-10^{33} yr if the mass scale M is in the range[2] 4×10^{14} - 2×10^{15} GeV. It is remarkable that M is independently estimated to be in this range from the observed low energy coupling constants for many theories leading to d = 6 operators.

A number of low dimension operators are listed in Table 1, along with typical processes that they initiate, their selection rules, and the mass scales for which they would lead to nucleon or nuclear life-times in the 10^{30} - 10^{33} yr range. Typical diagrams leading to some of these operators are shown in Figure 1.

For example, the d = 7 operators $qqq\ell^c\phi$ conserve B + L and (except for small mixing effects) lead to S = 1 final states. They can be generated by Higgs exchange[1] as in Figure 1b. Such decays can also be generated[7] by gauge boson mixing effects in the SO_{10} model (Figure 1c),

d	0	Process	ΔB, ΔL, ΔS	M (GeV)
6	$qqq\ell$	$p \to e^+\pi^o$	$\Delta B = \Delta L = -1$ $\Delta S = 0,1$	$4 \times 10^{14} - 2 \times 10^{15}$
7	$qqq\ell^c\phi$	$n \to e^-K^+$	$\Delta B = -\Delta L = -1$ $\Delta S = 1$	$2 \times 10^{10} - 10^{11}$
7	$qqq\ell^cD$	$n \to e^-\pi^+$	$\Delta B = -\Delta L = -1$	$4 \times 10^9 - 2 \times 10^{10}$
10	$qqq\ell^c\ell^c\ell^c\phi$	$n \to \nu\nu e^-\pi^+$	$\Delta B = -1/3\ \Delta L = -1$	$(3 - 7) \times 10^4$
11	$qqq\ell\ell\ell\phi^2$	$p \to e^+\nu^c\nu^c$	$\Delta B = +1/3\ \Delta L = -1$	$(2 - 4) \times 10^4$
9	$qqqqqq$	$n \leftrightarrow \bar{n}$ $pn \to \pi^+\pi^o$ $nn \to 2\pi^o$	$\Delta B = -2$ $\Delta L = 0$	$4 \times 10^5 - 10^6$
12	$qqqqqq\ell\ell$	$pp \to e^+e^+$ $H \leftrightarrow \bar{H}$	$\Delta B = \Delta L = -2$	>500

Table 1. Baryon number violating operators of dimension d. q, ℓ, φ, and D represent quarks, leptons, Higgs fields, and derivatives, respectively.

but the gauge boson mass scale ($M_{X'} \gtrsim 10^{14}$ GeV) is much too large to be interesting. The d = 10 operator can be generated by the exchange of three Higgs bosons (1d). It can lead to fermion number (F = 3B + L) conserving nucleon decays in the fractional-quark-charge versions of the Pati-Salam models.[4]

The d = 9 six quark operator can be rewritten as an effective di-neutron operator $\delta m\, n^T n$, where $\delta m \sim e^4\, m_p^{\ 6}/M^5$ (see 1e) is the neutron-antineutron transition mass. nn annihilations lead to a nuclear life-time[6] $\Gamma_{nuc} \sim a(\delta m)^2/m_p$. For $\tau_{nuc} > 10^{30}$ yr and $a \sim 10^{-2}$ (for wavefunction overlap effects) one requires $\delta m < 10^{-20}$ eV, leading to a free neutron-antineutron oscillation time $\tau_{n\bar{n}} = 1/\delta m > 10^5$ s.

The d = 12 operator can be written as a four-fermion operator $G_H\, p^T p\, e^T e$, where $G_H \sim e^6\, m_p^{\ 6}/M^8$ (See 1f). One has[8] $G_H < 10^{-20}\, G_F$ from nuclear stability and $G_H < 10^{-21} G_F$ from the absence of gamma rays from $\bar{H}H$ annihilation in interstellar space. Hence M > 500 GeV.

Two classes of baryon-number-violating models do not fit easily into the classification in Table 1. For example, there have been a number of models[9,10,6] in which the proton is made absolutely stable by imposing a globally conserved quantum number. This quantum number corresponds to baryon number for the ordinary fermions but not for a new class of "weird" quarks and leptons. Such theories therefore allow[11] a baryon asymmetry for the universe and predict the possibility of observing baryon number violation at accelerator energies[12] through such reactions as $e^+e^- \to Q\bar{Q} \to B\ell\ell\bar{E}^o\pi$'s where Q is a new heavy ($m \lesssim 100$ GeV) quark, and E^o is a stable massive ($m \sim 10$ GeV) neutrino.

In the integer-charge-quark versions of the Pati-Salam models[4] (e.g. $SU_4{}^4$) there is a $M_X \sim 10^4 - 10^6$ GeV leptoquark boson that can lead to quark decays such as $q \to \pi\nu$ with $\tau_q \sim 10^{-6} - 10^{-9}$s by mixing with the W. Proton decays into $3\nu\,\pi^+$, $3\nu\pi^+\pi^+\pi^-$, $\nu\nu e^-\pi^+\pi^+$, etc., with $\tau_p \sim 10^{29} - 10^{34}$ yr can then occur via the simultaneous decay of all three quarks.

II. d=6 Four-Fermion Interactions

A. Description

In a single family version of the Georgi-Glashow SU_5 model[13] the fifteen left-handed fermion fields are assigned to a 5* + 10 representation

$$5^*: \begin{pmatrix} \nu & \\ & d^c \\ e^- & \end{pmatrix}_L \qquad 10: \begin{pmatrix} & u & \\ e^+ & & u^c \\ & d & \end{pmatrix}_L \tag{3}$$

where the SU_2 doublets are arranged in columns. Proton decay can be mediated at tree level by the exchange of a color-antitriplet SU_2-doublet of gauge bosons, the X and Y, which have electric charges 4/3 and 1/3. A color-triplet SU_2-singlet Higgs boson H with charge -1/3 (which shares a 5 with the Higgs doublet of the standard model) can also mediate proton decay if sufficiently light ($m_H \lesssim 10^{10-11}$ GeV).

More generally, Weinberg[2] has classified the quantum numbers of all bosons which can mediate proton decay at tree level without violating $SU_3 \times SU_2 \times U_1$. In addition to bosons like the X and Y there can be a color-triplet SU_2-doublet of gauge bosons with charges 2/3 and -1/3. Such a doublet, called the X' and Y', occurs along with the X and Y in the SO_{10} model. There are also three types of color-triplet Higgs particles: two SU_2 singlets with charges -1/3 (like the H) and -4/3, and an SU_2-triplet with charges 2/3, -1/3, and -4/3. The importance of this classification is that all of these exchanges satisfy the selection rules $\Delta B = \Delta L$ and $\Delta B/\Delta S = -1, 0$, as well as a $\Delta I = 1/2$ rule for $\Delta S = 0$ decays.

As has already been described, these selection rules are more general than tree level exchange[1,2]. They hold for any theory in which $SU_2 \times U_1$ breaking effects (of order M_W/M) can be neglected. Weinberg[2] and Wilczek and Zee[1] have shown that there are only six d=6 operators relevant to nucleon decay (up to family indices). Two of them are

$$O_1 = O_{e^+_R} + O_{\nu^c_R}$$

$$O_2 = O_{e^+_L} \quad , \tag{4}$$

where (suppressing color indices)

$$O_{e^+_R} \equiv (\bar{u}^c_L \gamma^\mu u_L)(\bar{e}^+_R \gamma_\mu d_R)$$

$$O_{\nu_R{}^c} \equiv -(\bar{u}^c_L \gamma^\mu d_L)(\bar{\nu}_R{}^c \gamma_\mu d_R)$$

$$O_{e^+_L} \equiv (\bar{u}^c_L \gamma^\mu u_L)(\bar{e}^+_L \gamma_\mu d_L) \quad . \tag{5}$$

O_1 and O_2 can be generated by the exchange of gauge bosons. The X and Y contribute to both O_1 and O_2 while (X',Y') contribute only to O_1. There are also four scalar-tensor operators O_4, O_5, O_6, and O_7 which can be generated by the exchange of Higgs bosons. (There is a seventh operator $O_3 = O_{\nu_L{}^c}$ that is allowed if there is a light left-handed anti-neutrino). O_6 and O_7 are antisymmetric in family indices. Hence, for any theory involving only the first family $(u, u^c, d, d^c, \nu, e^+, e^-)_L$ the most general Lagrangian is

$$\mathcal{L} = 4G_1\, O_1/\sqrt{2} + 4G_2\, O_2/\sqrt{2} + 4G_4\, O_4/\sqrt{2} + 4G_5\, O_5/\sqrt{2} + \text{H.C.}, \tag{6}$$

up to corrections of order M_W/M. $\mathcal{L}$ in (6) satisfies $\Delta B = \Delta L$ and a $\Delta I = 1/2$ rule. If one also assumes that the interactions are mediated by gauge boson exchange, then $G_4 = G_5 = 0$. Different theories in this class are distinguished largely by the ratio $r \equiv G_2/G_1$ of effective couplings.[1] For example, the SU_5 and SO_{10} effective Lagrangians are

$$\mathcal{L}_{SU_5} = 4G\,[O_1 + 2\,O_2]\,/\sqrt{2} + \text{H.C.}$$

$$\mathcal{L}_{SO_{10}} = \mathcal{L}_{SU_5} + [4G' O_1/\sqrt{2} + H.C.] \ , \tag{7}$$

where $G/\sqrt{2} = g^2/8 M_X^2$, $G'/\sqrt{2} = g^2/8 M_{X'}^2$, and g is the gauge coupling constant. Hence $r = 2$ for SU_5 (anomalous dimension effects modify r to 2.1) and $r = 2M_{X'}^2/(M_X^2 + M_{X'}^2)$ for SO_{10}, which ranges from 0 to 2 as $M_{X'}/M_X$ goes from 0 to ∞.

In the presence of several families one must add interaction basis family indices to each of the four fermions in O_1, O_2, $O_4 - O_7$. It is difficult to draw general consequences beyond the $\Delta S/\Delta B = -1, 0$ rule. In a general SU_5 model with F families, for example, the representations are[6]

$$5^*: \begin{pmatrix} \nu_m \\ \\ e_m^- \end{pmatrix} \left(d^c A_R^{d\dagger}\right)_m \Bigg)_L \qquad 10: \left((A_L^{e^+} e^+)_m \quad \begin{matrix}(A_L^u u)_m \\ d_m\end{matrix} \quad (u^c A_R^{u\dagger})_m \right)_L , \tag{8}$$

where, for example, e_m^-, m=1,..., F is the mth mass eigenstate charged lepton. In (8), A_L^u is the F x F generalized Kobayashi-Maskawa matrix, which is in principle known from charged-current weak interactions. Baryon number violation, however, involves transitions between adjacent columns in (8). All predictions for proton decay depend on the totally arbitrary and unknown F x F mixing matrices A_R^d, $A_L^{e^+}$, and A_R^u. These unknown mixing effects turn out to be the largest uncertainty in proton decay in the standard models. It is not even a priori obvious that the dominant amplitude is for $p \to e^+\pi^o$ rather than $p \to \tau^+\pi^o$, for example. Fortunately, things simplify enormously in some specific models. For example, the minimal SU_5 model with all fermion masses generated by Higgs 5's leads to $A_L^{e^+} = A_R^d = I$ and $A_R^{u*} = A_L^u K^*$, where K is a diagonal matrix of (CP-violating) phases that are irrelevant to proton decay. For the two family version of this model, the effective Lagrangian is (suppressing color indices)[14]

$$\begin{aligned}\mathcal{L}_{SU_5} = \frac{4G}{\sqrt{2}} \{ &(\bar{u}_L^c \gamma^\mu u_L) \, [\bar{e}_R^+ \gamma_\mu d_R + \bar{\mu}_R^+ \gamma_\mu s_R \\ &+\{(1+c^2)\bar{e}_L^+ + sc\, \bar{\mu}_L^+\} \gamma_\mu d_L + \{(1+s^2)\bar{\mu}_L^+ + sc\, \bar{e}_L^+\}\gamma_\mu s_L] \\ &-[\bar{u}_L^c \gamma^\mu (c\, d_L + s\, s_L)][\bar{\nu}_{eR}^c \gamma_\mu d_R + \bar{\nu}_{\mu R}^c \gamma_\mu s_R]\} \ ,\end{aligned} \tag{9}$$

where $c = \cos\theta_c$, $s = \sin\theta_c$, and θ_c is the Cabibbo angle.

Unfortunately, this minimal SU_5 model makes the incorrect prediction $m_d/m_s = m_e/m_\mu \simeq 1/200$ (phenomenological estimates obtain[6] $\simeq 1/20$).

Wilczek and Zee[1] have introduced the Kinship Hypothesis, which states that baryon-number-violating interactions mainly involve transitions within the families (u, d, ν_e, e) or (c, s, ν_μ, μ) except for small (of order θ_c) mixing effects, as in the minimal SU_5 model.

B. Predictions for the Unification Mass[15]

The only theories for which the mass scale M_X can be predicted really cleanly are those for which an underlying gauge group G is spontaneously broken to a subgroup G_1 at M_X, with G_1 breaking to $SU_3^c \times U_1^{EM}$ at M_W. If there is a desert between M_X and M_W (i.e. no gauge boson, Higgs boson, or fermion thresholds) then M_X can be predicted from the observed low energy coupling constants if one knows G_1 and the spectrum and G_1 assignments of all of the fermions and light Higgs particles. (A knowledge of the underlying group G is not required). This is because the running coupling constants associated with G_1 will all approach the underlying coupling g_G at M_X^2 if their associated generators are properly normalized. (The normalization depends on the fermion content of the theory). If $G_1 = G_s = SU_3 \times SU_2 \times U_1$, for example, then the two independent ratios of running coupling constants will approach the asymptotic ratios

$$\sin^2\theta_W = \frac{e^2}{g^2} \rightarrow \frac{\Sigma t_3^{\,2}}{\Sigma\, q^2}$$

$$\frac{\alpha}{\alpha_s} = \frac{e^2}{g_s^{\,2}} \rightarrow \frac{\Sigma\, t_{c_3}^2}{\Sigma q^2} \tag{10}$$

for $Q^2 \rightarrow M_X^{\,2}$, where the normalization depends on the values of the electric charge q, third component of weak isospin t_3, and third component of color t_{c_3} of the fermions in the theory. For the standard model (F ordinary families), both asymptotic ratios are 3/8.

For $M_W^{\,2} < Q^2 < M_X^{\,2}$, the running coupling constant ratios are[16], to lowest order in the renormalization group equations (RGE) and treating thresholds as step functions,

$$\frac{\alpha(Q^2)}{\alpha_s(Q^2)} = \frac{3}{8}\left(1 - \frac{\alpha(Q^2)}{2\pi}\,(11 + \frac{n_H}{6})\,\ell n\,\frac{M_X^{\,2}}{Q^2}\right)$$

$$\sin^2\theta_W(Q^2) = \frac{3}{8}\left(1 - \frac{\alpha(Q^2)}{4\pi}\;\frac{(110-n_H)}{9}\,\ell n\,\frac{M_X^{\,2}}{Q^2}\right) \quad , \qquad (11)$$

where n_H is the number of light Higgs doublets. These lowest order results are independent of F. There are two independent estimates of M_X in (11). Their agreement can be considered a consistency check on this class of theories. For $n_H = 0$, for example, the consistency condition is $\sin^2\theta_W = 1/6 + 5/9\,\alpha_s$. An early estimate[14] utilized $\alpha^{-1} = 137$, $n_H = 0$, and the QCD parameter $\Lambda = 300$ MeV ($\alpha_s = 12\pi/25\,\ell n\,(Q^2/\Lambda^2)$), with the resulting predictions $M_X = 4 \times 10^{16}$ GeV and $\sin^2\theta_W(M_W^2) = 0.20$. Detailed studies of the corrections to (11) have been made by Goldman and Ross[17], Marciano[18], and others.[6] These include:
(a) incorporating the Q^2 dependence of α^{-1}:

$$\alpha^{-1}(M_W^{\,2}) \simeq \alpha^{-1}(0) - \frac{1}{3\pi}\sum_f q_F^{\,2}\,\ell n\,\frac{M_W^{\,2}}{m_f^{\,2}} \simeq 128.5 \quad , \qquad (12)$$

where the sum runs over the fermions f lighter than M_W. This reduces the prediction for M_X by an order of magnitude.
(b) using $n_H = 1$, which reduces M_X by a factor of 2.
(c) including two loop contributions to the RGE, which reduces M_X by $\simeq 4$ for F = 3 (and introduces a weak F dependence). Also, many authors[6] have analyzed the effects of thresholds at M_W and M_X. The result is that the estimate for M_X is not substantially altered.

There have now been several analyses[17-21] of these and smaller corrections, utilizing both $\overline{MS}$ (modified minimal subtraction) and MOM (momentum subtraction) renormalization schemes. The results agree with each other to within 1% in $\ell n\, M_X$ and 50% in M_X. The conclusion is that M_X depends approximately linearly on $\Lambda_{\overline{MS}}$, the QCD parameter evaluated in the two loop $\overline{MS}$ scheme. For F = 3, $n_H = 1$ one has

$$M_X \simeq 15 \times (1.5)^{\pm 1} \times 10^{14}\,\Lambda_{\overline{MS}} \rightarrow 6 \times (1.5)^{\pm 2} \times 10^{14}\ \text{GeV} \qquad (13)$$

where the final result uses the value[19,6] $\Lambda_{\overline{MS}} = 0.4 \times (1.5)^{\pm 1}$ GeV from deep inelastic scattering. The $(1.5)^{\pm 1}$ uncertainty in (13) is due to higher order corrections, thresholds, the t quark mass, and the masses of superheavy Higgs bosons.[22]

Using (13) one has the consistency condition[18,6]

$$\sin^2\theta_W(M_W^2) = 0.209 \mp 0.0025 \mp 0.0025 \qquad (14)$$

where the first uncertainty is theoretical and the second is from $\Lambda_{\overline{MS}}$. The experimental value[23] of $\sin^2\theta_W = 0.229 \pm 0.009$ ($\pm$ 0.005) is reduced by 0.010 - 0.015 by radiative corrections[24-26], in remarkable agreement with (14).

The prediction (13) holds for all theories in which (a) an arbitrary group breaks directly[22] to $SU_3 \times SU_2 \times U_1$ at M_X with a desert between M_X and M_W; (b) there are F = 3 families of ordinary fermions and $n_H = 1$ light complex Higgs doublets[27]; (c) $\Lambda_{\overline{MS}} = 0.4 \times 1.5^{\pm 1}$ GeV; and (d) 15 GeV < m_t < 50 GeV (The results are insensitive to m_t). That this wide class of models may be relevant to the real world is supported by the remarkable success of (14).

C. The Estimate of τ_p and τ_n.

A very simple estimate of the nucleon lifetime τ_N within SU_5-like models is $\tau_N \simeq M_X^4/\alpha_G^2 m_p^5$, where $\alpha_G \equiv g_G^2/4\pi \simeq 0.022$ is the underlying running coupling at M_X. For $M_X = 6 \times 10^{14}$ GeV this implies $\tau_N \simeq 8 \times 10^{30}$ yr. In fact, more detailed studies obtain the very similar value $\tau_p = 1.2 \times 10^{31\pm 2}$ yr.

The first step in a detailed estimate is to compute the anomalous enhancement factors[6] A^i to the coefficients G_i in (6). These factors are due to the exchange of $SU_3 \times SU_2 \times U_1$ bosons between the four external fermions. The leading log (e.g. $\alpha_g \ln(M_W/M_X)$) contributions to the A^i can be summed,[1,14] yielding $(A^1)^2 \simeq 11$ and $(A^2)^2 \simeq 12.5$. Also, r is renormalized by $A^2/A^1 \sim 1.06$ from its tree level value.[1]

The next step is to estimate the hadronic matrix elements of the O_i. I will discuss the results for the SU_5 model, but other gauge boson generated models are similar. (The general case including O_4 - O_7 has not been studied in detail). Essentially all estimates of the proton lifetime and branching ratios have utilized a valence approximation for the proton and have assumed that the dominant subprocess is $qq \to \ell^c q^c$, as in Figure 1a, with the third quark a spectator. (Quark decay processes $q \to \ell^c q^c q^c$ with two spectators, and quark fusion $qqq \to \ell^c$, which can occur in a nucleus, have been estimated to be unimportant[28]).

There have been many estimates of the nucleon lifetime and/or branching ratios. Most of the estimates (Ref. 7, 14, 17, 19, 28-30) use some form of SU_6 wave functions for the nucleon and mesons, as well as a phenomenological amplitude $\psi(0)$ for two quarks to overlap in the nucleon. Independent determinations include $10^3|\psi(0)|^2 = 1.1\ \text{GeV}^3$ (from Ω^- decay[31]), 4.4 GeV^3 (from S-wave Σ^+ decay[32]), and 11.5 GeV^3 (from P-wave Λ decay[33]). There are also three calculations utilizing MIT bag wave functions.[34-36]

The various estimates also differ in their treatments of the final state. There are several inclusive calculations[14,17,19,28] in which final state interactions are neglected (in the spirit of the parton model). Phase space is typically computed for the $\ell^c q^c q$ state assuming $M_u \simeq M_d \simeq m_p/3$, $M_s \simeq 500$ MeV for the constituent quark masses. There are also a number of exclusive calculations[29,34-36] in which the final $q^c q$ pair are projected onto single meson (π, ρ, η, ω, K) wave functions (e.g. bag or SU_6). The correct phase space is then computed for each exclusive final state. Such estimates actually obtain the partial rate for two body decays.

The results of seven calculations are given in Table 2. I have re-normalized the authors' results so that they use the same values for the anomalous enhancement factor $(A^i)^2 \simeq 12$, $M_X = 6 \times 10^{14}$ GeV, and (for the first four estimates) $|\psi(0)|^2 = 2 \times 10^{-3}\ \text{GeV}^3$. This value agrees with Ω and S-wave hyperon decay and is compatible with the bag model amplitudes used in the last three estimates. The p and n lifetimes scale roughly as M_X^4, except for small effects in A^i. The coefficients in $\tau_{p,n}(\text{yr}) \simeq a_{p,n} M_X^4$ (GeV) are also given in Table 2. The total lifetimes for the last four (exclusive) calculations involve factors of $\rho_{p,n}$, which are the branching ratios for two body decays.

I will take the order of magnitude spread (for $\rho_{p,n} \simeq 1$) of the values in Table 2 as an indication of the theoretical uncertainty in the hadronic matrix elements and the method of treating the final state interactions and phase space. Hence,

$$\tau_p(\text{yr}) \simeq (2.4\text{-}38) \times 10^{-29}\ M_X^4(\text{GeV}) = 4.8 \times 10^{32\pm1.3} \Lambda_{\overline{MS}}^4(\text{GeV}) \qquad (15)$$

where the $10^{\pm1.3}$ includes both the uncertainty in M_X ($\Lambda_{\overline{MS}}$) and in τ_p (M_X). In particular for $\Lambda_{\overline{MS}} = 0.4 \times 1.5^{\pm1}$ GeV, (15) implies

$$\tau_p(\text{yr}) = 1.2 \times 10^{(31\pm2)}\ \text{yr}, \qquad (16)$$

where the error now includes the quoted error in $\Lambda_{\overline{MS}}$ as well. τ_p is shown as a function of $\Lambda_{\overline{MS}}$ in Figure 2. Some recent estimates have

	$10^{29}\ a_p$	$10^{29}\ a_n$	For $M_X = 6 \times 10^{14}$ GeV τ_p (10^{30} yr)	τ_n (10^{30} yr)
JY[28]	3.7	4.3	4.8	5.6
EGNR[19]	4.8	---	6.2	---
GR[17]	2.4	3.7	3.1	4.8
GLOPR[29]	35 ρ_p	---	45 ρ_p	---
D[34]	38 ρ_p	50 ρ_n	50 ρ_p	65 ρ_n
G[36]	41 ρ_p	48 ρ_n	52 ρ_p	62 ρ_n
DGS[35]	2 ρ_p	---	2.6 ρ_p	---

Table 2. Lifetime estimates. The first three (last four) estimates are inclusive (exclusive). The last three calculations utilize MIT bag wave functions.

suggested that $\Lambda_{\overline{MS}}$ may be smaller than the value obtained from deep inelastic scattering. A value of 150 MeV, for example, yields $\tau_p \simeq 2.4 \times 10^{29\pm1.3}$ yr.

Equations (15) and (16) are for the SU_5 model, but other models within the general class described in the last section should give results well within the error bars in (16).

Two types of uncertainties are not included in the $10^{\pm 2}$ uncertainty in (16). (a) minor changes to the standard model (e.g. $n_H \neq 1$, $F \neq 3$, or a modification of the superheavy Higgs bosons or fermions) could affect τ_p, mainly through their effect on M_X. These could probably not change τ_p by more than another factor of $10^{\pm 2}$ without destroying the successful prediction of $\sin^2\theta_W$, however.[15] (b) Violations of the Kinship Hypothesis (for a non-minimal Higgs structure) could suppress τ_p without affecting $\sin^2\theta_W$. For example, the dominant amplitudes could be into energetically forbidden channels such as $\tau^+\pi^0$. It would be hard to suppress the decay rate by more than $\simeq \sin^2\theta_c \simeq 1/20$ by this mechanism, however (unless a symmetry is added to make the proton absolutely stable[9,10,6]).

The estimates in Table 2 give $\tau_n/\tau_p \gtrsim 1$. The ratio can be computed

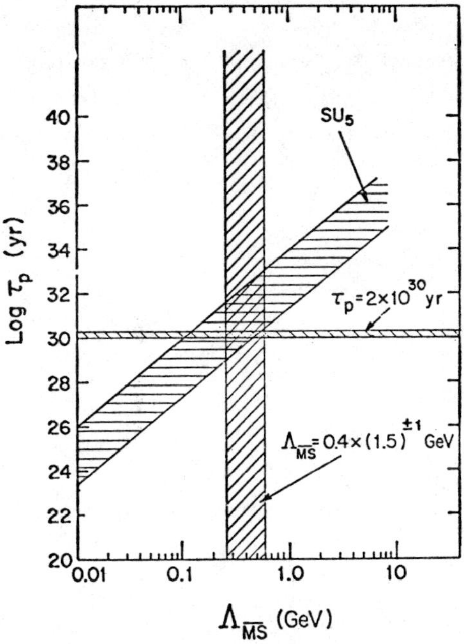

Figure 2. The SU_5 prediction for τ_p as a function of $\Lambda_{\overline{MS}}$.

directly using $\tau_n/\tau_p = \frac{1}{2} B(n \to e^+\pi^-)/B(p \to e^+\pi^o)$, which follows from the $\Delta I = 1/2$ rule. The branching ratio estimates described in the next section yield[6] $0.8 < \tau_n/\tau_p < 1.5$.

D. Predictions for Branching Ratios

It is useful to group the various symmetry and branching ratio predictions in decreasing order of generality.

All of the $d = 6$ operators (including the scalar-tensor operators O_4-O_7) satisfy the selection rules $\Delta B = \Delta L$ and $\Delta S/\Delta B = -1, 0$. Furthermore, the hadronic parts of O_1-O_7 transform as doublets under strong isospin ($\Delta I = \frac{1}{2}$). Hence, one has[1,2,7]

$$\Gamma(p \to e^+_{L,R}\, \pi^o) = \frac{1}{2}\Gamma(n \to e^+_{L,R}\, \pi^-)$$

$$\Gamma(p \to \nu^c\, \pi^+) = 2\Gamma(n \to \nu^c\pi^o) \quad . \tag{17}$$

Similar results hold for $\pi \to \rho$ and $e^+ \to \mu^+$. These three selection rules hold for arbitrary mixing between families. Any deviation would indicate the existence of $d > 6$ operators.

The hadronic parts of $O_{e^+_R}$ and $O_{\nu^c_R}$ are the two components of an isospin doublet. Hence, if one can ignore mixing between families (so that the two operators have the same coefficient)[1,2,7] then up to $O(m_e)$,

$$\Gamma(n \to e^+_R \pi^-) = \Gamma(p \to \nu^c \pi^+)$$

$$\Gamma(p \to e^+_R X) = \Gamma(n \to \nu^c X)$$

$$\Gamma(n \to e^+_R X) = \Gamma(p \to \nu^c X) \quad , \qquad (18)$$

with a similar result for $\pi \to \rho$. (18) also holds for O_4-O_7 but is broken by mixing.

Specializing now to the gauge boson induced operators O_1 and O_2, one has that $O_{e^+_R}$ and $O_{e^+_L}$ are related by a parity transformation. Hence

$$\Gamma(N \to e^+_L H_n) = r^2 \, \Gamma(N \to e^+_R H_n) \quad , \qquad (19)$$

where H_n is any non-strange hadronic state of definite parity. Hence, direct positrons will have a universal polarization $(1-r^2)/(1+r^2)$ for all decays involving $S = 0$ hadronic states, and a different universal polarization for $S = 1$ states. Similar statements apply to direct μ^+ decays. For the minimal SU_5 model in (9), the μ^+ polarization is -1 and $[1-(1+s^2)^2]/[1+(1+s^2)^2] \simeq -0.05$ for $S = 0$ and $S = 1$ states, respectively. These results hold for arbitrary mixing, but for O_1 and O_2 only.

Combining (17), (18), and (19), one has

$$\Gamma(p \to e^+\pi^o) = \frac{1}{2}\,\Gamma(n \to e^+\pi^-) = \frac{(1+r^2)}{2}\,\Gamma(p \to \nu^c\pi^+) = (1+r^2)\Gamma(n \to \nu^c\pi^o)$$

$$\Gamma(p \to e^+X) = (1+r^2)\Gamma(n \to \nu^c X)$$

$$\Gamma(n \to e^+X) = (1+r^2)\Gamma(p \to \nu^c X)$$

$$\Gamma(O^{16} \to e^+X) = (1+r^2)\Gamma(O^{16} \to \nu^c X)$$

$$\Gamma(p \to e^+X) \geq \frac{1}{2}\,(1+r^2)\Gamma(p \to \nu^c X) \quad , \qquad (20)$$

as well as $\pi \to \rho$. Relations (20) hold only for O_1 and O_2 in the absence of mixing. For SU_5, which has $r = 2$, (20) implies the desirable result that e^+ decays dominate over ν^c. Neutrinos are relatively more important for SO_{10}, which has $0 \le r \le 2$.

We have seen that the relative e^+/ν^c rates and the polarization of prompt muons would be very useful for discriminating between models and are insensitive to the details of the hadronic matrix elements. On the other hand, the relative rates for different hadronic states (i.e. the ratios $e^+\pi^o/e^+\rho^o/e^+\eta/e^+\omega$ and $\nu^c\pi^+/\nu^c\rho^+$) are independent of r and depend only on the hadronic wave functions. Hence, these ratios, although very important for designing and interpreting experiments, do not discriminate between different gauge mediated theories (they would presumably differentiate between vector and scalar exchange, but this question has not been carefully investigated).

A number of estimates of the proton and neutron branching ratios for the SU_5 model are shown in Tables 3 and 4. Results for other gauge models can be easily obtained by renormalizing the e^+/ν^c ratio using (20).

The estimates differ in the wave functions used (bag or SU_6), their treatment of phase space, and their mixing assumptions. However, Kane and Karl[30] have shown that the largest discrepancies are due to the method of projecting the q^c spin onto the meson wave function. They have therefore presented results for three models: a static (NR) model with the q^c at rest, a relativistic (R) model with the q^c mass $M = 0$, and an intermediate recoil (REC) model with $k/M \simeq 3/4$ for the q^c. The other major source of controversy is that several authors[34-36] have assumed that the pionic modes are suppressed by form factor effects due to the high recoil momentum; the other authors have argued that the recoil suppression is small. Donoghue and Karl[37] have shown within an oscillator model that the pion suppression may be large or small depending on which plausible set of oscillator parameters are used.

It is apparent from Tables 3 and 4 that the various calculations are in reasonable agreement, up to the theoretical uncertainties discussed above. In particular, the spread in the three models of Kane and Karl provides a reasonable estimate of the theoretical uncertainties, with the caveat that there may be additional recoil suppression of the pionic modes. Also, Karl and Lipkin[38] have argued that SU_6 violating hyperfine interactions will tend to enhance the ρ rate and suppress the π rate with respect to the results in Tables 3 and 4.

Mode	M^7 REC	GYOPR[29] NR	D^{34} R	G^{36} R	DGS[35] R	KK[30] NR	REC	R
$e^+\pi^o$	33	37	9	13	31	36	40	38
$e^+\rho^o$	17	2	21	20	21	2	7	11
$e^+\eta$	12	7	3	.1	5	7	1.5	0
$e^+\omega$	22	18	56	46	19	21	25	26
$\nu_e^c\pi^+$	9	15	3	5	11	14	16	15
$\nu_e^c\rho^+$	4	1	8	7	8	1.0	2.6	4
μ^+K^o	.3 - .5	19	--	7	.5	18	8	5
$\nu_\mu^c K^+$	--	0	--	.5	--	0	.2	.6

Table 3. Predictions for branching ratios for proton decay into the major two body modes in the SU_5 model.

Mode	M^7 REC	GYOPR[29] NR	D^{34} R	G^{36} R	KK[30] NR	REC	R
$\nu_e^c\pi^o$	8	8	2	3	8	7	7
$\nu_e^c\rho^o$	4	.5	5	4	.6	1.2	1.8
$\nu_e^c\eta$	3	1.5	1	--	1.5	--	--
$\nu_e^c\omega$	5	3.5	14	10	5	5	5
$e^+\pi^-$	50	74	23	32	79	72	68
$e^+\rho^-$	26	4	55	48	6	12	19
$\nu_\mu^c K^o$	--	10	--	2	1.1	3	0.6

Table 4. Predictions for neutron decay branching ratios in SU_5.

Little is known about multibody decay modes (which are ignored in Tables 3 and 4). Wise, Blankenbecler and Abbott[39] have estimated $\Gamma(p \to e^+\pi\pi(I))/\Gamma(p \to e^+\pi^o) = 0.24\ (1.5)$ for $I = 0\ (I = 1)$ from a current algebra model incorporating final state interactions. Also, the longer lifetime estimates in the exclusive calculations (Table 2) may hint at the importance of multibody decays.

Predictions for muons and S = 1 final states are much more model dependent. In addition to the large number (20-30%) of indirect muons expected from π^+ and K decay, there are two sources of prompt muons. μ^+ can be produced with S = 1 final states without mixing, from the elementary process $uu \to \mu^+ s^c$. The estimates in Table 3 yield $B(p\to\mu^+K^o) \simeq (5\text{-}20)\%$ in SU_5. The decay $n \to \mu^+K^o\pi^-$ cannot occur in valence approximation (but could occur at some level from sea effects). Also, μ^+ can be produced with S = 0 states through (model dependent) mixing effects. A similar statement holds for e^+ decays with S = 1 states. In the minimal SU_5 model in (9),

$$\frac{B(N\to\mu^+H_n)}{B(N\to e^+H_n)} = \frac{s^2c^2}{(1+c^2)^2+1} \simeq 0.01$$

$$\frac{B(N\to e^+H_s)}{B(N\to \mu^+H_s)} = \frac{s^2c^2}{(1+s^2)^2+1} \simeq 0.023\ , \tag{21}$$

where H_n (H_s) represent S = 0 (S = 1) hadronic states.

Higgs mediated decay could be important if the Higgs bosons are sufficiently light ($m_H \lesssim 10^{10\text{-}11}$ GeV). Little is known about their branching ratios (other than the general d = 6 selection rules), but one expects an enhancement of decays into heavy fermions. Machacek[7] has found that ν_τ^c and S = 1 modes are enhanced in some SU_5 and SO_{10} examples.

E. Nuclear Effects

All forseeable experiments involve nuclei rather than free nucleons. Nucleon decays within a nucleus are complicated by Fermi motion and also by the possibility that produced pions may be absorbed or elastically or charge-exchange scattered before getting out of the nucleus. For example, Sparrow has estimated[40] that only about half of the $p \to e^+\pi^o$ decays in water will lead to a back to back e^+ and π^o (and 40% of these are from the hydrogen atoms). Similarly, the secondary μ^+ rate should be reduced by $\simeq 2.5$ in water.

There may also be desirable effects. Dover et al.[41] have considered the process $p \to e^+ + (\text{virtual } \pi^o)$. The virtual π^o can be absorbed by another nucleon, producing an N, N*, or Δ. Such effects may enhance the decay rate in nuclei by $\simeq 2$.

Finally, $\Delta B = 2$ processes (Table 1) may be observable in nuclei.

III. Summary

The various general classes of baryon-number-violating interactions may be distinguished by their selection rules. The simplest (d = 6) class leads to $\Delta B = \Delta L$, $\Delta S/\Delta B = -1, 0$, and $\Delta I = 1/2$ for $S = 0$. A large subclass of these d = 6 theories (those with an $SU_3 \times SU_2 \times U_1$ invariant desert between M_X and M_W) lead to $M_X \sim 6 \times 10^{14}$ GeV and $\tau_p \sim 10^{31}$ yr, with large uncertainties. The successful prediction of $\sin^2\theta_W$ suggests that this class of theories may be relevant.

Within this class, the e^+/ν^c ratio and the polarization of prompt muons can effectively distinguish between models. The ratios $\pi/\rho/\omega/\eta$ are not useful for distinguishing between gauge models, but may be useful in determining whether the interaction is gauge boson or Higgs mediated. (The universality of the muon polarization is also useful in this respect). The largest uncertainties are in the mixing effects, which in turn depend on the detailed Higgs structure of the theory. The ratios μ^+/e^+ and $(S = 0)/(S = 1)$ depend sensitively on these mixings.

REFERENCES

1. F. Wilczek and A. Zee, Phys. Rev. Lett. 43 (1979) 1571, Phys. Lett. 88B (1979) 311.
2. S. Weinberg, Phys. Rev. Lett. 42 (1979) 850; 43 (1979) 1566; Phys. Rev. D22 (1980) 1694.
3. H. A. Weldon and A. Zee, Nucl. Phys. B173 (1980) 269.
4. J. C. Pati and A. Salam, First Workshop on Grand Unification, ed. P. H. Frampton, S. L. Glashow, and A. Yildiz (Math. Sci. Press, Brookline, 1980), p. 115, and references therein.
5. R. E. Marshak and R. N. Mohapatra, Orbis Scientiae 1980; R. N. Mohapatra, First Workshop on Grand Unification, p. 69.
6. For a complete set of references, see P. Langacker, SLAC-PUB 2544, to be published in Physics Reports.
7. M. Machacek, Nucl. Phys. B159 (1979) 37.
8. G. Feinberg, M. Goldhaber, and G. Steigman, Phys. Rev. D18 (1978) 1602.
9. M. Gell-Mann, P. Ramond, and R. Slansky, Rev. Mod. Phys. 50 (1978) 721.

REFERENCES - (Cont.)

10. P. Langacker, G. Segre, and H. A. Weldon, Phys. Lett. 73B (1978) 87 and Phys. Rev. D18 (1978) 552.

11. G. Segre and H. A. Weldon, Phys. Rev. Lett. 44 (1980) 1737.

12. P. Langacker, Z^0 Physics Workshop, Cornell, 1981, ed. M. E. Peskin and S.-H. H. Tye, pp. 95 and 467.

13. H. Georgi and S. L. Glashow, Phys. Rev. Lett. 32 (1974) 438.

14. A. J. Buras, J. Ellis, M. K. Gaillard, and D. V. Nanopoulos, Nucl. Phys. B135 (1978) 66; J. Ellis, M. K. Gaillard, and D. V. Nanopoulos, Phys. Lett. 88B (1979) 320.

15. For a more detailed discussion, see ref. 6 and 12.

16. H. Georgi, H. R. Quinn, and S. Weinberg, Phys. Rev. Lett. 33 (1974) 451.

17. T. J. Goldman and D. A. Ross, Phys. Lett. 84B (1979) 208 and Nucl. Phys. B171 (1980) 273.

18. W. J. Marciano, Phys. Rev. D20 (1979) 279 and Orbis Scientiae 1980.

19. J. Ellis, M. K. Gaillard, D. V. Nanopoulos and S. Rudaz, Nucl. Phys. B176 (1980) 61.

20. L. Hall, Nucl. Phys. B178 (1981) 75.

21. P. Binetruy and T. Schucker, Nucl. Phys. B178 (1981) 293, 307.

22. There is an additional uncertainty of ≃ 3 if the masses of superheavy fermions and Higgs bosons are allowed to differ significantly from M_X (Ref. 12).

23. J. E. Kim, P. Langacker, M. Levine, and H. H. Williams Rev. Mod. Phys. 53 (1981) 211.

24. W. J. Marciano and A. Sirlin, Phys. Rev. Lett. 46 (1981) 163.

25. S. Dawson, L. Hagelin, and L. Hall, HUTP-80/A090.

26. F. Antonelli and L. Maiani, INFN. 219.

27. M_X decreases by a factor of ≃ 1.8 for each additional light Higgs, while $\sin^2\theta_W$ increases by 0.004. The effects of F>3 are somewhat unclear. Fermion masses close to M_W apparently have little effect (W. J. Marciano, these proceedings) while lighter fermions tend to increase M_X and decrease $\sin^2\theta_W$ (Ref. 17, 19, 12).

28. C. Jarlskog and F. J. Yndurain, Nucl. Phys. B149 (1980) 29.

29. M. B. Gavela, A. Le Yaouanc, L. Oliver, O. Pene, and J. C. Raynal, Phys. Lett. 98B (1981) 51 and Phys. Rev. D23 (1981) 1580.

30. G. Kane and G. Karl, Phys. Rev. D22 (1980) 2808.

31. J. Finjord, Phys. Lett. 76B (1978) 116.

32. C. Schmid, Phys. Lett. 66B (1977) 353.

33. A. Le Yaouanc et al., Phys. Lett. 72B (1977) 53.

34. J. F. Donoghue, Phys. Lett. 92B (1980) 99.

REFERENCES - (Concl.)

35. A. M. Din, G. Girardi, and P. Sorba, Phys. Lett. 91B (1980) 77.

36. E. Golowich, Phys. Rev. D22 (1980) 1148.

37. J. F. Donoghue and G. Karl, U. Mass. preprint UM-HEP-144.

38. G. Karl and H. J. Lipkin, Phys. Rev. Lett. 45 (1980) 1223.

39. M. B. Wise, R. Blankenbecler, and L. F. Abbott, Phys. Rev. D23 (1981) 1591.

40. D. A. Sparrow, Phys. Rev. Lett. 44 (1980) 625.

41. C. B. Dover, M. Goldhaber, L. Trueman and L.-L. C. Wang, to be published.

SU(5) THEORY vs. EXPERIMENT

W. J. Marciano
Department of Physics and Astronomy
Northwestern University, Evanston, Illinois 60201
and
A. Sirlin
Department of Physics, New York University
New York, New York 10003

1. $SU(3)_c x SU(2)_L x U(1)$ Parameters

The "standard" $SU(3)_c x SU(2)_L x U(1)$ model of strong and electroweak interactions contains the following parameters: 1) It has three independent gauge couplings g_3, g_2, and g_1. The strong QCD coupling g_3 is often parametrized in terms of a mass scale $\Lambda_{\overline{MS}}$, while g_2 and g_1 are usually traded off for the electric charge e and the weak mixing angle θ_W via the relationships[1] $e^2 = g_2^2 \sin^2\theta_W$ and $\tan^2\theta_W = 3g_1^2/5g_2^2$. 2) There are two intermediate vector boson masses m_W and m_Z which are related by $m_W = m_Z \cos\theta_W$ and an arbitrary physical Higgs scalar mass m_ϕ. 3) Of its nine fermion masses m_e, m_μ, m_τ, m_d, m_s, m_b, m_u, m_c, and m_t, all except m_t have phenomenologically measured values. From e^+e^- annihilation experiments at PETRA one has the lower bound $m_t > 18 \sim 19$ GeV.[2] 4) There are three quark mixing angles θ_1, θ_2, θ_3, and a CP violating phase δ. In addition to the above, there may be neutrino masses, lepton mixing angles, new generations of fermions, etc. As experimentalists measure these parameters with higher precision, theorists seek natural relationships (such as $m_W = m_Z \cos\theta_W$) in an attempt to understand their origin. Such relationships are generally obtained by constraining the theory with additional symmetries. In that regard we now know how to correlate the three couplings g_3, g_2, and g_1 through grand unification, i.e. by embedding the "standard" model in a simple gauge group G.

Before reviewing the predictions of the SU(5) model [1,2] (and a class of similar grand unified theories (GUTS)), we begin by describing the present experimental status of $\alpha = e^2/4\pi$, $\Lambda_{\overline{MS}}$ and $\sin^2\theta_W$ with particular emphasis on our recent analysis of the effect of radiative corrections on the determination of $\sin^2\theta_W$ [3].

1.1 $\underline{\alpha}$: The best direct determination of the fine structure constant α presently comes from Josephson effect measurements which give [4]

$$\alpha^{-1} = 137.035987 \quad . \tag{1}$$

For later purposes, it is convenient to relate the long distance coupling α (defined at zero momentum transfer) with $\hat{\alpha}(m_W)$ which is defined by modified minimal subtraction $\overline{MS}$ in the dimensional regularization scheme with μ, the 't Hooft unit of mass [5], equal to m_W [6,7,8]

$$\hat{\alpha}^{-1}(m_W) = \alpha^{-1} - \frac{2}{3\pi}\sum_f Q_f^2 \ln(m_W/m_f) + 1/6\pi + \ldots \tag{2a}$$

where the sum is over all charged fermions (a color factor of three must be included for quarks), Q_f = electric charge, and the ... refer to higher order terms. For $m_W = 84.4$GeV, $m_u = m_d = m_s = 0.1$GeV (effective mass values consistent with a dispersive analysis of $e^+e^- \to$hadrons [7,9]) $m_c = 1.25$GeV, $m_b = 4.5$GeV, $m_t = 20$GeV and including QCD corrections (with $\Lambda_{\overline{MS}} = 0.4$GeV), we find

$$\hat{\alpha}^{-1}(m_W) = 127.49 \tag{2b}$$

(Our subsequent SU(5) analysis is independent of m_t in the leading log approximation.).

1.2 $\underline{\Lambda_{\overline{MS}}}$: The QCD coupling $\hat{\alpha}_3(\mu) = \hat{g}_3^2(\mu)/4\pi$ is also defined by modified minimal subtraction. It is often parametrized by a mass scale $\Lambda_{\overline{MS}}$ such that for $m_c < \mu < m_b$ (i.e. an effective 4 flavor theory) [10]

$$\hat{\alpha}_3(\mu) = \frac{12\pi}{25\ln(\mu^2/\Lambda^2_{\overline{MS}})}\left[1 - \frac{462}{625}\frac{\ln\ln(\mu^2/\Lambda^2_{\overline{MS}})}{\ln(\mu^2/\Lambda^2_{\overline{MS}})}\right] . \tag{3}$$

Using this expression, fits to scaling violations in electroproduction find $\Lambda_{\overline{MS}} \simeq 0.4$GeV. However, perturbative analyses of $e^+e^- \to$ hadronic jets and upsilon decay tend to give somewhat smaller values near $\Lambda_{\overline{MS}} \simeq 0.2$GeV. Cognizant of the uncertainty in the extraction of $\Lambda_{\overline{MS}}$ from present day low energy experiments, we will employ a conservative range $0.1\text{GeV} \leq \Lambda_{\overline{MS}} \leq 0.6\text{GeV}$.

For later purposes, we need to know the value of $\hat{\alpha}_3(m_W)$. It is obtained by integrating the β function equation

$$\mu\frac{\partial}{\partial\mu}\hat{\alpha}_3(\mu) = -\frac{1}{2\pi}(11-2N_F/3)\hat{\alpha}_3^2 - \frac{1}{8\pi^2}(102-38N_F/3)\hat{\alpha}_3^3 + \ldots \tag{4}$$

from $\mu = \sqrt{10}$ GeV (with $\hat{\alpha}_3(\sqrt{10})$ given by Eq. (3)) up to $\mu = m_W$, increasing the number of flavors N_F from 4 to 5 at $\mu = m_b$ and from 5 to 6 at $\mu = m_t$ (i.e. we employ a step function β). That procedure gives

$$\hat{\alpha}_3^{-1}(m_b) = \hat{\alpha}_3^{-1}(\sqrt{10}) + \frac{25}{6\pi}\ln(m_b/\sqrt{10}) + \frac{77}{50\pi}\ln\left(\frac{\hat{\alpha}_3(\sqrt{10})}{\hat{\alpha}_3(m_b)}\right) \tag{5a}$$

$$\hat{\alpha}_3^{-1}(m_t) = \hat{\alpha}_3^{-1}(m_b) + \frac{23}{6\pi}\ln(m_t/m_b) + \frac{29}{23\pi}\ln\left(\frac{\hat{\alpha}_3(m_b)}{\hat{\alpha}_3(m_t)}\right) \tag{5b}$$

$$\hat{\alpha}_3^{-1}(m_W) = \hat{\alpha}_3^{-1}(m_t) + \frac{7}{2\pi}\ln(m_W/m_t) + \frac{13}{14\pi}\ln\left(\frac{\hat{\alpha}_3(m_t)}{\hat{\alpha}_3(m_W)}\right) \tag{5c}$$

Using $m_b = 4.5 + 0.2$ GeV $= 4.7$GeV (the extra 0.2GeV takes into account b quark threshold effects which are neglected in extracting $\Lambda_{\overline{MS}}$ from experiments at $q^2 = -10\text{GeV}^2$), $m_t = 20$GeV and values of m_W consistent with our subsequent SU(5) analysis, we find for a given $\Lambda_{\overline{MS}}$ the following iterative solutions to Eq. (5) and values for $\hat{\alpha}^{-1}(m_W)$

$\Lambda_{\overline{MS}}$ (GeV)	m_W (GeV)	$\hat{\alpha}_3(\sqrt{10})$	$\hat{\alpha}_3(m_b)$	$\hat{\alpha}_3(m_t)$	$\hat{\alpha}_3(m_W)$	$\hat{\alpha}^{-1}(m_W)$
0.1	82.8	0.1732	0.1576	0.1217	0.1015	127.56
0.2	83.6	0.2106	0.1876	0.1385	0.1127	127.53
0.3	84.0	0.2423	0.2119	0.1510	0.1207	127.51
0.4	84.4	0.2721	0.2340	0.1616	0.1271	127.49
0.5	84.6	0.3019	0.2550	0.1710	0.1328	127.47
0.6	84.8	0.3325	0.2759	0.1797	0.1379	127.46

Table I

The above values of $\hat{\alpha}(m_W)$ and $\hat{\alpha}_3(m_W)$ will be used in section II to determine the SU(5) unification mass m_S.

1.3 $\sin^2\theta_W$: Presently the best experimental determination of $\sin^2\theta_W$ comes from deep-inelastic neutrino scattering via the measurement of $R_\nu \equiv \sigma(\nu_\mu + N \to \nu_\mu + X)/\sigma(\nu_\mu + N \to \mu^- + X)$. To obtain a precise value for $\sin^2\theta_W$, we have included the $O(\alpha)$ radiative corrections[3] to both the neutral and charged current cross-sections in R_ν [3,8,9,11,12]. Radiative corrections affect the extraction of $\sin^2\theta_W^{exp}$ from R_ν in the following ways: 1) the parameter ρ^2 which in lowest order is 1 for the $SU(2)_L \times U(1)$ model with Higgs isodoublets gets modified to [3]

$$\rho^2 = 1 - \frac{\alpha}{\pi}\Big[\ln\Big(\frac{m_Z^2}{2\langle -q^2\rangle}\Big) + 2 - \frac{3}{8s^4}\ln(c^2) + \frac{7}{8s^2} - \frac{1}{2c^4s^2}\Big(\frac{5}{2} - \frac{15s^2}{4} - \frac{s^4}{5} + \frac{14s^6}{9}\Big)$$

$$- \frac{3\xi}{8s^2}\Big(\frac{\ln(c^2/\xi)}{c^2-\xi} + \frac{1}{c^2}\frac{\ln(\xi)}{1-\xi}\Big) - \frac{3m_t^2}{8s^2m_W^2}\Big] + O(\alpha^2) \tag{6}$$

where $c^2 \equiv \cos^2\theta_W$, $s^2 \equiv \sin^2\theta_W$, $\xi \equiv m_\phi^2/m_Z^2$ and $\langle -q^2\rangle$ is the average momentum transfer. Taking the lowest order value $s^2 \simeq 0.23$ and using $\xi = 1$, $m_t = 20$GeV and $\langle -q^2\rangle = 20\text{GeV}^2$, we find [3]

$$\rho^2 = 0.983 \tag{7}$$

which is consistent with the experimental result [13] $\rho^2 = 0.998 \pm 0.050$.

Because ρ^2 and $\sin^2\theta_W^{exp}$ are correlated, a shift in ρ^2 away from 1 causes a change in the value of $\sin^2\theta_W^{exp}$ extracted from R_ν given by [3]

$$\delta s^2 = 0.49(\rho^2 - 1) = -0.0084 \ , \tag{8}$$

a $3 \sim 4\%$ reduction. 2) Analyses of R_ν in isoscalar targets usually assume equal u and d quark distribution functions (because of isospin symmetry). That equality is broken by radiative corrections. We find that this effect further reduces $\sin^2\theta_W^{exp}$ by [3,11,14]

$$\delta s^2 = -0.062\frac{\alpha}{2\pi}\left[\frac{4}{3}\ell n\left(\frac{2<-q^2>}{m^2}\right) + \frac{1}{3}\left(\pi^2 - \frac{19}{4}\right)\right] \simeq -0.0007 \tag{9}$$

where $m \simeq 0.3$GeV is the constituent target quark mass. 3) Hadronic energy cuts on the data which vary from experiment to experiment generally require the final state hadronic energy to be larger than some minimum value E_H^{cut}. Although adjustments are made to account for such cuts, they generally employ lowest order cross-section formulas. Therefore, each experiment must be separately corrected for the effect of radiative corrections on the cuts. Taking the CDHS experiment [15] for which E_H^{cut}=10GeV is an example, the effect of radiative corrections on this cut lowers $\sin^2\theta_W^{exp}$ by an additional [3]

$$\delta s^2 = -0.0035 \qquad (\text{For } E_H^{cut} = 10\text{GeV}) \ . \tag{10}$$

Applying the above mentioned effects of radiative corrections to five distinct measurements of R_ν (which are separately adjusted for charm threshold effects, sea quark effects, QCD corrections, etc. via the analysis in ref. 13), we obtain the corrected values for $\sin^2\theta_W^{(\nu_\mu;h)}(-20\text{GeV}^2)$ given in table II.

Collaboration	R_ν	E_H^{cut}	$\sin^2\theta_W^{(\nu_\mu;h)}$ (uncorrected)	$\sin^2\theta_W^{(\nu_\mu;h)}$ (corrected)
CDHS [15]	0.307 ± 0.008	10GeV	0.230 ± 0.013	0.217 ± 0.013
CHARM [16]	0.320 ± 0.010	2GeV	0.220 ± 0.015	0.211 ± 0.015
BEBC [17]	0.320 ± 0.030	15GeV	0.217 ± 0.045	0.203 ± 0.045
CITF [18]	0.280 ± 0.030	12GeV	0.272 ± 0.055	0.259 ± 0.055
HPWF [19]	0.30 ± 0.04	4GeV	0.274 ± 0.075	0.264 ± 0.075
	Weighted Average:		0.227 ± 0.010	0.216 ± 0.010

TABLE II

The corrected average value $\sin^2\theta_W^{(\nu_\mu;h)}(-20\text{GeV}^2) = 0.216 \pm 0.010$ given in Table II corresponds to an experimental definition of the weak mixing angle at $q^2 = -20\text{GeV}^2$. To obtain the corresponding value for a q^2 and experiment independent quantity such as $\sin^2\theta_W \equiv 1 - m_W^2/m_Z^2$ [9] or $\sin^2\hat{\theta}_W(m_W)$ defined by modified minimal subtraction [7,12], one uses

$$\begin{aligned} \sin^2\theta_W^{(\nu_\mu;h)}(-20\text{GeV}^2) &= \kappa^{(\nu_\mu;h)}(-20\text{GeV}^2)\sin^2\theta_W \\ &= \hat{\kappa}^{(\nu_\mu;h)}(-20\text{GeV}^2)\sin^2\hat{\theta}_W(m_W) \end{aligned} \tag{11a}$$

with [7,11] $\kappa^{(\nu_\mu;h)}(-20\text{GeV}^2) = 0.997$ and $\hat{\kappa}^{(\nu_\mu;h)}(-20\text{GeV}^2) = 1.005$. In that way we find [3]

$$\sin^2\theta_W = 0.217 \pm 0.010 \pm 0.004 \quad \text{(From } R_\nu\text{)} \tag{11b}$$

$$\sin^2\hat{\theta}_W(m_W) = 0.215 \pm 0.010 \pm 0.004 \quad \text{(From } R_\nu\text{)} \tag{11c}$$

where the ±0.004 is the overall theoretical uncertainty in this analysis.

In the case of the SLAC e-D asymmetry measurement [20], we found after including the dominant radiative corrections [12]

$$\sin^2\hat{\theta}_W(m_W) = 0.216 \pm 0.020 \quad \text{(e-D asymmetry)}\,. \tag{12}$$

Averaging the results in Eqs. (11c) and (12) we obtain

$$\sin^2\hat{\theta}_W(m_W) = 0.215 \pm 0.012 \quad \text{(Average)}\,. \tag{13}$$

This is the value that should be compared with predictions from grand unified theories.

From our results in Eqs. (11) - (13) and our previous calculations [6,9,12,21] of the radiative corrections to $W^\pm$ and Z^o mass formulas[4] we find

$$m_W = \frac{38.64\text{GeV}}{\sin\theta_W} = \frac{38.5\text{GeV}}{\sin\hat{\theta}_W(m_W)} = 83.0 \pm 2.4 \text{ GeV} \tag{14a}$$

$$m_Z = m_W/\cos\theta_W = 93.8 \pm 2.0 \text{ GeV}\ . \tag{14b}$$

(For changes in these predictions when $m_\phi \neq m_Z$ or $m_t > m_W$, see ref. 21.) We note that in total the radiative corrections to the experimental determination of $\sin^2\theta_W$ and the $W^\pm$, Z^o mass formulas increase m_W and m_Z by about 5% above lowest order predictions, a large measurable effect. Indeed, it is anticipated that the Z^o's mass will be measured to within $0.1 \sim 0.2$ GeV in high energy e^+e^- colliders; thereby probing higher order electroweak radiative corrections.

2. Grand Unified Theories

In this section we present the predictions of a class of grand unified theories (GUTS) generically denoted by G, the best known and simplest of which is the Georgi-Glashow SU(5) model [1]. We will assume that G contains only two symmetry breaking mass scales m_S (the super-heavy unification mass) and m_W, such that

$$G \underset{m_S}{\rightarrow} SU(3)_c \times SU(2)_L \times U(1) \underset{m_W}{\rightarrow} SU(3)_c \times U(1)$$

(This is certainly the case for the SU(5) model.) Taking G to be a simple Lie group requires $g_{G_o} = g_{3_o} = g_{2_o} = g_{1_o}$. In addition we assume charge assignments such that (as in the SU(5) model) $\sin^2\theta_W^o = e_o^2/g_{2_o}^2 = 3/8$. (Small subscript or superscript zero denotes bare unrenormalized parameters.) As a simplification, we assume that all physical Higgs scalars which are isodoublets under $SU(2)_L$ have mass m_W (we allow N_H such doublets) and all other physical scalars have mass m_S. (Our analysis is easily extended to models with several intermediate mass scales between m_W and m_S.)

2.1 SU(5) Analysis: Making the above assumptions, the effective low energy $SU(3)_c \times SU(2)_L \times U(1)$ couplings $\hat{\alpha}_i(m_W) = \hat{g}_i^2(m_W)/4\pi$, i=1,2,3 can be easily computed in terms of the grand unified coupling (at mass scale m_S) by using renormalization group techniques. Such calculations were first carried out in the pioneering work of Georgi, Quinn and Weinberg [2] and have since been refined and extended by others [6,22]. Because we employ modified minimal subtraction $\overline{MS}$, our approach is particularly simple; so we will describe it in detail.

The effective low energy couplings $\hat{\alpha}_i(m_W)$ i=1,2,3, $\hat{\alpha}(m_W)$ and $\sin^2\hat{\theta}_W(m_W)$ described in the first part of this talk are obtained by integrating the effective β function equations (super-heavy particles decouple and therefore do not contribute to the couplings' evolutions for $\mu < m_S$)

$$\mu\frac{\partial}{\partial\mu}\hat{\alpha}_i(\mu) = b_i\hat{\alpha}_i^2 + b_{ij}\hat{\alpha}_i^2\hat{\alpha}_j + \ldots, \quad i,j = 1,2,3 \tag{15a}$$

where

$$b_1 = -\frac{1}{2\pi}\left(-\frac{4}{3}n_g - \frac{1}{10}N_H\right)$$
$$b_2 = -\frac{1}{2\pi}\left(\frac{22}{3} - \frac{4}{3}n_g - \frac{1}{6}N_H\right) \tag{15b}$$
$$b_3 = -\frac{1}{2\pi}\left(11 - \frac{4}{3}n_g\right)$$

$$b_{ij} = -\frac{1}{8\pi^2}\begin{pmatrix} -\frac{19}{15}n_g & -\frac{3}{5}n_g & -\frac{44}{15}n_g \\ -\frac{1}{5}n_g & \frac{136}{3}-\frac{49}{3}n_g & -4\,n_g \\ -\frac{11}{30}n_g & -\frac{3}{2}n_g & 102-\frac{76}{3}n_g \end{pmatrix} \tag{15c}$$

(n_g = number of fermion generations) from $\mu = m_S$ to m_W using the boundary condition [12]

$$\hat{\alpha}_1^{-1}(m_S) = \hat{\alpha}_2^{-1}(m_S) - 1/6\pi = \hat{\alpha}_3^{-1}(m_S) - 1/4\pi = \alpha_i^{-1}(m_S) - 5/12\pi \tag{16}$$

($\alpha_i(m_S)$ is the $\overline{MS}$ coupling in the full SU(5) theory with $\mu = m_S$) and

$$\hat{\alpha}^{-1}(m_W) = \frac{5}{3}\hat{\alpha}_1^{-1}(m_W) + \hat{\alpha}_2^{-1}(m_W) \tag{17a}$$

$$\sin^2\hat{\theta}_W(m_W) = \hat{\alpha}(m_W)/\hat{\alpha}_2(m_W) \tag{17b}$$

Integrating Eq. (15) we find

$$\hat{\alpha}_i^{-1}(m_W) = \hat{\alpha}_i^{-1}(m_S) + b_i \ln(\frac{m_S}{m_W}) + \sum_{j=1}^{3}\frac{b_{ij}}{b_j}\ln(\frac{\hat{\alpha}_j(m_S)}{\hat{\alpha}_j(m_W)}), \qquad i=1,2,3 \tag{18}$$

then using the relationships in Eqs. (16) and (17) we obtain

$$\frac{\hat{\alpha}(m_W)}{\hat{\alpha}_3(m_W)} = \frac{3}{8}[1 - \frac{66+N_H}{6}\frac{\hat{\alpha}(m_W)}{\pi}\ln(\frac{m_S}{m_W}) + \frac{\hat{\alpha}(m_W)}{2\pi} + \frac{\hat{\alpha}(m_W)}{4\pi}\left\{\frac{272-\frac{176}{3}n_g}{11-\frac{4}{3}n_g}\ln(\frac{\hat{\alpha}_3(m_S)}{\hat{\alpha}_3(m_W)}) + \frac{-\frac{136}{3}+\frac{40}{3}n_g}{\frac{22}{3}-\frac{4}{3}n_g-\frac{1}{6}N_H}\ln(\frac{\hat{\alpha}_2(m_S)}{\hat{\alpha}_2(m_W)}) + \frac{\frac{4}{3}n_g}{-\frac{4}{3}n_g-\frac{1}{10}N_H}\ln(\frac{\hat{\alpha}_1(m_S)}{\hat{\alpha}_1(m_W)})\right\}] \tag{19}$$

and

$$\sin^2\hat{\theta}_W(m_W) = \frac{3}{8}[1 - \frac{110-N_H}{18}\frac{\hat{\alpha}(m_W)}{\pi}\ln(\frac{m_S}{m_W}) + \frac{5\hat{\alpha}(m_W)}{18\pi} + \frac{\hat{\alpha}(m_W)}{4\pi}\left\{\frac{-\frac{16}{9}n_g}{11-\frac{4}{3}n_g}\ln(\frac{\hat{\alpha}_3(m_S)}{\hat{\alpha}_3(m_W)}) + \frac{\frac{680}{9}-\frac{236}{9}n_g}{\frac{22}{3}-\frac{4}{3}n_g-\frac{1}{6}N_H}\ln(\frac{\hat{\alpha}_2(m_S)}{\hat{\alpha}_2(m_W)}) + \frac{\frac{16}{9}n_g}{-\frac{4}{3}n_g-\frac{1}{10}N_H}\ln(\frac{\hat{\alpha}_1(m_S)}{\hat{\alpha}_1(m_W)})\right\}] \tag{20}$$

The results in Eqs. (19) and (20) are very convenient for obtaining the SU(5) predictions and examining their dependence on N_H and n_g, as we shall see.

2.2 Predictions

One can solve either Eq. (19) or (20) to obtain a prediction for m_S. Although $\sin^2\hat{\theta}_W(m_W)$ is better determined experimentally than $\hat{\alpha}(m_W)/\hat{\alpha}_3(m_W)$ (see Eq. (13)), Eq. (19) is less sensitive than Eq. (20) to small variations in the input parameters and hence generally employed. Solving Eq. (19) for m_S/m_W we find

$$\frac{m_S}{m_W} = \exp\Big[\frac{6\pi}{66+N_H}\Big(\frac{1}{\hat{\alpha}(m_W)} - \frac{8}{3\hat{\alpha}_3(m_W)} + \frac{1}{2\pi} + \frac{1}{4\pi}\Big\{\frac{272-\frac{176}{3}n_g}{11-\frac{4}{3}n_g}\ell n\Big(\frac{\hat{\alpha}_3(m_S)}{\hat{\alpha}_3(m_W)}\Big) + \frac{-\frac{136}{3}+\frac{40}{3}n_g}{\frac{22}{3}-\frac{4}{3}n_g-\frac{1}{6}N_H}\ell n\Big(\frac{\hat{\alpha}_2(m_S)}{\hat{\alpha}_2(m_W)}\Big) + \frac{\frac{4}{3}n_g}{-\frac{4}{3}n_g-\frac{1}{10}N_H}\ell n\Big(\frac{\hat{\alpha}_1(m_S)}{\hat{\alpha}_1(m_W)}\Big)\Big\}\Big)\Big] . \quad (21)$$

For $n_g=3$, $\hat{\alpha}_i(m_S)=0.0243$ and the values in table I (i.e. using $\hat{\alpha}_2(m_W)=\hat{\alpha}(m_W)/\sin^2\hat{\theta}_W(m_W) \simeq 0.037$ and $\hat{\alpha}_1(m_W) \simeq 0.0165$), we find

$$\frac{m_S}{m_W} = 6.5\times 10^{12}\times 10^{-0.19(N_H-1)}\times(\Lambda_{\overline{MS}}/0.4\text{GeV})^{1.04} \quad (22)$$

which when inserted into Eq. (20) yields the SU(5) prediction

$$\sin^2\hat{\theta}_W(m_W) = 0.2083+0.004(N_H-1)+0.006\ell n(0.4\text{GeV}/\Lambda_{\overline{MS}}) . \quad (23)$$

For $N_H \geq 1$ and $0.1\text{GeV} \leq \Lambda_{\overline{MS}} \leq 0.6\text{GeV}$, the SU(5) prediction in Eq. (23) is in excellent agreement with the experimental average $\sin^2\hat{\theta}_W(m_W) = 0.215 \pm 0.012$ found after we included radiative corrections (see Eq. (13)); a tremendous success for the SU(5) model.

Using $m_W = 38.5\text{GeV}/\sin\hat{\theta}_W(m_W)$ and the SU(5) estimate for the proton lifetime[5] (with $n_g = 3$) [22,23]

$$\tau_p \simeq 10^{-28\pm1}(m_S \text{ in GeV})^4 \text{ yr.} \quad (24)$$

we find from Eq. (22)

$$\tau_p \simeq 9\times 10^{30\pm1}\times 10^{-0.76(N_H-1)}\times(\Lambda_{\overline{MS}}/0.4\text{GeV})^{4.16} \text{ yr.} \quad (25)$$

Note that each increase in N_H by one Higgs doublet decreases the proton lifetime by a factor of 0.17. Given the present experimental bound [24]

$$\tau_p^{exp} \gtrsim 10^{30} \text{ yr.} \quad (26)$$

one sees that there cannot be too many light Higgs doublets in the SU(5) model, i.e. from Eq. (25) $N_H \lesssim 3$. For the minimum case $N_H=1$, we have illustrated the SU(5) model's predictions as a function of $\Lambda_{\overline{MS}}$ in table III

$\Lambda_{\overline{MS}}$ (GeV)	$\sin^2\hat{\theta}_W(m_W)$	m_W (GeV)	m_Z (GeV)	m_S (GeV)	τ_p (yr)
0.1	0.216	82.8	93.6	1.4×10^{14}	$4 \times 10^{28\pm1}$
0.2	0.212	83.6	94.2	2.7×10^{14}	$5 \times 10^{29\pm1}$
0.3	0.210	84.0	94.6	4.1×10^{14}	$3 \times 10^{30\pm1}$
0.4	0.208	84.4	94.9	5.5×10^{14}	$9 \times 10^{30\pm1}$
0.5	0.207	84.6	95.1	7.0×10^{14}	$2 \times 10^{31\pm1}$
0.6	0.206	84.8	95.3	8.5×10^{14}	$5 \times 10^{31\pm1}$

TABLE III

If any quantity in table III is precisely measured (m_Z will be determined to within $0.1 \sim 0.2$ GeV by proposed e^+e^- colliders), then within the framework of SU(5), all the others are known. We note that the SU(5) model favors a proton lifetime in the range $10^{30} \sim 10^{32}$ yrs. which is within easy reach of the ongoing and proposed experiments discussed at this meeting.

3. A Fourth Generation?

Is there a fourth generation of fermions? Some large GUTS based on SO(N) groups require an even number of generations [25], thereby suggesting at least a fourth generation. In addition, mass sum rules derived in some dynamical symmetry breaking schemes call for fermion masses $\gtrsim m_W$ [26], implying the existence of new heavy fermions. Furthermore, there are no arguments against a fourth generation[6]; hence we consider it a viable possibility. Indeed, its introduction may eventually help shed light on the origin of CP violation and provide us with some basic understanding of the quark masses and mixing angles.

In this section we describe the effect of a fourth generation on the SU(5) predictions for $\sin^2\hat{\theta}_W(m_W)$ and τ_p.[7] Calling the fourth generation members ν_E, E, U and D, we assume $m_E \simeq 33$GeV and $m_D \simeq m_W$. (These values simplify our analysis and seem like reasonable guesses; the final results are rather insensitive to this specific assumption.) Our results do not depend on m_U (in leading log); however, one also expects $m_U \simeq m_W$. For these mass values $\hat{\alpha}_3(m_W)$ is unchanged while $\hat{\alpha}^{-1}(m_W)$ decreases by 0.20 and (for n_g=4) $\hat{\alpha}_i(m_S)$ increases from 0.0243 to 0.0287. Using these values and n_g=4 in Eq. (21), we find

$$\frac{m_S}{m_W} = 7.9 \times 10^{12} \times 10^{-0.19(N_H-1)} \times (\Lambda_{\overline{MS}}/0.4\text{GeV})^{1.04} \tag{27}$$

Comparing this with Eq. (22) one sees that the unification mass m_S is increased by about 20%. Next, inserting this expression into Eq. (20) yields

$$\sin^2\hat{\theta}_W(m_W) = 0.2079 + 0.004(N_H-1) + 0.006\ln(0.4\text{GeV}/\Lambda_{\overline{MS}}) \tag{28}$$

a slight decrease of 0.0004 compared with the 3 generation prediction in Eq. (23), i.e. only a 0.2% reduction.

To obtain the proton lifetime prediction appropriate for the case of four generations, we must modify the usual SU(5) results in the following ways. Since $\hat{\alpha}_i(m_S)$ is larger for four generations, this effect directly reduces τ_p (which is proportional to $\hat{\alpha}_i^{-2}(m_S)$) by a factor of 0.72. Also, the one loop enhancements [27] to the decay rate formulas are increased, which further reduces τ_p by a factor of 0.84. In total, we find for four generations the prediction in Eq. (24) becomes

$$\tau_p \simeq 0.6 \times 10^{-28\pm1} \times (m_S \text{ in GeV})^4 \text{ yr.} \tag{29}$$

Inserting the value of m_S obtained from Eq. (27) in this formula gives

$$\tau_p \simeq 1.2 \times 10^{31\pm1} \times 10^{-0.76(N_H-1)} \times (\Lambda_{\overline{MS}}/0.4\text{GeV})^{4.16} \text{ yr.} \tag{30}$$

Comparing this result with the 3 generation prediction in Eq. (25), we find that τ_p has increased overall by about 30%, not a very significant change. In summary, going from 3 to 4 generations causes fairly insignificant modifications in the SU(5) model's predictions.

The basic idea of grand unification is very appealing. The simplest model based on SU(5) has scored an important success in predicting a value for $\sin^2\hat{\theta}_W(m_W)$ which is in excellent agreement with recent experimental findings (after radiative corrections are included). It makes an additional dramatic prediction that the proton will decay with a lifetime in the range $10^{30} \sim 10^{32}$ yrs. If correct, such decays will be seen by the planned experiments within the coming year (or may have already been seen). An incredible discovery may be awaiting us.

Endnotes

1. Our $g_1 = \sqrt{5/3}\, g'$, where g' is the usual U(1) coupling.
2. A recent analysis of $K_L \to \mu^+\mu^-$ and the K_L-K_S mass difference in the six quark model by A. Buras [28] finds $m_t < 33$GeV.
3. Leading log radiative corrections to R_ν have been studied by the authors of ref. 29.
4. Radiative corrections to the $W^\pm$ and Z^o masses have also been analyzed in ref. 30.
5. A detailed review of proton decay in GUTS has been given by P. Langacker [23].
6. Helium production within the framework of big bang cosmology suggests the bound $N_\nu \leq 3 \sim 4$ for the number of distinct neutrino species [31]. This bound is consistent with 4 generations.
7. The effect of a fourth generation on m_S and $\sin^2\hat{\theta}_W(m_W)$ has been independently examined by M. Fischler and C. Hill. Their results agree with ours (private communication).

References

[1] H. Georgi and S. Glashow, Phys. Rev. Lett. 32, 438 (1974).

[2] H. Georgi, H. Quinn and S. Weinberg, Phys. Rev. Lett. 33, 451 (1974).

[3] A. Sirlin and W. Marciano, "Radiative Corrections to $\nu_\mu + N \to \mu^- + X$ and their Effect on the Determination of ρ^2 and $\sin^2\theta_W$", NYU preprint TR3/81, April 1981.

[4] T. Kinoshita, in New Frontiers in High-Energy Physics edited by A. Perlmutter and L. Scott (Plenum Press, New York 1978) p 127.

[5] G. 't Hooft, Nucl. Phys. B61, 455 (1973).

[6] W. Marciano, Phys. Rev. D20, 274 (1979).

[7] W. Marciano and A. Sirlin, Phys. Rev. Lett. 46, 163 (1981).

[8] W. Marciano, in Particles and Fields-1979, edited by B. Margolis and D.G. Stairs, AIP Conference Proceedings No. 59 (American Institute of Physics, New York, 1980) p 373.

[9] A. Sirlin, Phys. Rev. D22, 971 (1980).

[10] A. Buras, Rev. Mod. Phys. 52, 199 (1980).

[11] W. Marciano and A. Sirlin, Phys. Rev. D22, 2695 (1980).

[12] W. Marciano and A. Sirlin, "SU(5) predictions: τ_p, $\sin^2\theta_W^{exp}$, m_W and m_Z" to be published in the proceedings of the VPI workshop on Weak Interactions as Probes of Unification (1980).

[13] J. Kim, P. Langacker, M. Levine and H. Williams, Rev. Mod. Phys. 53, 211 (1978); I. Liede and M. Roos, Nucl. Phys. B167, 397 (1980).

[14] J. Kiskis, Phys. Rev. D8, 2129 (1973); A. DeRújula, R. Petronzio and A. Savoy-Navarro, Nucl. Phys. B154, 394 (1979).

[15] CDHS Collab. C. Geweniger, Proc. Intern. Conf. Neutrino' 79 (Bergen 1979), eds. A. Haatuft and C. Jarlskog, Vol. 2 (Bergen, Norway, 1979) p 392.

[16] CHARM Collab., M. Jonker et al., Phys. Lett. 99B, 265 (1981).

[17] BEBC ABCLOS Collab., M. Deden et al., Nucl. Phys. B149, 1 (1979).

[18] CITF Collab., F. Merritt et al., Phys. Rev. D17, 2199 (1978).

[19] HPWF Collab., P. Wanderer et al., Phys. Rev. D17, 1679 (1978).

[20] C. Prescott et al., Phys. Lett. 77B, 347 (1978); 84B, 524 (1979).

[21] W. Marciano and A. Sirlin, in Proceedings of the Cornell Z^o Theory Workshop, edited by M. Peskin and S-H. Tye, 1981, p40.

[22] T. Goldman and D. Ross, Phys. Lett. 84B, 208 (1979); Nucl. Phys. B171, 273 (1980); W. Marciano, in "Orbis Scientiae, Recent Developments in High Energy Physics", Jan. 1980 edited by A. Perlmutter and L. Scott (plenum, New York) p121; S. Weinberg, Phys. Lett. 91B, 51 (1980); L. Hall, Harvard University preprint (1980); P. Binetray and T. Schucker, CERN preprints (1979) & (1980) N. Chang, A. Das and J. Perez-Mercader, CCNY preprint (1979); I. Antoniadis, C. Bouchiat and J. Iliopoulos, Phys. Lett. 97B, 367 (1980); C. Llewellyn Smith, G. Ross and J. Wheater, Nucl. Phys. B177, 263 (1981); J. Ellis et al. CERN preprint TH-2833 (1980).

[23] P. Langacker, "Grand Unified Theories and Proton Decay", SLAC preprint 2544 (1980).

[24] J. Learned, F. Reines and A. Soni, Phys. Rev. Lett. 43, 907 (1979).

[25] A. Zee, Univ. of Penn. preprint UPR-0151T (1980).

[26] A. Carter and H. Pagels, Phys. Rev. Lett. 25, 43 (1979).

[27] A. Buras, J. Ellis, M.K. Gaillard and D. Nanopoulos, Nucl. Phys. B135, 66 (1978); F. Wilczek and A. Zee, Phys. Rev. Lett. 43, 1571 (1979); J. Ellis, M.K. Gaillard and D. Nanopoulos, CERN preprint TH.2749 (1980).

[28] A. Buras, Phys. Rev. Lett. 46, 1354 (1981).

[29] F. Antonelli and L. Maiani, Univ. of Rome preprint (1980); S. Dawson, J. Hagelin and L. Hall, Harvard Univ preprint (1980).

[30] M. Veltman, Phys. Lett. 91B, 95 (1980); Erratum M. Green and M. Veltman, Nucl. Phys. B169, 137 (1980); F. Antonelli, M. Consoli and G. Corbo, Phys. Lett. 91B, 90 (1980).

[31] See, M. Turner, in Proceedings of this Conference.

EFFECTIVE LAGRANGIAN AND EFFECTIVE PARAMETERS IN GRAND UNIFIED THEORIES†

York-Peng Yao

Department of Physics, University of Michigan
Ann Arbor, Michigan 48109

I. Introduction

This conference is devoted to discussing a very appealing idea that particles interact with a universal strength at the most fundamental level[1]. The observed strengths in various interactions are apparently not the same, because there exist mass hierarchies and because different effective coupling constants suffer different radiative corrections and evolve separately.

Let us accept this premise of universality in a unification scheme, which is supposed to occur at extremely high energy. Naturally one wonders whether one can obtain a low energy effective theory to describe the more common physical processes.

In mathematics, let the grand unification Lagrangian be $\mathcal{L}(\phi)$, in which some of the fields describe heavy particles, ϕ_{heavy}, while the others describe light particles, ϕ_{light}. Let us suppose that the energy of reaction is way below the thresholds of producing heavy particles. The question is whether we can reduce $\mathcal{L}(\phi)$ to $\mathcal{L}_{eff}(\phi_{light})$, which can be used to construct all the relevant matrix elements.

It seems that an answer can be readily given, because the generating functional is

$$e^{iW} = \int d\phi \; e^{i\int d^4x \, (\mathcal{L}+J\phi)} , \tag{I.1}$$

in which J is the source function, the variation of which can produce an external ϕ.

If an external ϕ_{heavy} cannot be made, we can just turn off J_{heavy} and 'integrate out' ϕ_{heavy}, i.e.

†Work supported partially by the U.S. Department of Energy

$$\int d\phi_{\text{heavy}}\, e^{i\int d^4x\,\mathcal{L}} = e^{i\int d^4x\,\mathcal{L}_{\text{eff}}} . \tag{I.2}$$

A formal effective Lagrangian is thus obtained. There are in fact many problems in this approach when one tries to simplify $\mathcal{L}_{eff}$, they have been discussed by Weinberg, Ovrut, Schnitzer, Hagiwara, Nakazawa, Hall, Weisberger, etc.[2] I would like to focus on some of them, which are caused by quantum effects:

(1) We should ask: What does one mean by 'integrating out' the heavy fields? There are complications at the quantum level, because the light and the heavy fields mix at every order in a loop expansion. We must make sure that there is a sharp separation of the heavy sector from the light sector in order to know what to 'integrate out'. This was a problem first pointed out by Gildener[3] and I shall call it an issue in the gauge hierarchy.

(2) Is there a limit such that $\mathcal{L}_{eff}$ is in fact local and renormalizable? I take this as a physical requirement that such a limit should exist if grand unification is valid. The reason is that our present wealth of physics is well described by the standard model $(SU(3))_{color} \times (SU(2) \times U(1))_{electroweak}$. However, it is by no means obvious that a local Lagrangian will emerge right after the functional integration, even if we restrict ourselves to a certain order of accuracy in external momenta relative to a heavy mass scale (M). The reason is simple: external momenta correspond to the derivative ∂ acting on ϕ_{light} in Eq. (I.2), but not every ∂ is external. Thus, it is very risky to make cavalierly the approximation

$$M \gg \partial \tag{I.3}$$

in $\mathcal{L}_{eff}$. This approximation assumes that the ultraviolet cutoff and the large mass limit are interchangeable. They are not.

We may ask: why do we trouble ourselves to find a local renormalizable effective Lagrangian, $\mathcal{L}_{eff}$? Why don't we just use $\mathcal{L}$ to calculate? The answer has to do with simplicity and with necessity. It is obvious that a local $\mathcal{L}_{eff}$ is simpler in structure since it has only light fields. (As Veltman has informed us, there are more than five hundred vertices in the full Lagrangian of SU(5).) More importantly, it is a necessity to have $\mathcal{L}_{eff}$, because the effective expansion parameter is not just the universal coupling constant g, but also $g^2 \ln M/\mu$, in which μ is the energy scale at which we do our everyday physics. This logarithm is not a small number. We must devise a way to sum powers of these logarithms to perform an improved perturbation expansion. $\mathcal{L}_{eff}$ serves just that purpose.

I am going to describe to you the work by Kazama, Unger and myself.[4]

We work within the framework of gauge theories with spontaneously broken symmetry. Consider a symmetry G, which is broken because one of the scalar particles acquires a big vacuum expectaton value v_1. There may ensue stages of symmetry breaking, but let us assume for simplicity that the next stage is that another scalar particle acquires a vacuum expectation value v_2, which is much smaller, i.e.

$$v_1 \gg v_2. \tag{I.4}$$

Thus, the successive symmetry reductions are

$$\begin{aligned} G &\underset{v_1}{\rightarrow} (SU(3))_{color} \times (SU(2)\times U(1))_{electroweak} \\ &\underset{v_2}{\rightarrow} (SU(3))_{color} \times U(1)_{EM}. \end{aligned} \tag{I.5}$$

We have obtained the following all order results:

(1) Light particles have small masses to all orders without the need of order by order fine tuning. Radiative effects give only logarithm corrections, i.e.

$$m_{light} \sim v_2 \,(1+ \sum_{loop} a\,(\ln v_1/\mu)^n) \tag{I.6}$$

We must emphasize that the naturality question, why $v_1 \gg v_2$, is not addressed by us.

(2) There exists an effective local renormalizable Lagrangian, which can be used to reproduce all the light particle matrix elements when $p_i \ll v_1$ to $\mathcal{O}(1)$ accuracy. Specifically, let $\mathcal{L}(g,v_1,v_2,\alpha,\mu)$ be the

Lagrangian of the complete theory, where g is a set of coupling constants, α is a gauge parameter, and μ is a dimensionful scale introduced in dimensional regularization. Through $\mathcal{L}$, we construct Γ^n, which is the amputated Green's function with n light external particles, and which is one particle irreducible with respect to light internal lines. Similarly, let $\mathcal{L}^*(g^*,v^*,\alpha^*,\mu)$ be the reduced effective Lagrangian, which is obtained from $\mathcal{L}$ by discarding all terms with heavy fields and rescaling some of the coupling constants, v_2, and the gauge parameters properly. Let Γ^{*n} be the amputated 1PI Green's function with the same n light particles, which is calculated according to $\mathcal{L}^*$. We have shown that

$$\Gamma^n = Z^{n/2}\,\Gamma^{*n} + 0\,(\frac{1}{v_1^2}) \tag{I.7}$$

where Z is a finite wave function renormalization constant.

(3) There is a set of renormalization group equations, which relate (g,v_1,v_2,α) to the effective parameters (g^*,v^*,α^*). The procedure to accomplish this is straightforward and is done within the minimal subtraction scheme.

II. Effective Lagrangian

There are several steps we take to arrive at an effective Lagrangian $\mathcal{L}^*$.

(1) It is expedient that we should deal directly with the renormalized tree vacuum expectation values v_1 and v_2. In other words, the perturbation series will be built with v_1 and v_2 as input parameters. If we did not do so, we would find that in the ith loop order

$$\begin{aligned}(v_1)^i &= (1+a_1g^2)(v_1)^{i-1} + a_2g^2\,(v_2)^{i-1},\\ (v_2)^i &= b_2g^2(v_1)^{i-1} + (1+b_1g^2)(v_2)^{i-1}.\end{aligned} \tag{II.1}$$

This was a source of confusion. If we argued that $v_1^{true} >> v_2^{true}$, we would need to do very fine tuning at each order to achieve it. In fact, it would seem that the natural solution should be

$$v_2^{true}/v_1^{true} \sim O(g^2) \,, \tag{II.2}$$

which would deal a mortal blow to the development of grand unified theories - protons would have decayed in no time!

What happens is that in such a way of doing the calculation, the tadpoles are non-vanishing, because we are not at the true minimum of the potential. These tadpoles also have corrections of $O(v_1)$. When we calculate the masses of the light fields, correctly incorporating all the terms we find that $O(v_1)$ terms cancel, so that light particles remain light.

(2) As we said earlier, heavy and light states mix further in each order. We shall see that one light particle irreducible Green's functions automatically take this mixing into account to the order of accuracy we want.

Let us illustrate how this works in a simple case when we have just one heavy scalar which mixes with one light scalar. The mass matrix of this subsystem is

$$\mathcal{M}^2 = \begin{pmatrix} M_H^2 + \Sigma_{HH} & \Sigma_{hH} \\ \Sigma_{hH} & m_h^2 + \Sigma_{hh} \end{pmatrix} . \tag{II.3}$$

The orthogonal matrix which diagonalizes $\mathcal{M}^2$ is

$$\mathcal{O} = \begin{pmatrix} \cos\delta & \sin\delta \\ -\sin\delta & \cos\delta \end{pmatrix} \tag{II.4}$$

with

$$\sin\delta = -\frac{\Sigma_{hH}}{M_H^2 + \Sigma_{HH}} + \mathcal{O}\left(\frac{1}{v_1^3}\right) \tag{II.5}$$

in which we have used the following power counting results

$$\Sigma_{hh} \cong \mathcal{O}(v_2^2)$$
$$\Sigma_{hH} \cong \mathcal{O}(v_1 v_2) \qquad \text{(II.6)}$$
$$\Sigma_{HH} \cong \mathcal{O}(v_1^2) \quad .$$

Then, the true light state is

$$h^{true} = h \cos\delta + H \sin\delta$$
$$\cong h - \Sigma_{hH} \frac{1}{M_H^2 + \Sigma_{HH}} H + \mathcal{O}\left(\frac{1}{v_1^3}\right) \qquad \text{(II.7)}$$

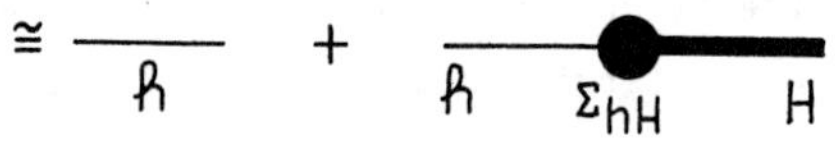

but this is precisely what one light particle irreducible Green's functions will generate for us. The heavy particles occur only internally and one can show that it is sufficient to approximate

$$\langle H^{true} H^{true} \rangle = \langle HH \rangle + \mathcal{O}(1/v_1^2), \qquad \text{(II.8)}$$

in which the $\mathcal{O}(1/v_1^2)$ piece does not couple with vertices $\sim v_1$.

(3) We can show that given any diagram or subdiagram with heavy internal lines, the corresponding integral can be rearranged so that the $\mathcal{O}(1)$ terms all have the heavy lines shrunken into vertices with no more than 4 light lines entering and/or leaving. The rest are negligible. This establishes the local renormalizable nature of the effective vertices. In fact, we can identify what these effective vertices are in relation to the light Lagrangian. However, a more economical way to show what resulting local theory ensues is via BRS identities.[5]

To illustrate this part, it is best to give examples. Consider a three point function

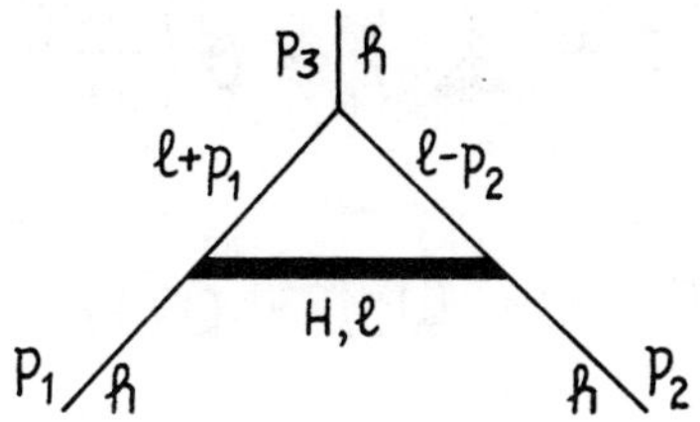

$$I = M^2\int d^4\ell \; \frac{1}{(\ell+p_1)^2+m^2} \; \frac{1}{(\ell-p_2)^2+m^2} \; \frac{1}{\ell^2+M^2}$$

We define:

shrinking a (sub) graph to a point

$\equiv$ setting all external momenta which go into/out of this graph to zero

$\equiv$ localization

There are two (sub) graphs which contains a heavy line

1. 

$$(\text{shrink operator } \tau_1)\; \frac{1}{\ell^2+M^2} = \frac{1}{M^2}$$

2. 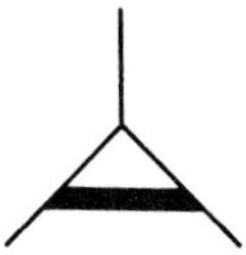

$$(\text{shrink operator } \tau_2)\; \frac{1}{\ell^2+M^2} \; \frac{1}{(\ell+p_1)^2+m^2} \; \frac{1}{(\ell^2-p_2)^2+m^2}$$

$$= \frac{1}{\ell^2+M^2} \; \frac{1}{\ell^2+m^2} \; \frac{1}{\ell^2+m^2}$$

Then, we have an identity

$I = (1-\tau_2)(1-\tau_1)\,\big(I \sim$  $\big) \sim \mathcal{O}\left(\frac{1}{V_1^2}\right)$

$+\,(1-\tau_2)\,\big(\tau_1 I \sim$  $\big) \sim \mathcal{O}(1)$

$+\,\big(\tau_2 I \sim$  $\big) \sim \mathcal{O}(1)$

Note that the resulting integrals are automatically renormalized. Only the second and third terms which have been localized contribute to order 1. The first term is negligible.

The reason that the first term is negligible is because

$$(1-\tau_1)\,\frac{1}{\ell^2+M^2} = \frac{1}{\ell^2+M^2} - \frac{1}{M^2} \tag{II.9}$$

improves the infrared behavior of the integrand to $\sim -\ell^2/M^4$, so that the range of integration $\ell^2 \ll M^2$ gives no contribution of $\mathcal{O}(1)$.

Also

$$\begin{aligned}(1-\tau_2)&\left[(1-\tau_1)\,\frac{1}{\ell^2+M^2}\,\frac{1}{(\ell+p_1)^2+m^2}\,\frac{1}{(\ell-p_2)^2+m^2}\right]\\ &= \left(\frac{1}{(\ell+p_1)^2+m^2}\,\frac{1}{(\ell-p_2)^2+m^2} - \frac{1}{(\ell^2+m^2)^2}\right)\cdot\\ &\quad\cdot\left(\frac{1}{\ell^2+M^2} - \frac{1}{M^2}\right)\end{aligned} \tag{II.10}$$

improves the ultraviolet behavior $\ell^2 \sim M^2$ of the integrand, so that there is no contribution of $\mathcal{O}(1)$ from there either. Thus, the first term can be dropped.

Let us summarize what we have learned in this example:

(i) We have identities to operate on the integrand in such a way that

(iia) $\mathcal{O}(1)$ terms correspond to having all the heavy lines shrunken, a localization process. This also identifies effective coupling constants as appropriate zero momentum Green's functions of the full theory.

(iib) $\mathcal{O}(1)$ terms are automatically renormalized, if and when it is necessary.

(iic)The remainders in which some of the heavy lines are not shrunken are sufficiently oversubtracted that they are negligible.

In general, (iia) and (iic) can be organized by a generalization of the oversubtraction algebraic identity due to Zimmerman.[6]

(4) I would like to briefly discuss power counting which is necessary to prove that the terms we said we could discard can indeed be neglected when we work to $\mathcal{O}(1)$ accuracy.

Because of the appearance of large scale (v_1) factors in vertices, large mass effects can come from any region of an internal integration. Let us artifically divide a momentum integration ℓ into two regions. We call $\ell^2 > M^2$ the ultraviolet region, and $\ell^2 < M^2$ the infrared. In the general situation when several integrations are involved, we must consider the overlap of these regions.

I shall confine myself to discussing the very special situation in which all sub-regions of integration are in the ultraviolet. Discussion of the general situation can be found in the paper by Kazama and myself.[4]

I assume in the following that the grand unifying theory we deal with has a reflection symmetry: $\phi_{\text{light scalar}} \to -\phi_{\text{light scalar}}$ and $v_2 \to -v_2$ simultaneously. We can demand this as a symmetry in SU(5). I shall motivate here that the proper 1LPI local induced vertices in which there is at least one heavy line can have at most $\mathcal{O}$ (ℓn (v_1/μ)) dependence. This disucssion also bears on the mass hierarchy problem.

Now, from dimensional analysis, we see that any (sub-)diagram with n external lines can depend on v_1 with a power not stronger than

$$d=4-n.$$

This is also the maximal power of v_1 any induced vertices can have, where the external lines are all light. Thus, if $n>4$, we can immediately discard such a subgraph. When $n=4$, we have $d=0$, which is equivalent to saying that the dependence is at most $\mathcal{O}(\ell n\ v_1/\mu)$.

When $n=3$, $d=1$, we might have $v_1\mathcal{O}(\ell n\ v_1/\mu)$ behavior. However, the reflection symmetry easily reduces the dependence to $v_2\ \mathcal{O}(\ell n(v_1/\mu))$ for scalar-scalar-scalar and scalar-vector-vector vertices. The dependence is $\mathcal{O}(\ell n\ v_1/\mu)$ for vector-vector-vector and vector-scalar-scalar vertices.

For $n=2$, we are dealing with (light mass)^2. Here, $d=2$ and we might have $v_1{}^2\mathcal{O}(\ell n\ v_1/\mu)$. This is indeed the case when individual diagrams are calculated. We must use BRS identities, which relate proper three point Green's functions to proper two point Green's functions. We can show that in fact (light mass)^2 can have only $v_2^2\ \mathcal{O}(\ell n\ v_1/\mu)$ dependence.

(5) We can show that the relevant BRS identities in the full theory (when $v_1 \gg |p_i|$, v_2) go over to the spontaneously broken BRS identities of the system left behind. In a realistic grand unification model, the residual symmetry implied by these BRS identities will be $SU(3) \times SU(2) \times U(1)$. Besides, only v_2 appears explicitly in these identities. Since this is the only scale which is left to give mass to the low energy system, all particles can have only light mass $\sim v_2$. This is a confirmation of the existence and the stability of a mass hierarchy. The $\ell n\ v_1/\mu$ powers appear as various finite renormalization constants which we can have when we solve these identities to obtain the effective Lagrangian. They are, on the other hand, completely determined when we compare the effective action of the full theory with that of the effective theory.

III. Effective Parameters

Let us turn to the renormalization group equations, which will be used to sum the dependence on $g^2\ell n\ v_1/\mu$ of the effective parameters. We shall stay in the low energy region to do our physics. This means that the scale parameter μ will eventually be set to a small value ($=m_W$).

We shall see that the evaluations of the effective parameters in the present approach completely avoids the issue of threshold crossing. One may anticipate this, since the effective parameters are process-independent universal quantities.

The derivation of the renormalization group equations is very simple: when we work with the full theory, we have the renormalization group equation

$$\left(\mu \frac{\partial}{\partial\mu} + \beta_g \frac{\partial}{\partial g} + \gamma_{v_1} v_1 \frac{\partial}{\partial v_1} + \gamma_{v_2} v_2 \frac{\partial}{\partial v_2} + \gamma_\alpha \alpha \frac{\partial}{\partial\alpha} - n\gamma\right) \Gamma^n = 0 \tag{III.1}$$

with

$$\beta_g = \mu \frac{d}{d\mu} g$$

$$\gamma_{v_1} = \mu \frac{d}{d\mu} \ell n\, v_1 \tag{III.2}$$

etc.

where Γ^n is the one light particle irreducible amputated Green's function with n light external lines. I have dropped terms due to running of the scalar self couplings and Yukawa couplings.

On the other hand, when we deal with the effective theory at low energy, we have

$$\left(\mu \frac{\partial}{\partial\mu} + \beta^*_{g^*} \frac{\partial}{\partial g^*} + \gamma^*_{v^*} v^* \frac{\partial}{\partial v^*} + \gamma^*_{\alpha^*}\alpha^* \frac{\partial}{\partial\alpha^*} - n\gamma^*\right) \Gamma^{*n} = 0 \tag{III.3}$$

We put the decoupling result

$$\Gamma^n = Z^{n/2}\, \Gamma^{*n} + \mathcal{O}\left(\frac{m^2}{M^2}\right) \tag{III.4}$$

into the renormalization group equation of the full theory, Eq. (III.1), and demand that the resulting equation should agree with that of the effective theory, Eq. (III.2). For this to hold, we have, in particular, the relation[7]

$$\beta^*_{g*} = \left(\mu \frac{\partial}{\partial\mu} + \beta_g \frac{\partial}{\partial g} + \gamma_{v_1} \frac{\partial}{\partial \ln v_1} + \gamma_\alpha \frac{\partial}{\partial \ln \alpha}\right) g^* \qquad \text{(III.5)}$$

Let us apply this last equation to the minimal SU(5) model.[8] We shall assume that there are three fermion families and that there are one 24-plet and one 5-plet of scalars. We shall perform this calculation to the two loop accuracy. Although the final results agree with those obtained before via other methods, our present approach is simpler.

We use[9]

$$\beta = a\, g^3 + b\, g^5$$

with

$$16\pi^2\, a = \frac{-52}{3} + \frac{4}{3}\, n_f$$

$$(16\pi^2)^2\, b = -\frac{850}{3} + \frac{324}{5} + \frac{698}{15}\, n_f \qquad \text{(III.6)}$$

and

$$\beta^*_i = a^*_i\, g^{*3}_i + \sum_j g^{*3}_i\, b^*_{ij}\, g^{*2}_j \qquad i,j=1,2,3$$

with

$$16\pi^2\, a^*_1 = \frac{1}{10}\, n_s + \frac{4}{3}\, n_f$$

$$16\pi^2\, a^*_2 = \frac{-22}{3} + \frac{1}{6}\, n_s + \frac{4}{3}\, n_f$$

$$16\pi^2\, a^*_3 = -11 + \frac{4}{3}\, n_f$$

$$(16\pi^2)^2\, b^*_{ij}$$

$$= \begin{pmatrix} 0 & 0 & 0 \\ 0 & \frac{-136}{3} & 0 \\ 0 & 0 & -102 \end{pmatrix} + \begin{pmatrix} \frac{9}{50} & \frac{9}{10} & 0 \\ \frac{3}{10} & \frac{13}{6} & 0 \\ 0 & 0 & 0 \end{pmatrix} n_s$$

$$+ \begin{pmatrix} \frac{19}{15} & \frac{3}{5} & \frac{44}{15} \\ \frac{1}{5} & \frac{49}{3} & 4 \\ \frac{11}{30} & \frac{3}{2} & \frac{76}{3} \end{pmatrix} n_f \qquad \text{(III.7)}$$

Wherever there are three numbers, the first is due to gauge bosons (plus their associated ghosts and Goldstone scalars), the second is due to physical scalars, while the third is due to the fermions. n_f is the number of fermion families (=3) and n_s is the number of light scalar doublets (=1).

The solutions of Eq. (III.5) are supplemented by the boundary conditions

$$g^*_i = g(1+g^2\rho_i) \qquad \text{(III.8)}$$

which are obtained perturbatively to one loop order at $\mu=v_1$ via Eq. (III.4). They are

$$16\pi^2\,\rho_1 = \frac{-35}{4}\,\ell n\left(\frac{25}{8}\,g^2\right) + \frac{1}{30}\,\ell n\left(\frac{-5}{2}\,\beta\right) + \frac{5}{6}$$

$$16\pi^2\,\rho_2 = \frac{-21}{4}\,\ell n\left(\frac{25}{8}\,g^2\right) + \frac{1}{6}\,\ell n\,(10b) + \frac{1}{2}$$

and

$$16\pi^2\rho_3 = \frac{-7}{2}\,\ell n\left(\frac{25}{8}\,g^2\right) + \frac{1}{4}\,\ell n\left(\frac{5}{2}\,b\right) + \frac{1}{12}\,\ell n\left(\frac{-5}{2}\,\beta\right) + \frac{1}{3} \qquad \text{(III.9)}$$

where b is a 4-scalar coupling of the 24-plet and β is a crossed coupling of the 24-and the 5-plets. Note that the boundary conditions of Eq. (III.9) are independent of gauge, as we have explicitly checked.

The solutions are

$$\frac{1}{(g_i^*(t))^2} = \frac{1}{(\bar{g}(T))^2} + 2\, a_i^* \, T$$

$$+ \sum_j \frac{b_{ij}^*}{a_j^*} \ln\left(1 + 2\, a_j^* \, \bar{g}^{\,2}\, T\right) \tag{III.10}$$

$$- 2\, \rho_i(\bar{g}\,)$$

where

$$t = \ln v_1/\mu$$

$$T = \ln \bar{v}_1/\mu$$

$$\mu \frac{d}{d\mu} \ln \bar{v}_1 + \gamma_{v_1}\,(\bar{v}_1, \bar{g}, \bar{\alpha}) = 0 \tag{III.11}$$

and

$$\frac{1}{(\bar{g}(t))^2} = \frac{1}{g^2} - 2\, at + \frac{b}{a}\, \ln\left(1 + 2\, a\, g^2 t\right) \tag{III.12}$$

is the two loop running SU(5) gauge coupling, which can be identified as g_M, the grand unification coupling constant defined at the grand unification scale. Our results are consistent with those given by Weinberg and Hall.

We can relate v_1 to the physical mass of the x-boson, which satisfies the equation

$$\mu \frac{d}{d\mu} M_{x\ \mathrm{physical}} = 0 \quad , \tag{III.13}$$

which is a consequence of the independence of $M_x|_{\mathrm{physical}}$ of the renormalization scale. To the one loop order,

$$M_{x\ \mathrm{physical}} = \sqrt{\frac{25\, \bar{g}^{\,2}\, \bar{v}^{\,2}}{8}} \left[1 + \frac{\bar{g}^{\,2}}{32\pi^2}\left(\frac{185}{18} + \frac{5\pi\sqrt{3}}{2} - \frac{20}{9}\, n_f\right)\right] \tag{III.14}$$

Our numerical values for $\sin^2\theta_W$ and $M_x|_{\mathrm{physical}}$ agree completely with those of Marciano and Sirlin.[10] Thus, using as inputs

$$6.7 < \alpha_s^{-1} (\mu = m_W) < 11.2$$

and

$$126.8 < \alpha_{em}^{-1} < 127.8 \qquad \text{(III.15)}$$

we have

$$4.09 \times 10^{13}\ \text{GeV} < M|_{\text{physical}} < 1.36 \times 10^{15}\ \text{GeV}$$

and

$$.204 < \sin^2\theta_W < .222 \qquad \text{(III.16)}$$

The on-shell correction to the x-boson mass is only 2%. It should be taken into account only when we calculate g_i's to three loop order.

REFERENCES

[1] J.C. Pati and A. Salam, Phys. Rev. D8, 1240 (1973); H. Georgi and S.L. Glashow, Phys. Rev. 32, 438 (1974); H. Georgi, H.R. Quinn and S. Weinberg, Phys. Rev. Letts. 33, 451 (1974).

[2] S. Weinberg, Phys. Lett 91B, 51 (1980), B. Ovrut and H. Schnitzer, Phys. Rev. D21, 3369 (1980); D22, 2518 (1980) and Brandeis Preprints, T. Hagiwara and N. Nakazawa, Harvard Preprint HUTP 80/A012, L. Hall, Nucl. Phys. B178, 75 (1981), W. Weisberger Stonybrook Preprint ITP-SB-81-6 (1981).

[3] E. Gildener, Phys. Rev. D14, 1667 (1976).

[4] Y. Kazama, D. Unger and Y.-P. Yao UM HE 80-36; Y. Kazama and Y.-P. Yao Fermilab-Pub 81/18 THY.

[5] C. Becchi, A. Rouet and R. Stora, CPT Report, Marseille, 1974 (unpublished) and Ann. Phys. (NY) 98, 287 (1976).

[6] W. Zimmermann in Lectures on Elementary Particles and Quantum Field Theory, edited by S. Deser et al (MIT Press, Cambridge Mass., 1971) Vol. I., p. 397.

[7] This relation was first derived by E. Witten Nucl. Phys B122 109 (1977).

[8] D.G. Unger and Y.-P. Yao, University of Michigan Preprint UM-HE 81-30.

[9] D.R.T. Jones, Nucl. Phys. B75, 531 (1974); W. Caswell, Phys. Rev. Lett. 33, 244 (1974); T. Goldman and D. Ross, Phys. Lett. 84B, 208 (1979); D.R.T. Jones supplied the general formula for the 2 loop β function including scalars.

[10] W. Marciano and A. Sirlin, Phys. Rev. Lett. 46, 163 (1981) and this volume. See also D. Ross, Nucl. Phys. B140, 1 (1978); T. Goldman and D. Ross, Phys. Lett. 84B, 208 (1979) and Nucl. Phys. B171, 273 (1980); W. Marciano, Phys. Rev. D20, 274 (1979); Orbis Scientiae 1980; L. Hall, Ref. 2; P. Binetruy and J. Schucker, Nucl. Phys. B178, 293 and 307 (1981); N. Chang, A. Das and J. Perez-Mercader, Phys. Lett. 39B, 137 (1980) and Phys. Rev. D22, 1414 (1980); C. Llewellyn Smith, G. Ross and J. Wheater, Nucl. Phys. B177 273 (1981); I. Antoniadis, C. Roiesnel and C. Kounnas, Preprint LPTENS 81/4, March 1981.

THRESHOLDS IN PARTICLE PHYSICS

M. Veltman
Institute for Theoretical Physics
University of Utrecht
The Netherlands

Experimentally, particle physics is more or less known to about 18 GeV, the present limit on the mass of the top quark. Theoretically everything seems well described by the standard model of strong, e.m. and weak interactions [1].

However, the standard model is unlikely to be true indefinitely, as a function of energy, since it does not include the theory of gravitation. Thus there is a limit, 10^{19} GeV, where the model is incomplete. This aside from the fact that there are other unsatisfactory features such as the large number of free parameters; also we really do not understand the remarkable family structure, nor do we know why nature has chosen the groups SU_3, SU_2 and U_1. We may put things as follows. There is a cut-off Λ above which the present structure either disappears or is amended, and we would like to explore the various possibilities concerning the magnitude of Λ.

To be strict, the present experimental situation is in fact well described by the four-fermion theory plus electromagnetism. No one has as yet found any evidence for vector bosons. The starting point of any investigation must therefore necessarily be the four-fermion theory. Several authors [2] have investigated the limit of validity of this model, and very recently Abbott and Fahri [3], and Fritzsch and Mandelbaum [4] have proposed more or less explicit models.

There are basically two ways in which the four-fermion theory may break down at higher energies. First, it may indeed be that vector bosons exist, and that indeed the standard model, or something similar, is true. The authors mentioned have derived upper bounds on the W-masses for this case. Basically there are two inputs to this problem: (i) the strength and structure of the neutral currents, and (ii) the maximally allowed magnitude of the radiative corrections. The limits for the vector boson masses come out to be of order 150 GeV and 250 GeV.

Secondly, it may turn out that the basic fermions, quarks and leptons, are composite, and that this structure shows up at about 300

GeV. Let us explore this possibility.

One of the first arguments concerning this possibility relates to the family problem. We now believe in three, possibly four, but no more families with the structure and mass pattern as the observed ones. It seems therefore that this family structure stops well before 200 GeV. It is then quite natural to believe that quarks and leptons are composites, with an "ionization" energy of the order of 300 GeV [5], which is incidentally the "Fermi length".

Toying with this idea one very soon arrives at the following basic structure. Assume the existence of one basic family, made up from a high and a low quark, and a lepton with its neutrino. Furthermore there is a scalar particle. The families are obtained by binding the basic family with one, two or three scalars.

The model of Fritzsch and Mandelbaum assumes the existence of a quark-like coloured doublet, except that the charges are $\pm 1/2$, and two scalars, behaving like a 3 and a $\bar{3}$ with respect to color and charges $-1/6$ and $+1/2$. Some new hypercolor force binds scalars and fermions to give rise to a family; other families are supposed to be excitations involving some extra bosonic object. The binding force is due to a new type of confining interactions called hypercolor. The extra bosonic object is identified with some hypercolor gluon. The usual vector bosons of weak interactions are more like resonances, comparable to the well-known ρ of strong interactions.

The model of Abbott and Fahri uses for the basic fermions precisely a family structure as indicated above. The scalar is called H and has much in common with the Higgs of the standard model. The families are bound states of the left-handed basic fermions and one H; additional families require other scalars. 't Hooft's [6] work on anomalies is quoted to explain the correct number of nearly mass zero fermions. The confining force is precisely the usual $SU(2)_L$; thus in this theory the usual vector bosons are massless. Masses arise through Yukawa couplings.

It is an advantage of these schemes that no spontaneous symmetry breakdown is needed, i.e. no vacuum expectation values and cosmological constants. As such they are very attractive. In these schemes the usual vector bosons become wide resonances supposedly located at about 100 GeV.

At this moment the lack of knowledge of physics around 100 GeV makes the models somewhat arbitrary. It is not very clear if they are in trouble with respect to the radiative corrections generated. After all, the confining interactions are strong, and in the first instance one would expect radiative corrections of order one. This could upset for instance CVC, i.e. the close experimental agreement between the vector coupling constants of μ-decay and β-decay, taking into account the radiative corrections as computed by Sirlin [7] in the standard model.

Experiment will tell us rather soon if something of this type is true. If indeed leptons and quarks are confined objects with a structure of the order of 300 GeV, and if there is a confining strong interaction, then at energies of this order of magnitude all these particles start behaving like hadrons involved in some strong interaction. Massive production of muons and electrons is then to be expected. Probably this would become visible in the $p\bar{p}$ experiments of the near future.

The main difficulty with these models is in fact the absence of neutral, strangeness changing currents, i.e. the GIM mechanism. Consider the very first model cited. In this model we would have:

electron	$\bar{L}H$
muon	$\bar{L}HH$
down quark	ℓH
strange quark	ℓHH

The following process would be possible:

$$e^- + s \to \mu^- + d$$

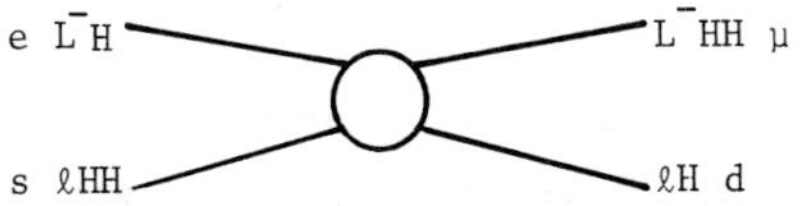

This is a simple exchange process. It would give rise to the decay $K \to \pi\mu e$. No particular kinematics seems to inhibit this possibility. The presently known limits on this decay suggest that the structure must be smaller than 300 GeV, presumably of the order of a few TeV.

Recently the connection between infrared and ultraviolet theories has been elaborated [8]. The main conclusions are:

(i) If the distance between the infrared and ultraviolet is sufficiently large then the infra-red theory must be renormalizable.

(ii) If symmetry persists over some range then the symmetry must be valid indefinitely.

These "theorems" apply if we assume that the next threshold is higher than 300 GeV, say 1000 GeV or higher. Let us first consider the case that the next threshold is at 1000 GeV. Then we must assume that the standard model, including vector-boson masses of about 90 GeV is valid to 1000 GeV. Above this threshold the particle spectrum may or may not change, i.e. quarks and leptons may or may not be composite. If quarks and leptons are composite then the same difficulty as mentioned before (unwanted currents) applies. More modest, one may assume that only the Higgs (and the longitudinal vector bosons) are composite. Due to the insensitivity of weak interactions to the Higgs system (the screening theorem) this possibility will be very difficult to exclude for a long time to come.

The forces necessary to obtain bound states are commonly called technicolor. These theories have been discussed very extensively in recent times, and we will not consider them here [9]. There is however one aspect that needs consideration, and that is naturalness.

In the standard model there is precisely one quadratic divergence, and if the standard model is an effective theory with a cut-off then the magnitude of this divergence is of the order of the cut-off energy Λ squared. Now this divergence is part of the vector boson mass which is of the order of 100 GeV. Assuming naturalness this states that $\alpha\Lambda^2$ is of the order of $(100\ \text{GeV})^2$, which gives the usual technicolor value Λ 1000 GeV. If indeed technicolor starts at 1000 GeV then this is just fine.

However, if there is no technicolor at 1000 GeV then we have here a problem with respect to naturalness. The only solution at this time seems to be that things are arranged in such a way that the coefficient of the quadratic divergence is zero [8]. This coefficient is zero if:

$$\sum_{\text{fermions}} \frac{m_f^2}{M^2} = \frac{3}{2} + \frac{3}{4}\frac{M_o^2}{M^2} + \frac{3}{4}\frac{m_H^2}{M^2} .$$

Can this equality be true? Now, first of all, the masses appearing in this equation (the fermion masses m_f, the Higgs mass m_H and the charged and neutral vector boson masses M and M_o) are the masses in the bare Lagrangian. The observed masses differ by radiative corrections, involving $\ln(\Lambda/M)$. They have not yet been computed.

Aside from this it seems that some heavy fermions are needed. For instance, fitting the top mass to this equation, assuming a relatively light Higgs gives $m_t \sim 69$ GeV. This may be too high. The point is that in the present model the top quark mass influences K-decay and the K_L-K_s mass difference. This constrains the mixing angles, and a limit on the top quark mass can be obtained. Recently such a reasoning and calculation has been done by A. Buras [10], and he quotes the value 26±7 GeV.

If this is true then probably a fourth generation is necessary to keep the W-mass down to 100 GeV.

A number of questions concerning the existence of further generations and Higgs masses may be resolved if both the neutral vector boson mass as well as the weak mixing angle are measured accurately. Precision of the order of 0.3% is perhaps possible with the SLAC collider and/or CERN's LEP. This precision is just enough to limit the Higgs mass to below 1000 GeV. Also, every very heavy generation contributes about 100 MeV to M_o; thus the sensitivity is to three generations. The feature that the radiative corrections to the W masses remain finite even if the fermion masses become very large is a very exciting aspect of these radiative corrections. Everything should be done to measure at least two quantities to the highest possible precision. One is for sure: M_o will be measured up to 0.1%.

In this way we may expect that within the next decade at least certain things will become clear. The $\Lambda = 300$ GeV models will be confirmed or ruled out; the naturalness of the standard model will be better understood in case of a Λ far above 300 GeV.

Let us now consider the case for which Λ is really very high, say 10^{17}GeV. If that is so then SU_5 unification will be reached before Λ is visible. How could this situation be understood?

Actually, one may make a very amusing observation. In the SU_5 theory there are two Higgs sectors, being the 100 GeV and the 10^{15} GeV systems. There are now two quadratic divergencies, namely one associated with each Higgs system. It is now conceivable that due to a vanishing coefficient for the low Higgs system there is really only one quadratic divergence, namely for the other Higgs sector. But this is fine, perhaps that is the reason that the X and Y vector boson and that Higgs system are so heavy!

In this way we arrive at the following point of view. There is a super symmetric theory valid above 10^{17} GeV. A renormalizable zero-mass subset of the SU_5 type is part of that theory. Subsequently there is some small breaking of the theory giving rise to masses of the order of

100 GeV; next quadratic divergencies lift the X, Y bosons etc. to 10^{15} GeV. It seems an entirely natural situation.

There is of course in the scheme a very big problem as to the details of the theory between 10^{15} and 10^{19} GeV, as well as the origin of the 100 GeV mass scale. No one has come up with any ideas on that so far.

References:

[1] S. Bludman, Il Nuovo Cim. 9 (1958) 433.
S. Glashow, Nucl. Phys. 22 (1961) 579.
S. Weinberg, Phys. Rev. Lett. 19 (1967) 1264.
S. Glashow, J. Iliopoulos and L. Maiani, Phys. Rev. D2 (1970) 1285.
C. Bouchiat, J. Iliopoulos and Ph. Meyer, Phys. Let. 42B (1972) 91.
[2] J. Bjorken, Phys. Rev. D19 (1979) 335.
P. Hung and J. Sakurai, Nucl. Phys. B143 (1978) 81.
M. Veltman, Varenna lectures 1977.
[3] L. Abbott and E. Fahri, CERN preprints Jan. 1981 and March 1981.
[4] H. Fritzsch and G. Mandelbaum, Univers. of München preprint, Febr. 1981.
[5] M. Veltman, talk at the Int. Conf. on Photons and Leptons, Fermilab., Aug. 23-29, 1979.
[6] G. 't Hooft, Cargèse lectures 1979.
[7] A. Sirlin, Nucl. Phys. 71B (1974) 24.
See also M. Veltman, ref. 2.
[8] M. Veltman, Michigan preprint Oct. 1980; to be published in Acta Phys. Pol.
[9] S. Weinberg, Phys. Rev. D13 (1976) 974.
L. Susskind, Phys. Rev. D20 (1979) 2619.
[10] A. Buras, Fermilab preprint, March 1981.

THE TeV PICTURE

Stuart Raby
Stanford Linear Accelerator Center
Stanford, California

In this talk we shall discuss a not-so-standard model of strong and electroweak interactions. We will discuss some direct consequences of this model and finally we will speculate on the outlook for the future.

Let us now briefly describe the parameters of the standard model so that we will know exactly what we shall be missing in the not-so standard model which follows. There are three sectors: gauge, fermion and Higgs.

(a) Gauge sector: we have the three gauge groups $SU(3)_C \otimes SU(2)_L \otimes U(1)_Y$ and their respective gauge couplings g_3, g_2, g_1. $SU(3)_C$ is asymptotically free and thus g_3 gets strong at low energies and sets the scale of strong hadronic interactions. g_1 and g_2 are electroweak parameters which are directly related to the measured parameters α and $\sin^2\theta_W$ (discussed previously by Langacker and Marciano). Finally there are the gauge bosons—8 gluons, photon, $W^{\pm}$ and Z^o where the last two, of course, have yet to be directly observed. Note that the $W^{\pm}$ and Z^o masses are determined by parameters in the Higgs sector.

(b) Fermion sector: there are three generations of quarks and leptons (assuming the top quark exists). Their gauge couplings are determined by their standard charge assignments. However at least 9 masses and 4 mixing angles of this sector are given by 13 arbitrary parameters in the Higgs sector (namely Yukawa couplings).

Finally when I talk about these masses, I am referring to the so-called current algebra masses which are essentially local mass terms in the effective low energy standard model Lagrangian. This is to distinguish them from their dynamical masses obtained as a result of QCD which I shall refer to later.

(c) Higgs sector: we have the standard Higgs doublet $\Phi = \begin{pmatrix} \phi^+ \\ \phi^o \end{pmatrix}$ with Y = +1. Its self interactions are described by the scalar potential $V(|\Phi|) = -\mu^2|\Phi|^2 + \lambda|\Phi|^4$. We thus have two additional parameters making a total of 18 — 15 arbitrary constants in addition to the 3 gauge couplings — which successfully parametrize the low-energy physics.

The two Higgs couplings λ and μ determine: (1) the $W^{\pm}$ mass via the relations $G_F^{-\frac{1}{2}} \sim \langle\phi_o\rangle \sim \mu/\sqrt{\lambda}$ and $G_F/\sqrt{2} = g_2^2/8M_W^2$; (2) the neutral current parameter $\rho = M_W^2/M_Z^2 \cos^2\theta_W = 1$ + small corrections, and (3) the physical Higgs boson mass $\mu_h \sim \sqrt{\lambda}\, G_F^{-\frac{1}{2}}$. Hence we see that two Higgs parameters determine the above 3 measurable parameters. Consequently one of them, (2), is a natural relation having to do with symmetries which I shall discuss shortly. It is the soft violation of these symmetries which allows for the small corrections to ρ as discussed originally by Veltman [1] (see Marciano, these talks). Finally the quark and lepton masses are given in terms of arbitrary Yukawa couplings; i.e., $m_{q,\ell} \sim g_Y\langle\phi^o\rangle$. Since $\langle\phi^o\rangle \sim 250$ GeV as determined from G_F, then g_Y is of order 10^{-5} for the lightest generation of quarks and leptons.

We know that this effective Lagrangian for the standard model, accurately describes the low energy world as borne out by experiment, modulo a few particles which have yet to be discovered. What are the remaining open questions. There are marked regularities in the fermion mass matrix which have no explanation — the generation hierarchy, up-down symmetry breaking masses, Cabibbo and CP violating angles. All of these are determined in terms of small and arbitrary dimensionless parameters $g_Y \sim 10^{-5}$-10^{-1}. In addition there is no explanation for the huge gauge hierarchy associated with the small dimensionless ratio $G_F^{-\frac{1}{2}}/m_{p\ell} \sim 10^{-17}$. What we would like to do is to remove the Higgs sector from the theory and replace it by a more fundamental component. (By more fundamental, I mean that the 15 extra parameters in the Higgs sector are in principle determinable by a few parameters in the new sector).

For the moment, let us just remove the Higgs sector from the theory and imagine what the physics would be like before we replace it with any other component. We note that the leptons would be massless. However the quarks are still massive as a result of the strong QCD forces. They have their so-called constituent masses.

What about the W's and Z^o? Note that they are in fact massive with a mass $M_W = g_2 f_\pi/2 = M_Z \cos\theta_W$. Let us consider this in more detail. Consider u and d quarks only for simplicity. We have a left-handed doublet and two right-handed singlets

$$\begin{pmatrix} u \\ d \end{pmatrix} \quad \frac{\bar{u}}{d} \quad . \tag{1}$$

If we turn off the weak interactions, i.e., $g_1 = g_2 = 0$, the Lagrangian is invariant under a global chiral symmetry $SU(2)_L \otimes SU(2)_R$. However, as we all know, this symmetry is <u>not</u> a symmetry of the QCD vacuum. The condensates $\langle\bar{u}u\rangle = \langle\bar{d}d\rangle \sim (3f_\pi)^3$ form and spontaneously break $SU(2)_L \otimes SU(2)_R$ down to the remaining symmetry $SU(2)_{\text{Isospin}}$. As a result of this spontaneous symmetry breaking, we obtain 3 massless Nambu-Goldstone bosons $\pi^\pm$, π^o.

Technically we say that the 3 axial vector currents create these massless states out of the vacuum with a characteristic decay constant $f_\pi = 93$ MeV; i.e., we have

$$\left\langle 0 \middle| J^i_{\mu 5} \middle| \pi^j \right\rangle \propto f_\pi \, \delta_{ij} \, q_\mu . \tag{2}$$

Note that as a result of $SU(2)_{\text{Isospin}}$ all the pions have the same f_π. This result is very important as we shall soon see.

Now we turn on the weak interactions. The pions couple directly to the electroweak gauge bosons W^i_μ ($i = 1, 2, 3$) and B_μ via the currents $J^i_{\mu L} = \frac{1}{2}(J^i_{\mu\text{vector}} - J^i_{\mu 5})$ and $Y_\mu = J^3_{\mu 5}$ + vectorial pieces. Hence the pions are eaten, leading directly to the gauge mass squared matrix

$$\frac{g_2^2 f_\pi^2}{4} \left(\begin{array}{cc|cc} 1 & 0 & 0 & 0 \\ 0 & 1 & 0 & 0 \\ \hline 0 & 0 & 1 & -g_1/g_2 \\ 0 & 0 & -g_1/g_2 & g_1^2/g_2^2 \end{array} \right) \begin{array}{c} W^1_\mu \\ W^2_\mu \\ W^3_\mu \\ B_\mu \end{array} . \tag{3}$$

As a result we obtain

$$M_W = \frac{g_2 f_\pi}{2} , \qquad \tan\theta_W = g_1/g_2 \tag{4}$$

and the relation $M_W = M_Z \cos\theta_W$ follows directly from the symmetry $SU(2)_{Isospin}$ which required all the f_π's to be equal.

The moral of our short monologue is twofold: (1) the non-Abelian gauge symmetry QCD has provided naturally light scalars whose scale ($f_\pi \sim \Lambda_{QCD}$) is determined by a logarithmically varying coupling constant, and (2) the relation $M_W = M_Z \cos\theta_W$ can be natural in such a scenario.

We are thus lead to consider the following not-so-standard model - preliminary version as discussed originally by Weinberg and Susskind [2]. It includes a gauge sector and fermion sector only.

(a) Gauge sector: we have the gauge group $G_T \otimes SU(3)_c \otimes SU(2)_L \otimes U(1)_Y$ containing the four parameters g_T, g_3, g_2 and g_1, respectively. The gauge group G_T where T stands for Technicolor (or Hypercolor) is assumed to be asymptotically free. Thus g_T becomes strong at a scale $\Lambda_T >> \Lambda_{QCD}$ which sets the scale for all strong Technihadronic physics.

(b) Fermion sector: we assume the usual 3 generations of quarks and leptons. In addition we suppose that there is at least one left-handed $SU(2)_L$ doublet of Technifermions and 2 right-handed singlets with hypercharge assignments Y = 0, ±1 such that the theory is anomaly free.

$$\begin{pmatrix} N \\ E \end{pmatrix} \qquad \begin{matrix} \bar{N} \\ \bar{E} \end{matrix} \quad . \tag{5}$$
$$Y = 0 \qquad Y = \mp 1$$

If we turn off the weak interactions ($g_1 = g_2 = 0$), we have:

(a) $SU(2)_L \otimes SU(2)_R$ global symmetry in the Technifermion sector;

(b) when g_T becomes strong we assume that the following strong interaction condensates form

$$\langle \bar{N}N \rangle = \langle \bar{E}E \rangle \sim \left(3F_T\right)^3 \quad , \tag{6}$$

(c) hence $SU(2)_L \otimes SU(2)_R$ is spontaneously broken to $SU(2)_I$, and

(d) as a result, 3 Nambu-Goldstone bosons $\pi_T^{\pm}$, π_T^3 are formed, with a decay constant $F_T \sim \Lambda_T$. Finally, if we now turn on the weak interactions, we obtain

$$M_W = \frac{g_2 F_T}{2} = M_Z \cos\theta_W \tag{7}$$

which determines the Technicolor scale F_T to be ~250 GeV. In general if there are several Techni-doublets with their right-handed singlet partners we obtain

$$M_W = \frac{g_2 F_T}{2} \sqrt{N_{TD}} \tag{8}$$

where N_{TD} is the number of Techni-doublets and thus we have

$$F_T \sim 250\ \text{GeV}/\sqrt{N_{TD}} \quad . \tag{9}$$

This is a preliminary version, however, since it is easy to see that quarks and leptons remain massless. There is no mechanism whereby they may flip their chirality. To remedy this difficulty, the following authors: Weinberg, Dimopoulos and Susskind, Eichten and Lane, have suggested unifying G_T with some symmetry of the ordinary generations; i.e., $SU(3)_c$ or generation symmetry, etc. What this means is that we put some Technifermions (Q) and ordinary fermions (q) together in a single representation of a larger group, referred to as Extended Technicolor (ETC) (or Sideways color). This group ETC must then break down at a scale V_{ETC} leaving only TC which then gets strong as before. If left- and right-handed couplings exist (not necessarily vectorial) then we can flip the chirality of an ordinary quark by letting it feel the spontaneously generated mass of the Technifermion. If V_{ETC} is much greater than F_T, then in the low energy world we may describe this ETC interaction by the following effective four-Fermi interaction

$$\frac{1}{V_{ETC}^2} \quad \bar{Q}^* \sigma_\mu \bar{q}\ Q^* \sigma^\mu q \tag{10}$$

where Q and $\bar{Q}$ are left-handed Techniquarks and q and $\bar{q}$ are left-handed ordinary quarks. Upon Fierz transforming Eq. (10) we obtain

$$\frac{1}{V_{ETC}^2} \quad \bar{q}q\ \bar{Q}^* Q^* \tag{11}$$

or an effective current quark mass

$$m_q = \frac{\langle \bar{Q}Q \rangle}{V_{ETC}^2} \quad . \tag{12}$$

Note that all strong TC corrections are implicit in the condensate $\langle\bar{Q}Q\rangle$. In addition the scale of $\langle\bar{Q}Q\rangle$ as determined by the weak interactions is of order ~600 GeV (for $N_{TD} = 1$) or ~300 GeV (for $N_{TD} = 4$). Thus the scale V_{ETC} effectively determines the quark and lepton masses.

In general in order to give mass to all the quarks and leptons, one seems to require a complete family of Technifermions [4]. In the minimal scenario we have the following Technifermions

$$\begin{pmatrix} U \\ D \end{pmatrix} \quad \begin{matrix} \bar{U} \\ \bar{D} \end{matrix} \qquad \begin{pmatrix} N \\ E \end{pmatrix} \quad \begin{matrix} \bar{N} \\ \bar{E} \end{matrix} \qquad . \tag{13}$$

In the limit $g_1 = g_2 = g_3 = 0$ there is an $SU(8)_L \otimes SU(8)_R$ global symmetry of the Techni-Lagrangian. When g_T gets strong, the condensates $\langle\bar{U}U\rangle = \langle\bar{D}D\rangle = \langle\bar{N}N\rangle = \langle\bar{E}E\rangle$ are assumed to form which spontaneously breaks $SU(8)_L \otimes SU(8)_R$ down to $SU(8)_{vector}$. As a result, 63 Nambu-Goldstone bosons are produced. Three of these are eaten by the W's and Z^0 when the weak interactions are turned on. The other 60 remain as physical pseudo-Nambu-Goldstone bosons. Many authors [4-6] have discussed the following properties of these very interesting states.

(a) Spectrum: they obtain their mass from interactions which are weak at the scale F_T. They are all very light on this scale and will thus be the leading signals for TC. In addition their mass may be calculated reliably using standard current algebra and Dashen's theorem. The results are listed in Table I. Note that the lightest states, the neutral axions a_T^o and a_T^3 do not receive any mass from $SU(3)_C \otimes SU(2)_L \otimes U(1)_Y$ forces. In evaluating their mass we have assumed that there is a Pati-Salam interaction which gives them a maximum mass of 2-3 GeV since lepto-quark gauge bosons are constrained by limits on the reaction $K_L \to \mu^+ e^-$ to be heavier than 310 TeV [6].

(b) Production: the production cross sections and decay rates may be found in the literature cited [4-6]. It suffices to remark here that the neutral states η_T^a and a_T^o may be produced singly in the reactions pp or $\bar{p}p$ going to η_T^a + anything or a_T^o + anything. For example in a 1 TeV on 1 TeV beam at the Tevatron one would expect about 440 η_T^a events/10^7 sec assuming $\mathcal{L} = 10^{30}/cm^2 \cdot sec$. The charged axions should be seen in the reaction $e^+e^- \to a_T^+ + a_T^-$, especially if one sits on the Z^0 (see Lane [5]). The dominant decay modes for all these states are via Yukawa couplings to the heaviest fermions allowable.

TABLE I

STATE	COLOR	CHARGE	MASS (GeV)	NAME
$\bar{U}U-\bar{D}D+\bar{N}N-\bar{E}E$	1	0	0	Technipions (eaten
$\bar{U}D+\bar{N}E$	1	-1	0	by gauge bosons)
$\bar{U}(\lambda^a/2)U+\bar{D}(\lambda^a/2)D$	8	0	$240\sqrt{4/N}$	Colored Technieta η_T^a
$\bar{U}(\lambda^a/2)U-\bar{D}(\lambda^a/2)D$	8	0	$240\sqrt{4/N}$	Colored Technipions
$\bar{U}(\lambda^a/2)D$	8	-1	$240\sqrt{4/N}$	π_T^{a3}, $\pi_T^{a\pm}$
$\bar{E}U$	3	5/3	$165\sqrt{4/N}$	
$(1/\sqrt{2})(\bar{E}D-\bar{N}U)$	3	2/3	$160\sqrt{4/N}$	Techni-
$\bar{N}D$	3	-1/3	$155\sqrt{4/N}$	leptoquarks
$(1/\sqrt{2})(\bar{E}D+\bar{N}U)$	3	2/3	$160\sqrt{4/N}$	
$\bar{U}U+\bar{D}D-3(\bar{N}N+\bar{E}E)$	1	0	2-3	Paraxion a_T^o
$\bar{U}U-\bar{D}D+\bar{E}E-\bar{N}N$	1	0	2-3	Axion a_T^3
$\bar{U}D-\bar{N}E$	1	-1	8-10	Charged Axion $a_T^{\pm}$

Generation hierarchy: Dimopoulos, Raby and Susskind [7] have discussed the possibility that a large group, e.g., SU(N), may sequentially spontaneously break itself down at different scales (so-called Tumbling). This could lead to the following possible explanation of the generation hierarchy. Assume that the ETC group is $SU(N+3)_{ETC}$ which sequentially breaks down to

$$SU(N+3)_{ETC} \overset{V_u}{\rightarrow} SU(N+2) \overset{V_c}{\rightarrow} SU(N+1) \overset{V_t}{\rightarrow} SU(N)_{TC} \quad (14)$$

at a scale $V_u > V_c > V_t$, where $SU(N)_{TC}$ is the remaining TC group. We assume the fermions transform in fundamental representations of ETC as for example

$$\text{ETC} \updownarrow \underbrace{\begin{pmatrix} U \\ t \\ c \\ u \end{pmatrix} \quad \begin{pmatrix} D \\ b \\ s \\ d \end{pmatrix}}_{SU(2)_L} \quad \begin{pmatrix} \bar{U} \\ \bar{t} \\ \bar{c} \\ \bar{u} \end{pmatrix} \quad \begin{pmatrix} \bar{D} \\ \bar{b} \\ \bar{s} \\ \bar{d} \end{pmatrix} \quad (15)$$

where U, D, $\bar{U}$, $\bar{D}$ are Techniquarks. When $SU(N)_{TC}$ gets strong we expect the condensates $\langle\bar{U}U\rangle = \langle\bar{D}D\rangle$ to form. This then results in the following quark mass relations

$$
\begin{aligned}
m_t &\sim \frac{\langle\bar{U}U\rangle}{V_t^2} \\
m_c &\sim \frac{\langle\bar{U}U\rangle}{V_c^2} \\
m_u &\sim \frac{\langle\bar{U}U\rangle}{V_u^2}
\end{aligned}
\tag{16}
$$

or $m_t > m_c > m_u$ since $V_u > V_c > V_t$. We thus obtain a generation hierarchy as a direct result of a presumed gauge hierarchy. In addition we note that such a gauge hierarchy is necessary in order to suppress dangerous neutral current processes, which we discuss next.

Generation changing neutral currents: there are of course many unavoidable experimental consequences of a local generation symmetry. These have been discussed by several authors [8].

The $\Delta G = 2$ processes are the most dangerous. By using the relations of Eq. (16) to evaluate the scales, V_t, V_c, V_u one can predict the rates for $K_0 \to \bar{K}_0$ or $D_0 \to \bar{D}_0$ modulo Cabibbo-like mixing angles in the up or down quark sectors. One finds that any s-d mixing angle must be less than 10^{-2} in order to be consistent with the $K_L - K_S$ mass difference. Theoretically this can be arranged if the ETC group in the down quark sector is vectorial. We are then forced to have essentially the entire Cabibbo angle in the up quark sector. As a result the process $D_0 \to \bar{D}_0$ should be seen at a rate which is close to the present experimental limits. $\Delta G = 1$ processes such as $\mu N \to eN$ (where N is any nucleus) or $\mu \to e\bar{e}e$ should be seen soon if the μ-e mixing angle is of order one. The $\Delta G = 0$ process $K^+ \to \pi^+\mu^+e^-$ should be seen at a rate which is just below the experimenal upper bound, i.e., $\Gamma(K^+ \to \pi^+\mu^+e^-)/\Gamma(K^+ \to \pi^o\nu_\mu\mu^+) \leq 1.5 \times 10^{-7}$ [9]. The amplitude for the analogous process $K_L \to \mu^+e^-$ vanishes if ETC is vectorial in the down quark sector.

Finally we should note that, as discussed, ETC in the down quark is expected to be vectorial. As a consequence, the Yukawa couplings of pseudo-Nambu-Goldstone bosons to the down quark sector should be parity conserving and of course pseudoscalar. This should not be true in the up quark sector, however, since ETC cannot be purely vectorial there. Thus the parity of the pseudos cannot be measured via their up quark decay modes.

To summarize, the TeV picture looks as follows:

$$
\begin{array}{lcl}
 & & a_T^{\pm},\ a_T^{3}\ \ a_T^{o} \\
 & & W^{\pm},\ Z^{o} \\
100\ \text{GeV} & - & \\
 & & \text{Technileptoquarks} \\
 & & \pi_T^{\pm a},\ \pi_T^{3a},\ \eta_T^{a} \\
 & & \\
 & & \sigma_T \\
1\ \text{TeV} & - & \rho_T,\ W_T,\ \phi_T,\ \rho_T^{a} \\
 & & \text{Technibaryons} \\
 & & \text{---------------}V_{t,b} \\
 & & \\
10\ \text{TeV} & - & \text{---------------}V_{c,s} \\
 & & \\
100\ \text{TeV} & - & \text{---------------}V_{u,d}
\end{array}
\qquad (17)
$$

Technicolor and GUTS: Frampton [10] has asked the question whether or not there can exist a grand unified Technicolor model. For example he considered, among others, the case $SU(N) \supset SU(n)_{TC} \otimes SU(3)_C \otimes SU(2)_L \otimes U(1)_Y$. He required the resulting theory satisfy 4 criteria:

(1) SU(N) is anomally free;

(2) $SU(n)_{TC}$ is asymptotically free;

(3) g_T grows faster than g_3, and

(4) there are at least two complete generations of quarks and leptons.

He was not able to find a set of fermion representations which satisfied all these assumptions. Although this result is not at the level of a rigorous no-go theorem it nevertheless seems extremely plausible that there are no grand unified Technicolor theories.

Let us thus conclude by discussing a few of the alternatives to grand unification. (1) It is extremely possible that there is a grand proliferation of new scales and new groups as one goes up in energy. (2) Grand unification did not appear possible since there are too many elementary fermions [10]. This might suggest a composite structure of quarks and leptons at a TeV scale. (3) Elementary scalars are necessary, and even possible if we include supersymmetry to keep them naturally light down to a TeV scale, (see the discussions by Dimopoulos, Georgi and Srednicki, this conference). Finally there is always the fourth possibility: none of the above. It is clear, however, that the TeV picture is certain to provide answers to some very fundamental questions.

This work was supported by the Department of Energy, contract DE-AC03-76SF00515.

References

[1] M. Veltman, Nucl. Phys. B123, 89 (1977).

[2] S. Weinberg, Phys. Rev. D13, 974 (1976), D19, 1277 (1979); L. Susskind, Phys. Rev. D20, 2619 (1979).

[3] S. Weinberg, ref. [2]; S. Dimopoulos and L. Susskind, Nucl. Phys. B155, 237 (1979); E. Eichten and K. Lane, Phys. Lett. 90B, 125 (1980).

[4] S. Dimopoulos, Nucl. Phys. B168, 93 (1980); E. Eichten and K. Lane, ref. [3]; E. Farhi and L. Susskind, SLAC preprint SLAC-PUB-2361, (1979); S. Dimopoulos, S. Raby and L. Susskind, "Technicolor," talk at McGill APS Conference, Nov. 1979.

[5] M. E. Peskin, Nucl. Phys. B175, 197 (1980); J. Preskill, Nucl. Phys. B177, 21 (1981); M. A. B. Bég, H. D. Politzer and P. Ramond, Phys. Rev. Lett. 43, 1701 (1979); J. Ellis, M. K. Gaillard, D. V. Nanopoulos and P. Sikivie, Nucl. Phys. B182, 529 (1981); S. Chadha and M. E. Peskin, CERN preprints TH. 3023 and 3038 (1981); F. Hayot and O. Napoly, Zeitschrift f. Phys. C7, 229 (1981); E. Farhi and L. Susskind, Phys. Rep., to be published; and K. Lane, Ohio State University preprint DOE/ER/01545-306.

[6] S. Dimopoulos, G. L. Kane and S. Raby, University of Michigan preprint UM-HE-80-22 (1980).

[7] S. Dimopoulos, S. Raby and L. Susskind, Nucl. Phys. B169, 373 (1980).

[8] G. L. Kane and R. Thun, Phys. Lett. 94B, 513 (1980); R. N. Cahn and H. Harari, LBL Berkeley preprint LBL-10823 (1980); and S. Dimopoulos and J. Ellis, Nucl. Phys. B182, 505 (1981).

[9] A. Diamant-Berger et al., Phys. Lett. 62B, 485 (1976).

[10] P. Frampton, Harvard preprint HUTP-79/B006 (1979).

COMPLEXIONS OF BARYON AND LEPTON NUMBER VIOLATIONS WITHIN MAXIMAL SYMMETRIES

Jogesh C. Pati
Center for Theoretical Physics
Department of Physics and Astronomy
University of Maryland
College Park, Maryland 20742

ABSTRACT

Scenarios for filling the grand plateau between 10^2 and 10^{15} GeV with elementary or composite structures are outlined. Some of the associated experimental consequences, which would be of importance for future high energy accelerators and cosmic ray physics, are listed. Observing that grand unification based on maximal symmetries permit intermediate mass-scales as low as about 10^3-10^5 GeV, the origin of baryon, lepton and fermion number-violations within maximally gauged symmetries and the complexions of these violations arising within such symmetries are noted. In particular it is noted that maximal symmetries permit not only (B-L)-conserving proton-decay mode ($p \to e^+\pi^o$ etc.), but also (B-L)-violating modes, i.e. $p \to e^-$+pions, $p \to 3\ell$+pions and $p \to 3\bar{\ell}$+pions, as well as n-$\bar{n}$-oscillations and neutrinoless double β-decay. This is in contrast to nonmaximal symmetries such as SU(5) and SO(10). where the possible existence of these modes is not permitted. The relative rates of the various B,L-violating processes depend on the pattern of spontaneous breakdown of the maximal symmetry. For the special case that the maximal symmetry breaks through a single mass scale to $SU(2)xU(1)xSU(3)^{col}$, the standard result that there exists a grand "desert" and that the (B-L)-conserving modes ($p \to e^+\pi^o$,etc.) should dominate, with $\sin^2\theta_W \approx 0.21$ and grand unification mass $\approx 10^{15}$ GeV, emerges. In general, though, maximal symmetries would permit a richer complexion of phenomena with intermediate mass-scales and no desert. Searches for (B-L) violating proton-decay modes, n-$\bar{n}$ oscillations and neutrinoless double β-decay can serve as probes for the possible existence of the intermediate mass scales.

I. FILLING THE GRAND PLATEAU WITH ELEMENTARY OR COMPOSITE STRUCTURES

There are a number of experiments requiring only low or modest energies $\lesssim$ 100 GeV, which pertain to basic questions. Some of these are listed in Table I. These include (i) an improvement in the measurement of the electric dipole moment of the neutron, (ii) an improvement on the limit on V+A-interactions, (iii) an improvement on the limits on rare processes such as $\mu \to e\gamma$ and $K_L \to \bar{\mu}e$, (iv) a search for neutrino-oscillations, (v) improvements in the measurements of the masses of ν_e and ν_μ, (vi) a search for fractional charges, and (vii) a search for primordial magnetic monopoles of masses as light perhaps as 100 GeV.

Specially significant among these key low energy experiments are the ones which pertain to searches for baryon and lepton number-non-conservation. These include the searches for (1) proton-decay and decay-modes of differing selection rules (see Table I), (2)n-$\bar{n}$-oscillations and (3) neutrinoless double β-decay. In particular the discovery of protons decaying into leptons and/or antileptons would serve as a test for the basic hypothesis that quarks and leptons are ultimately members of <u>one family</u>[1-4] -- in the context of a gauge unification of basic particles and their forces.

The investigations of these key low energy experiments will hopefully be exhausted, however, before the end of this century. Looking ahead one may therefore ask: what sort of <u>new physics</u> could one look forward to exploring in the high energy regime? In practice, accelerator-energy will be limited to at most 100 TeV for the conceivable future. Likewise cosmic ray-energies are also limited for all practical purposes to about 100 TeV. Thus the only relevant theorists' question concerning future high energy-experiments is: Can one legitimately anticipate, on the basis of a certain theoretical picture, to observe truly new physics involving some novel phenomena in the energy range of order 1 to 100 TeV?

To answer this question, we note that there are at least two mileposts in the scale of masses signifying new physics. They are:

TABLE I

KEY LOW ENERGY EXPERIMENTS

(1) Search for proton-decay and alternative decay modes*

(i) $\Delta(B-L)=0$, $\Delta F=-4$: $p\to e^+\pi^o$, μ^+K^o etc.

(ii) $\Delta(B+L)=0$, $\Delta F=-2$: $p\to e^-\pi^+\pi^+$ etc.

(iii) $\Delta(B+\frac{L}{3})=0$, $\Delta F=0$: $p\to e^-\nu\nu\pi^+\pi^+$ etc.

(iv) $\Delta(B-\frac{L}{3})=0$, $\Delta F=-6$: $p\to e^-\bar{\nu}\bar{\nu}$ etc.

(2) Search for n-$\bar{n}$-oscillations ($\Delta B=2$, $\Delta L=0$)

(3) Search for neutrinoless double β-decay ($\Delta L=2$, $\Delta B=0$)

(4) Searches for rare processes: $\mu\to e\gamma$, $K_L\to\bar{\mu}e$

(5) Improving measurements of masses of $\nu_{e,\mu,\tau}$;

(6) Search for Neutrino-Oscillations

(7) Search for exotic fractional charges.

(8) Improving measurement of e.d.m. of neutron to 10^{-27} ecm.

(9) Search for V+A//Left-right symmetry

(10) Search for second Z^o through forward-backward asymmetry in $e^-e^+\to\mu^-\mu^+$, $q\bar{q}$ at PETRA, PEP, and LEP energies

---

*Here F stands for fermion number: $F = B_q+L = 3B+L$ (see text).

(a) the masses of $W^{\pm}$ and Z^{o} gauge-particles of order 10^{2} GeV,

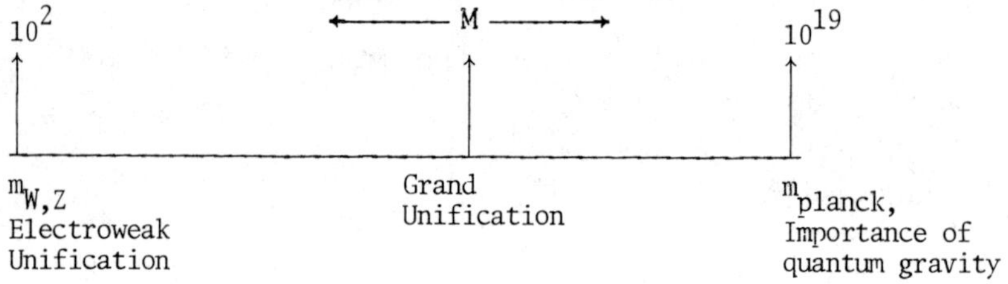

Fig. 1.

representing an energy-scale at which electroweak unification is achieved, and (b) the Planck mass $\sim 10^{19}$ GeV, at which quantum gravity would presumably be important. If one assumes grand unification of quarks and leptons and of their forces,[1-4] one would expect that there exists at least one more mass-scale M, signifying the energy-scale at which weak, electromagnetic as well as strong interactions would exhibit their unity, as also quarks and leptons. The scale M must naturally far exceed the scale of electroweak unification of order 10^{2} GeV; in a class of models M is as high as about 10^{15}-10^{16} GeV;[5] though in general it can be as low as even 10^{4}-10^{5} GeV [6] consistent with the observed value of $\sin^{2}\theta_{W}$.

If M is of order 10^{15}-10^{16} GeV and if there does not exist any intermediate mass-scale between 10^{2} and 10^{15} GeV, as is the case for SU(5) and effectively for[7] SO(10), there would emerge the grand plateau or the so-called grand "desert" of antidiscovery. Within this scenario, one would not expect to discover any new physics beyond the discoveries of $W^{\pm}$ Z^{o}-particles and possibly of a few new flavors. Such a scenario, for all we know, can be true, but no doubt it will be very dull.

It has been consistently stressed by Salam and myself that the hypothesis of grand unification is not synonymous with the existence of a grand "desert". The desert may "bloom" or the plateau may be filled with relatively low mass-structures either because (a) the unification-mass scale M itself is relatively low $\sim 10^{4}$-10^{5} GeV;[6] or (b) there are intermediate mass-scales[8] (IMS) of order 10^{4}-10^{5} and 10^{8}-10^{10} GeV even though the unification mass M is still of order 10^{15} GeV; this situation arises within maximal symmetries, which I shall discuss; or because (c) quarks, leptons, Higgs and perhaps the gauge bosons including W, Z and gluons are composites of a more elementary set of objects,[9-11] which we shall call "preons," and that their sizes are of order 10^{-17}-10^{-18} cm, the inverse sizes being of order 1-10 TeV.

For the first two cases (a) and (b), the spirit of grand unification of quarks and leptons holds in a conventional sense with the basic gauge bosons and fermions being treated as elementary at least up to the momentum-scale of the order of the unification mass M, or distance-scale $\gtrsim 1/M$. The question of the spin-0 Higgs particles being composites with sizes bigger than 1/M can still be left open for these two cases. If the Higgs are composites with sizes $r \gg 1/M$, one may expect to see manifestations of new kinds of quarks, spin-1 gauge bosons and spin-0 scalar particles exhibiting a new force at an energy-scale $\sim 1/r \ll M$ following the ideas of technicolor.[12]

For the third case, with quarks, leptons and gauge bosons (i.e. W, Z, gluons etc.) having inverse sizes M_o less than the "would be" grand unification mass M, on the other hand, grand unification in the conventional sense would not apply at the level of quarks and leptons. This is because we would expect the composites to "dissociate" into their constituents and the constituent dynamics to begin to manifest once the collision-momenta exceeds the inverse size M_o of the composites. Thus we expect that the effective local quark-lepton gauge symmetry would be limited in this case to only such symmetries, which hold at relatively low momenta $\lesssim M_o$. With M_o of order $1 \sim 100$ TeV, such a symmetry may be either $SU(2)_L \times U(1) \times SU(3)^{col}$, or for example, the subunification symmetry[2] $SU(2)_L \times SU(2)_R \times SU(4)^{col}_{L+R}$, which incorporates left-right symmetry as well as quark-lepton unification; but we would not expect either of these symmetries to be part of a bigger grand unified symmetry G.

The effective coupling constants of $SU(2)_L \times U(1) \times SU(3)^{col}$ or of $SU(2)_L \times SU(2)_R \times SU(4)^{col}_{L+R}$ would still be related to each other within this picture through the dynamics of the substructures (preons), which (presumably) is governed by a single basic coupling constant. By the same token, protons will be expected to decay into leptons and/or antileptons, as quarks and leptons are still related to each other through their preonic structures. But one can no longer apply perturbative renormalisation group equations for the effective coupling constants of $SU(2)_L \times U(1) \times SU(3)^c$ beyond a momentum-scale of order M_o, and in this sense the familiar merging of three forces into one would not apply.

From the observational point of view, there thus appear to be three alternative scenarios, which are summarized below:

Scenario I: There exists a Grand "Desert" beyond 10^2 GeV,

Grand Unification Mass $M \sim 10^{15}$ GeV; Intermediate mass-scales (IMS) either do not exist or far exceed 100 TeV; quarks, leptons, gauge and Higgs-bosons behave as elementary "point" like particles up to momenta $\gtrsim M$.

Observational Possibilities: No new basic physics is expected to be discovered in the energy range ~ 1 to 100 TeV.

Scenario II: "Desert" blooms through (a) light Intermediate mass-scales (IMS) $\sim 10^4$-10^5 GeV with unification-mass M being $\sim 10^{15}$ GeV, or (b) Light unification-mass $M \sim 10^4$-10^5 GeV, and/or (c) compositeness of Higgs with sizes $\sim (1 \text{ TeV})^{-1}$.

Here quarks, leptons and gauge bosons behave as elementary point-like particles up to momenta $\gtrsim M$ and thus conventional picture of grand unification still applies, even though Higgs may reveal their compositeness at momenta $< M$. Existence of light Intermediate mass-scales may be signalled by observation of B-L violating proton-decay modes (e.g. $p \to 3\ell$ or $3\bar{\ell}$ + mesons) or n-$\bar{n}$ oscillations (see later).

Observational Possibilities for Scenario II:

(1) Exotic Quark-Lepton Interactions: The light IMS or light unification mass $M \sim 10^4$-10^5 GeV can represent the masses of leptoquark gauge bosons X's as in $SU(4)^{col}$ (Ref. 2). The exchange of such gauge bosons would lead to a new class of quark-lepton-interactions: $q\bar{q} \to \ell\bar{\ell}$ which would begin to compete favorably with electromagnetic contributions for CM energies exceeding $M_X/10$. Such exotic quark-lepton interactions would thus appear in the form of enhanced lepton-pair production in $\bar{p}p$- and pp-collisions for CM energies exceeding 1 to 10 TeV. If seen, they would constitute a direct proof of the hypothesis of quark-lepton unification.[1,2]

(2) Right-handed V+A-Interactions: Light IMS may also represent masses of "right handed" gauge bosons $W_R^{\pm}$ coupled to V+A-currents. Within maximal symmetries (see later), $W_R^{\pm}$ can be as light as about 1 TeV. Direct production of W_R^{+} in a $\bar{p}p$-machine with its subsequent decay to a fermion pair via V+A-current would demonstrate left-right symmetry of the basic interactions.[13]

(3) Majorana-Neutrino ν_R: Light IMS may also represent the mass-scale of the right handed Majorana neutrino ν_R, which arises within L$\leftrightarrow$R symmetric theories.[14] High energy machines in the 1-100 TeV range may thus not only lead to the discovery of $W_R^{\pm}$, but also the discovery of the Majorana heavy neutrino ν_R, which would arise as a decay-product of $W_R^+ \to e_L^+ + \nu_R$.

(4) "TechniColored" Hadrons and Techniforce: Independent of the observational possibilities listed above, one can expect to see technihadrons (fermions, gauge bosons and spin-0 particles) and techniforce in the energy-range $\sim$1 to 100 TeV, if Higgs are composites with size $\sim$ (1 TeV)$^{-1}$. [Note, within the present scenario quarks, leptons and the associated gauge bosons are still assumed to be elementary, as in a technicolor-theory. The observational possibilities become more "colored," if this assumption does not hold (see below).]

Scenario III: "Desert" blooms through compositeness of quarks, leptons, gauge and Higgs bosons with sizes $r_o \sim 10^{-17}$-10^{-19} cm (inverse sizes $M_o \equiv 1/r_o \sim 1$ to 100 TeV).

Here the quark-lepton gauge symmetry may effectively be $SU(2)_L \times U(1) \times SU(3)^{col}$ or the subunification-symmetry $SU(2)_L \times SU(2)_R \times SU(4)^{col}$, but the conventional grand unification of quarks and leptons[1-4] in terms of a simple or semisimple group (with components being related by discrete symmetry) would not hold even effectively, as the "would be" grand unification mass M now exceeds the inverse size M_o of the composites. The force which binds preons--the constituents of quarks and leptons--can now play the role of technicolor force. The technihadrons can now be composed of the same substructures (preons) as the normal quarks and leptons; this provides an economical basis[11] for the idea of technicolor.

Observational Possibilities for Scenario III:

(1) Power Form Factors for quarks and leptons; Gross Scaling Violations: Deep inelastic ep or μp-scattering and high energy lepton-lepton scatterings would reveal power electromagnetic and weak form-factors for quarks and leptons of the form $\sim[M_o^2/(Q^2+M_o^2)]^n$, n=1 or 2 with $M_o \sim 1$ to 100 TeV. This would be reflected by gross-scaling violations in ep or μp-structure functions, which cannot be accounted for by the familiar logarithmic scaling violations of QCD.

(2) Appearance of Rescaling for $Q \gg M_o$; Revelations of Preon-Dynamics and the New Preon-Binding Force: Scaling may be restored with structure functions described in terms of preon-distribution functions once relevant energy-momentum variables far exceed the inverse size M_o of quarks and leptons. This in the first place would be a sure demonstration of the existence of preonic substructure. Studies of this nature would furthermore reveal the detailed nature of preons, the new force which binds preons and the nature of its dynamics.

(3) Change in R; Departures from Drell-Yan Cross Sections: Likewise the R-parameter for e^-e^+-annihilation would pass through a period in which it would exhibit gross-scaling violation; it may eventually "settle" at a value, which is appropriate for preons rather than quarks and leptons. This would happen when the momentum Q far exceeds M_o.

In a similar fashion the charged lepton pair-production in pp or $\bar{p}$p-collisions would exhibit sharp departures from Drell-Yan like cross sections once the invariant mass of lepton-pair approaches M_o.

(4) Enhanced Lepton-Multiplicity: Once the center of mass energy in pp or $\bar{p}$p-collisions exceeds the inverse size M_o of quarks and leptons, we expect that preon-pairs would be produced. Even if the preons are confined permanently or semipermanently at "large" distances (like quarks) due to the preon-binding force, the preon-antipreon pairs can be produced at short distances, which can create additional pairs; these would recombine to make quarks and leptons in comparable numbers, which are neutral with respect to the binding force. Thus the lepton-pair multiplicity will increase substantially for energy-momenta exceeding M_o.[14a] Recall that lepton-pair multiplicity will also increase if the exchange of leptoquark gauge boson becomes prominent. A study of the dependence of enhanced lepton-pair production on the invariant mass of the pair can clearly distinguish between enhanced lepton-pair production due to elementary lepto-quark exchange on the one hand and compositeness of quarks and leptons on the other.

(5) If the effective quark-lepton gauge symmetry is as big as $SU(2)_L \times SU(2)_R \times SU(4)^{col}$, we would expect to see in addition manifestations of leptoquark gauge bosons as well as right handed gauge particles $W_R^{\pm}$ and possibly also Majorana right-handed neutrino ν_R on the 1-100 TeV range in a manner described under II.

(6) With Higgs being composites with sizes $\sim$ (1 TeV)$^{-1}$ we would expect to see manifestations of technihadrons as in II except that the

preon binding force can now play the role of technicolor force.[11]

Remark: If quarks and leptons are composites of size $\sim 10^{-17}$-10^{-19} cm, there may exist several layers of increasing elementarity --preons, pre-preons and pre-pre-preons--within quarks and leptons with the last but one layer having a size of the order of inverse Planck mass $\sim 10^{-33}$ cm. In this case there would be no long extending "desert" in physics between the two mileposts exhibited in Fig. 1.

The observational possibilities which may arise within scenario II and/or III, are summarized in Fig. 3.

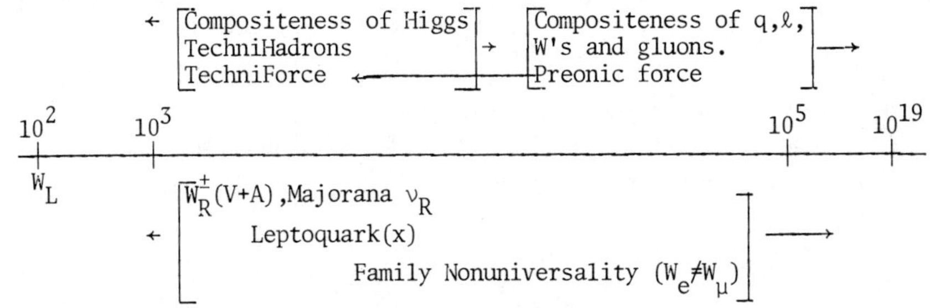

Fig. 2. Blooming of the "Desert" with Elementary or Composite Structures Following Scenario II and/or III.

Needless to say, the scope for discoveries of fundamental new physics in the energy range 1 to 100 TeV is immense if either there exists intermediate mass-scales, or if quarks and leptons are composites with sizes $\sim 10^{-17}$-10^{-19} cm. or both.

From now on I turn my attention exclusively to grand unified symmetries which permit intermediate mass-scales. For these purposes, quarks and leptons would be assumed to be elementary at least up to probe-momenta of the order of the unification-mass. The question of compositeness of quarks and leptons is discussed elsewhere.[11] As a rule intermediate mass-scales as low as 10^4-10^5 GeV arise within grand unified symmetries, which are "maximal" in the space of basic fermions.[3,15] Examples of such symmetries are SU(16) and its family-extensions (see later). Such symmetries, depending upon the pattern of symmetry-breaking, permit not only the co-existence of the four types of decay modes of proton[3] listed in Table I, but also the co-existence of (B-L)-conserving proton-decay (i.e. $p \to e^+\pi^0$ etc.) with neutron-oscillations.[15] This is in contrast to nonmaximal symmetries such as SU(5) and SO(10), where such possible co-existences and low intermediate gauge-masses are not permitted. In what follows, I outline certain typical features of maximal symmetries, the alternative patterns of spontaneous breakdown of such symmetries and the complexions of baryon-lepton number violating processes, associated with such patterns.

II. MAXIMAL SYMMETRIES AND THEIR SPONTANEOUS DESCENTS

As the name suggests, "Maximal" symmetry[3] corresponds to gauging all fermionic degrees of freedom with fermions consisting of quarks and leptons. Thus with n two-component left-handed fermions F_L and n two-component right-handed fermions F_R, which may be replaced by the left-handed charge conjugate fields $(F^C)_L$, the maximal symmetry is SU(2n). As an example, for a single family consisting of eight left-handed fermions (i.e. 6 quarks + 2 leptons) plus the corresponding charge conjugate fields, the maximal symmetry is SU(16). For the electron-family the multiplet is:

$$F_e = \{(u,d)_{r,y,b},(\nu,e^-)\,|\,(d^c,u^c)_{r,y,b},e^c\nu^c\}_L \ . \qquad (1)$$

Such symmetries generate triangle anomalies, which are avoided by postulating that there exists a conjugate <u>mirror</u> set of fermions[16] $F^m_{L,R}$, which couples to the gauge mesons through the helicity flip coupling (i.e. $F_{L,R} \leftrightarrow F^m_{R,L}$). Thus by maximally gauged symmetries, we mean gauge symmetries which are maximal in the space of fermions up to the mirror discrete symmetry.

The discrete mirror symmetry applies to the gauge interactions only. The gauge invariant Yukawa coupling constant of the basic fermions $F_{L,R}$ with scalar field can differ[17] from that of the mirror fermions. This would in general assign differing diagonal mass terms to basic versus mirror fermions. A direct invariant mass-mixing between basic and mirror fermions can be avoided, if so desired, by imposing the discrete symmetry $F_{L,R} \leftrightarrow F_{L,R}$; $F^m_{L,R} \leftrightarrow - F^m_{L,R}$. Since the diagonal mass terms of the mirror fermions violate weak $SU(2)_L$, like those of the basic fermions, it follows that <u>the mirror fermions masses can not be superheavy</u>. With the strength of Yukawa coupling being limited above by the electric charge e, the mirror fermion-masses cannot be much heavier than about 100-150 GeV. <u>The LEP and high energy $\bar{p}p$-machines may thus</u>

be very suitable to test the mirror-hypothesis.[16] The characteristic signature of mirror fermions is their helicity flip-coupling compared to the basic fermions. This would reflect itself in V+A rather than V-A coupling for weak decays of mirror heavy leptons. If the mirror fermions do not mix at all, or mix very little, with the basic fermions, there would also exist stable or semistable heavy mirror-matter.[16]

For maximal gauging of three families (e, μ and τ) one would need to gauge $[SU(16)]^3$ or the still extended symmetry SU(48). As expressed elsewhere,[9,11] such gigantic symmetries should be viewed only as effective gauge symmetries with the associated gauge particles, quarks, leptons, mirror fermions and Higgs bosons being composites of a more economical system of preons. Spontaneous symmetry breaking can permit the descent of SU(48) or $[SU(16)]^3$ to observed low energy symmetry $SU(2)_L \times U(1) \times SU(3)_c$ via the diagonal symmetry $SU(16)_{e+\mu+\tau}$. In what follows, I shall for simplicity use SU(16) as a language for maximal symmetry, though it will ultimately be viewed as part of an extended maximal symmetry such as SU(48) or $[SU(16)]^3$.

A possible chain of extending the observed low energy symmetry $SU(2)_L \times U(1) \times SU(3)^{col}$ towards maximally gauged symmetries is outlined in Fig. 3.

$$[SU(16)]^3$$
$$\uparrow$$
$$SU(16)_{e+\mu+\tau}$$
$$\uparrow$$
$$SU(8) \times SU(8)$$
$$\uparrow$$
$$SU(2)_L \times SU(2)_R \times SU(4)^{col}_{L+R}$$
$$\uparrow$$
$$SU(2)_L \times SU(2)_R \times U(1)^{B-L}_{L+R} \times SU(3)^{col}$$
$$\uparrow$$
$$SU(2)_L \times U(1) \times SU(3)^{col}$$

Fig. 3. Towards a Maximal Gauging.

The chain exhibited here is based upon the assumption that Nature is intrinsically left-right symmetric;[13] the symmetry breaks spontaneously such that low energy weak interactions utilize primarily left-handed particles. This chain inevitably passes through the subunification-symmetry[2]

$$SU(2)_L \times SU(2)_R \times SU(4)^{col}_{L+R}$$

which embodies three attractive features: (i) left-right symmetry, (ii) quantisation of electric charge, as well as (iii) quark-lepton unification with lepton-number being the fourth color. This is why, we regard $SU(2)_L \times SU(2)_R \times SU(4)^{col}$ as the master-symmetry on which grand unification may be based. A few desirable intermediate symmetries, which are not exhibited in Fig. 3, are[16] $[SU(4)]^4$ and[18] $[SU(6)]^4$ based on two and three families respectively.

Motivations for "Maximal" Gauging

Grand unification would be achieved by gauging a simple or appropriate semisimple subgroup of the maximal one family-symmetry SU(16), such as SU(8) x SU(8) or SO(10), or even the "left-right asymmetric" SU(5). The motivations for gauging a "maximal" symmetry as opposed to a nonmaximal one are essentially twofold:

(i) Aesthetics: We believe that in the context of gauge unifications of basic particles and their forces, quantum numbers like baryon number B and lepton number L, or their linear combinations B-L and fermion number $F \equiv 3B+L$, are best defined in terms of a nonabelian local symmetry G just like electric charge. One of the characteristic features of "maximal" gauging is to ensure that baryon and lepton numbers or equivalently their linear combinations B-L and $F = 3B+L$ are among the generators of the local "maximal" symmetry G; they are thus conserved in the basic gauge Lagrangian; they are violated spontaneously and inevitably as the gauge particles coupled to the linear combinations of B and L acquire masses (see elaboration below). By contrast, for the case of nonmaximal gauging, B, L and F are violated, in general, explicitly in the basic gauge Lagrangian; that is the case for example for SU(5) and SO(10).

Furthermore "maximal" symmetry is essentially unique. Modulo mirror-discrete symmetry, it treats every fermion on the same footing in that it gauges all fermionic degrees of freedom. By contrast, the choice of a nonmaximal symmetry is rather arbitrary and in general several bonafide degrees of freedom are suppressed within nonmaximal symmetries.

(ii) Pragmatic: Maximal symmetries permit the possible existence of intermediate mass-scales and thereby observability of signals of grand unification at modest energies $\sim$10-100 TeV; simultaneously they allow richer complexions for baryon and lepton-number violations at lower energies. These possibilities do not arise within nonmaximal symmetries such as SU(5) and SO(10).

The Origin of B, L, F Violations Within Maximal Symmetries:

As mentioned above baryon and lepton-numbers, or equivalently B-L and fermion number F = 3B+L, are among the generators of the maximal local symmetry such as SU(16). This may be seen simply as follows.

First, as soon as we gauge[2] $SU(4)^{col}$, which is the "maximal" symmetry in the space of three quarks plus one lepton of the same flavor (q_r, q_y, q_b, ℓ), B-L becomes a generator of the local symmetry. The 15th generator of $SU(4)^{col}$ is associated with a massless gauge particle V_μ^{15} with the fermionic coupling:

$$g\, V_\mu^{15} [J_\mu^{15}]_{fermions} = g\, V_\mu^{15} \{ \sum_{r,y,b} \bar{q}_i \gamma_\mu q_i - 3\bar{\ell}\gamma_\mu \ell \} (2\sqrt{6})^{-1} \,. \qquad (2)$$

We see that J_μ^{15} generates a charge which is proportional to B_q-3L=3(B-L). [Here B_q denotes quark-number, which is 3B.]

Next if fermions and antifermions of the same helicity are put in one multiplet and the maximal symmetry of this multiplet is gauged,[3] as in SU(16), the traceless generator which is proportional to +1 for all fermions and -1 for all antifermions, representing fermion-number, gets included among the generators of the nonabelian local symmetry. The said generator is thus associated with an additional massless gauge particle V_μ^F whose coupling to fermions is given by:

$$g\, V_\mu^F [J_\mu^F]_{fermions} = g\, V_\mu^F (1/4)\; [(\Sigma \bar{q}\gamma_\mu q + \bar{\ell}\gamma_\mu \ell)_L - (q \to q^c, \ell \to \ell^c)] \qquad (3)$$

In 1955 Lee and Yang[19] showed that if the conservation of baryon number (which was generally assumed to hold at that time) was due to a massless gauge particle coupled to baryon number, then the effective coupling generated through the exchange of such a massless particle must be much weaker than gravitational in order that the limits from Eötvos type experiments may be satisfied. It is easy to see that as long as the Eötvos type experiments involve different kinds of matter with varying baryon to lepton-ratios, the same conclusion would follow for the effective coupling of a massless gauge particle coupled to any linear combination of baryon and lepton numbers.

So long as one did not wish to postulate the grand unification hypothesis unifying quarks and leptons, however, one did not have to introduce a massless spin-1 particle coupled to any linear combination of baryon and lepton-numbers in the first place; and in the second place if one did introduce such a particle, one was free to associate with it an arbitrarily small effective coupling. Now the crucial point is that within the grand unification-hypothesis, with a maximal gauging of the quark-lepton-symmetry, massless spin-1 gauge particles coupled to linear combinations of baryon and lepton-numbers would have to be introduced,[2,3] as we saw before. Furthermore, within this hypothesis, the effective coupling of a gauge particle can not be arbitrarily weak, since all effective gauge interactions are of order $\alpha \approx 1/137$ at the grand unification mass-scale. There is then just one choice left in order for the hypothesis to be consistent with Eötvos-type experiments. Such spin-1 gauge particles coupled to linear combinations of B and L must acquire mass spontaneously. It then follows that the associated charges (for the illustrative example B-L and F) must be violated spontaneously.

Instead of gauging the maximal symmetry, one might have gauged a subgroup of the maximal symmetry still retaining the requirement that quarks and leptons are in the same irreducible multiplet of the subgroup. In this case one need not have any generator within the gauged subgroup representing a linear combination of baryon and lepton numbers; but then these quantum numbers would be violated explicitly by the basic gauge interactions. An analogous situation is in fact what happens in some realistic models of grand unification such as SU(5) and SO(10), both of which are subgroups of the maximal one family symmetry SU(16).

Thus the violations of B and L may well be explicit. This would be the case if there was a good reason why Nature would choose a specific subgroup of the maximal symmetry. The point of the remark made, however, is that even if baryon and lepton numbers are conserved in the basic lagrangian (and this holds automatically if we gauge the "maximal" symmetry of the quark-lepton-multiplet), either B or L or both must be violated spontaneously as the gauge particles of the quark-lepton symmetry acquire masses; this is in order that the theory may be compatible with limits from Eötvos type experiments. This line of reasoning had led in 1973 to a questioning of baryon and lepton-number conservations with the prediction that the lightest baryon -- the proton -- must decay.[2,19]

To see more explicitly the manner in which B, L, F-violations arise within a maximal symmetry such as SU(16), note that SU(16) generates 255 gauge particles. Of these, (63+63+1) = 127 belong to the subgroup SU(8) x SU(8) x $U(1)_F$, where the two SU(8)'s operate on fermions and antifermions respectively; $U(1)_F$ denotes fermion number. These 127 gauge particles include the familiar W's, gluons, leptoquarks, X's, V_{B-L} and V_F all carrying fermion number zero. The remaining (8x8) + (8x8) = 128 gauge particles take fermions to antifermions and vice versa and carry fermion number ±2. The different types of gauge particles are listed in Fig. 4.

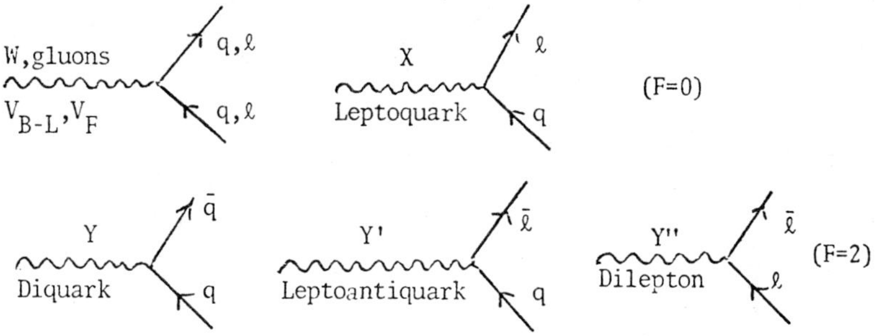

Fig. 4 Gauge particles of a maximal symmetry

Note that each of the gauge particles prior to SSB carry definite baryon, lepton and fermion numbers. Furthermore, among the gauge particles there exist two (i.e. V_{B-L} and V_F), which couple to B-L and fermions number F respectively. Following the argument presented above, it follows that both B-L and F must be violated spontaneously as the associated massless spin-1 particles acquire masses. They are violated by two distinct mechanisms:

(i) Gauge mixing: Gauge particles carrying different B, L and F get mixed through SSB; this induces violations of B, L and F. In particular the mixing of Y with $\bar{Y}'$ will induce proton-decay into antilepton while that of Y with $\bar{X}$ will induce proton decay into lepton as illustrated in Figs. 5(a) and (b) respectively.

Mixings	Selection Rules	Transitions
$Y \leftrightarrow \bar{Y}'$	$\Delta(B-L)=0; \Delta F=-4$	$p \to \bar{\ell} +$ mesons
$X \leftrightarrow \bar{X}$	$\Delta(B+L)=0; \Delta F=-2$	$p \to \ell +$ mesons

The mixing of Y and $\bar{Y}'$ occurs as SU(16) descends spontaneously to $SU(2)_L \times SU(2)_R \times SU(4)^c$, while that of Y with $\bar{X}$ occurs at a lower stage of SSB when $SU(2)_L \times U(1)$ breaks to $U(1)_{em}$. Both these mixings can arise through vacuum expectation values of a Higgs multiplet of the type $\Omega^{\{AB\}}_{\{CD\}} \sim \{16x16\}_{symm} \times \{16^*x16^*\}_{symm}$ (see Ref. 15 for details).

The amplitudes for the corresponding processes will depend upon the mixing masses versus the diagonal masses, which in turn depend upon the route of spontaneous descent (see later):

$$\mathrm{Amp}(p \to e^+\pi^0)_{\Delta F=-4} \approx g^2(\Delta^2_{YY'}/m^2_Y m^2_{Y'}) \tag{4}$$

$$\mathrm{Amp}(p \to e^-\pi^+\pi^+)_{\Delta F=-2} \approx g^2(\Delta^2_{XY}/m^2_X m^2_Y) \tag{5}$$

For the simplest pattern of SSB, $m^2_Y = m^2_{Y'} = \Delta^2_{YY'}$; and thus the amplitude for $p \to e^+\pi^0$ reduces to (g^2/m^2_Y). The mixing $(\mathrm{mass})^2$ Δ^2_{XY} must be proportional to a scale representing $SU(2)_L$-breaking. Thus $\Delta^2_{XY} \lesssim M m_{W_L}$, where M represents the heavier of the two masses: m_X and m_Y.

(ii) Spontaneously Induced Yukawa Transitions: Alternatively the violations of B, L and F may also arise through spontaneously induced Yukawa-transitions of the type:

$$q \to \ell + \phi, \quad q \to \bar{\ell} + \phi', \quad q \to \bar{q} + \phi'' \tag{6}$$

Two typical mechanisms for the first and the third transitions are exhibited in Figs. 6a and 7a. The effective Yukawa-transition $q \to \ell + \phi$ shown in Fig. 6a, taken in third order followed by an invariant $\lambda\phi^4$ scalar quartic coupling, subject to $\langle\phi\rangle \neq 0$, induces proton-decay into three leptons plus mesons (see Fig. 6b)

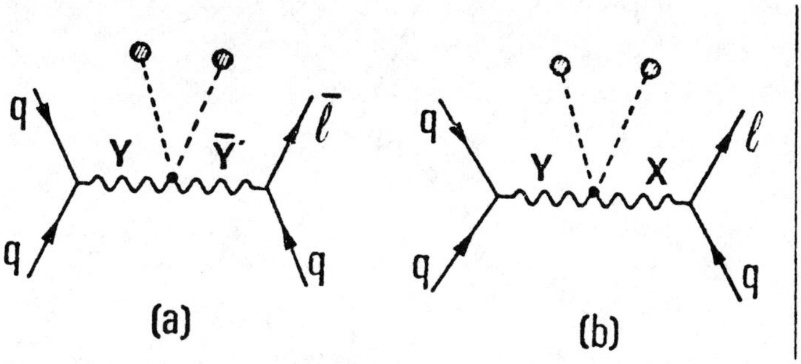

Fig. 5a,b: Spontaneously Induced Gauge Mixings leading to $3q \to \bar{\ell}$ and $3q \to \ell$-transitions. For definition of the canonical gauge fields, see Fig. 2.

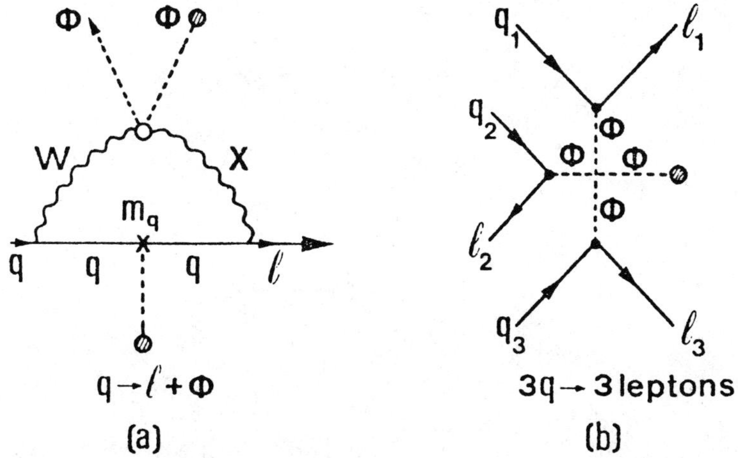

Fig. 6a,b: Spontaneously induced $q \to \ell + \phi$ and $3q \to 3\ell$ transitions.

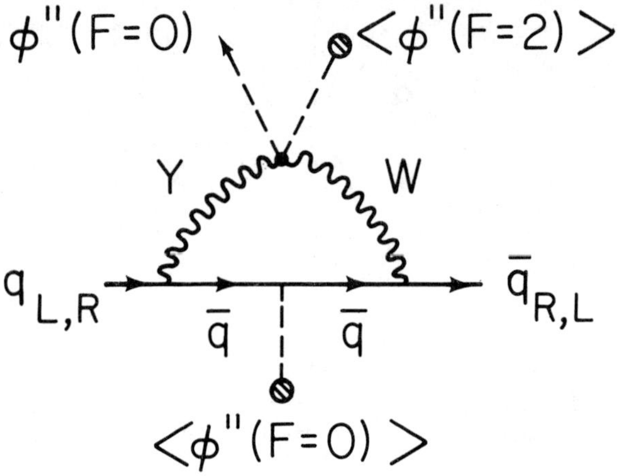

Fig. 7a. A typical diagram for the induced Yukawa-transition $q \to \bar{q} + \phi''$ (F=0), which violates fermion number by two units.

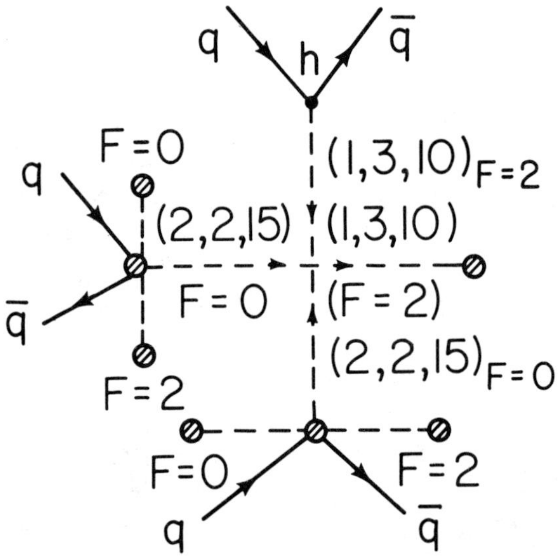

Fig. 7b. The effective transition $3q \to 3\bar{q}$ which induces n-n̄ oscillations. The intermediate scalar fields denoted by dotted lines all belong to the single multiplet $\underline{136}$ of SU(16). They are specified in terms of their transformation properties with respect to the subsymmetry $SU(2)_L \times SU(2)_R \times SU(4)^c \times U(1)_F \subset SU(16)$, see text. Only one of the three Yukawa vertices, labelled by h, is SU(16) invariant, the other two are not.

$$\text{proton} \to 3 \text{ leptons} + \text{mesons} \quad (\Delta(B + \tfrac{L}{3}) = 0; \Delta F = 0) \; . \tag{7}$$

Likewise the effective transitions $q \to \bar{\ell} + \phi'$ and $q \to \bar{q} + \phi''$ taken in third order induce proton decay into three antileptons plus mesons and n-$\bar{n}$ oscillations respectively (see Fig. 7b).

$$p \to e^{+}\bar{\nu}\bar{\nu} \text{ etc.} \quad (\Delta(B - \tfrac{L}{3}) = 0; \Delta F = -6) \; . \tag{8}$$

$$n \leftrightarrow \bar{n} \quad (\Delta B = \pm 2, \Delta L = 0, \Delta F = \pm 6) \; . \tag{9}$$

Details of the processes (6)-(9), including the types of Higgs multiplets which are relevant for these processes, may be found in Refs. 20 and 15.

Two common features of both mechanisms (i) and (ii), are worth noting:

(a) They utilise only spontaneous rather than explicit violations of B, L and F.

(b) None of these mechanisms is tied in any way to the nature of quark-charges. They hold for quark-charges being either integral or fractional.

The Subgroups SU(5) and SO(10)

Instead of gauging the maximal symmetry SU(16), one might have gauged[4,7] one of its subgroups SU(5) or SO(10). The subgroup SU(5) contains neither B-L nor F among its generators; while SO(10) contains B-L but not F. Both lead to explicit violations of B,L as well as F. This comes about as follows: While Y and Y' coupled to $\bar{q}^c\gamma_\mu q$ and $\bar{\ell}^c\gamma_\mu\ell$-currents are distinct gauge particles within the maximal symmetry SU(16) (as shown in Fig. 4), for the subgroups SO(10) and SU(5) only the symmetric combinations $Y_S \equiv (Y+\bar{Y}')/\sqrt{2}$ are present; the antisymmetric combinations $Y_a \equiv (Y-\bar{Y}')/\sqrt{2}$ do not appear. The exchange of Y_S in the second order of gauge interactions induces transitions of the type $3q \to \bar{\ell}$ (i.e. $p \to e^{+}\pi^{o}$ etc.) leading to explicit violations of B, L and F.

These two cases, spontaneous versus explicit violations of B, L and F may agree in their predictions for proton-decay and for that matter in all low energy-properties. Nevertheless they differ from each other conceptually as well as in their physical consequences at high temperatures. Spontaneous violation would disappear at high temperatures exceeding the masses of the relevant gauge particles, while explicit violation would acquire its maximum strength at such temperatures, where the gauge particles would be massless. This distinction plays its most obvious role in the early stage of the universe.[21]

Route of Spontaneous Descent of SU(16)

It is now of interest to see the alternative routes of spontaneous descent of SU(16) down to the low energy symmetry $SU(2)_L \times U(1) \times SU(3)_C$. The three most obvious routes are:[15]

(I) via $SU(8) \times SU(8) \times U(1)_F$, where the two SU(8)'s operate in the spaces of the eight fermions and the either antifermions, respectively, and $U(1)_F$ represents fermion number.

(II) via SO(10) with respect to which the 16-plet remains irreducible.

(III) via $SU(12)_q \times SU(4)_\ell \times U(1)_{|B|-|L|}$, where $SU(12)_q$ operates on six quarks plus six antiquarks, $SU(4)_\ell$ on two leptons plus two antileptons and $U(1)_{|B|-|L|}$ represents $|B|-|L|$.

These three alternative routes together with the Higgs multiplets, whose VEV could induce spontaneous descent of SU(16) via these routes, are exhibited in Fig. 8. Vacuum expectation values of three scalar multiplets suffice to give the desired descents: the adjoint 255, the fourth rank tensorial field $\phi^{\{AB\}}_{\{CD\}} \sim \{16\text{x}16\}_{symm} \times \{16^*\text{x}16^*\}_{symm}$ and 136 = $(16\text{x}16)_{symm}$. The multiplet 136 is used to give masses to fermions as well as to gauge bosons.

VEV of the adjoint 255 by itself can take SU(16) into $SU(8) \times SU(8) \times U(1)_F$; while that of the tensorial field $\phi^{\{AB\}}_{\{CD\}}$ can break SU(16) in general straight to $SU(2)_L \times SU(2)_R \times SU(4)^C$ and for a special case[15] to SO(10); simultaneously it breaks B, L and F. Thus if $<255>$ far exceeds $<\phi^{\{AB\}}_{\{CD\}}>$, the descent via $SU(8) \times SU(8) \times U(1)_F$ would be prominent. In this case it turns out[15] that the proton may decay via a variety of modes: (i) $p \to e^+ + \pi^0$ etc. ($\Delta F=-4, \Delta(B-L)=0$); (ii) $p \to e^- + \pi^+\pi^+$ etc. ($\Delta F=-2, \Delta(B+L)=0$); and (iii) $p \to e^- + \nu\nu + \pi^+\pi^+$ etc. ($\Delta F=0, \Delta(B+L/3)=0$). Here the $\Delta F=-2$ and $\Delta F=0$ modes can even far supercede the $\Delta F=-4$ mode.

The relevant characteristic mass-scales for the three processes turn out to be $\approx (10^{12}\text{-}10^{14})$ GeV, $(10^{8}\text{-}1\text{-}^{11})$GeV and $(10^{4}\text{-}10^{5})$GeV

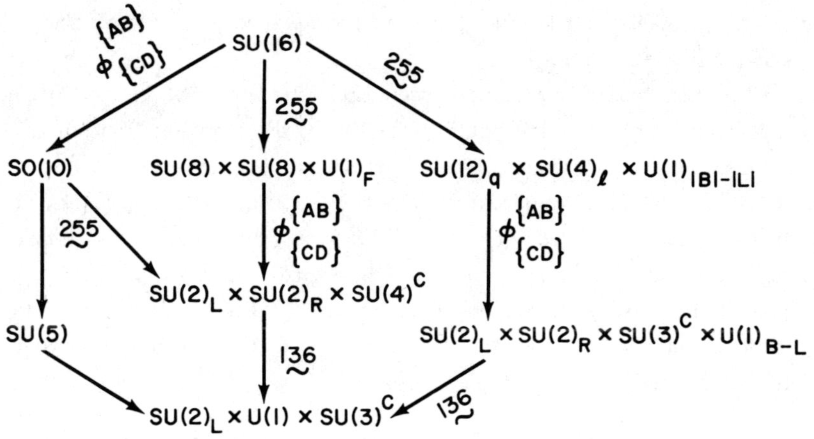

Fig. 8. Alternative routes of SSB of SU(16). Here $\phi\{AB\}_{\{CD\}}$, $\underset{\sim}{255}$ and $\underset{\sim}{136}$ denote scalar multiplets transforming like $(16 \times 16)_{symm} \times (16^* \times 16^*)_{symm}$, adjoint and $(16 \times 16)_{symm}$ - representations of SU(16) respectively.

respectively, the corresponding proton-lifetime being $10^{31\pm2}$ years (see Ref. 15) for details).

If on the other hand $\langle\phi^{\{AB\}}_{\{CD\}}\rangle$ far exceeds $\langle\underline{255}\rangle$ and in particular the special SO(10)-chain materialises, then only the $\Delta F=-4$ mode (i.e. $p \to e^+ + \pi^o$ etc.) would be prominent. Thus, as mentioned before, predictions of SU(5) and SO(10) can emerge as special cases of the "maximal" symmetry SU(16).

The third route involving descent via $SU(12)_q \times SU(4)_\ell \times U(1)$ can result from an alternative pattern of VEV of the adjoint $\underline{255}$ with the subsequent breakdowns taking place through $\langle\phi^{\{AB\}}_{\{CD\}}\rangle$ and $\langle\underline{136}\rangle$. In contrast to the first two routes, the third route separates quarks from leptons at the first stage of SSB and thus assigns the heaviest masses to the lepto-antiquark Y' as well as the leptoquark X (see Fig. 4); the diquark Y in this case can, however, be much lighter than Y' and X, as it acquires a mass only at the second stage of SSB. We now observe that for a gauge mass-pattern given by

$$\begin{aligned} m_X &\approx m_{Y'} \approx 10^{14}\ \text{GeV} \\ m_Y &\approx \Delta_{YY'} \sim 10^4\text{-}10^5\ \text{GeV} \end{aligned} \tag{10}$$

which is permissible for the third route, $\Delta B=2$, $\Delta L=0$ n-$\bar{n}$ oscillations as well as proton-decay of the $\Delta F=-4$ variety (i.e. $p \to e^+\pi^o$ etc.) can co-exist, each with measurable strength. One obtains the neutron-oscillation period for free neutrons of order 10^5-10^9 sec. with a deuteron lifetime of order 10^{30}-10^{38} years. (The $\Delta B=2$, $\Delta L=0$ decay of deuteron into pions was considered sometime ago in Ref. 3.) The proton lifetime with these mass patterns is $\approx 10^{30.5\pm2}$ years. What is worth noting is that proton does not decay unduly rapidly <u>despite</u> the ultralight mass of the diquark Y($m_Y \approx 10^4$ GeV). This is something which was possible only because of the spontaneous nature of the B, L, F violations occurring within a maximal symmetry (for which Y-$\bar{Y}'$ mixing vanishes as $m_Y \to 0$).

In contrast to the case presented here, it may be noted that Mohapatra and Marshak,[22] working within the subunification symmetry $SU(2)_L \times SU(2)_R \times SU(4)^c$, proposed in Ref. 2, have constructed a model which permits n-$\bar{n}$-oscillation, but not proton-decay. Most grand unification models (e.g. these based on SU(5) or SO(10) -- at least in their simplest version) permit proton-decay but not n-$\bar{n}$-oscillation.

The case arising here within the maximal symmetry SU(16) permits both.

Finally, it may be observed that in this third route of spontaneous descent, not only the lepto-antiquark Y' but also the leptoquark X becomes superheavy ($m_X \sim m_{Y'} \sim 10^{14}$ GeV). This has the consequence that the ΔF=0, ΔF=-2 and ΔF=-6 proton-decay modes, i.e. p → 3 leptons + mesons, p → ℓ + mesons and p → 3$\bar{\ell}$ + mesons, will all be strongly suppressed since these processes involve exchange of either the leptoquark X or the leptoantiquark Y' (see e.g. Figs. 6a, 6b and 5b).

The selection rules for B, L, F violations associated with the three alternative routes of spontaneous descent and the corresponding characteristic mass-scales are summarised in Table II.

Table II

Route of Spontaneous Descent	Selection Rules*	Characteristic Mass Scales
I SU(16)	$\Delta F=-4(p\to e^+\pi^o)$--Yes	10^{12}-10^{14} GeV
→SU(8)xSU(8)xU(1)$_F$	$\Delta F=-2(p\to e^-\pi^+\pi^+)$--Yes	10^8-10^{11} GeV
→SU(2)xSU(2)xSU(4)	$\Delta F=0(p\to e^-\nu\nu\pi^+\pi^+)$-Yes	10^4-10^5 GeV
$<255> >> <\phi^{\{AB\}}_{\{CD\}}>$	$\Delta F=-6(p\to 3\bar{\ell})$--No	
	$\Delta B=2(n\leftrightarrow\bar{n}$ oscill.)No	
II SU(16)→SU(2)xSU(2)xSU(4)	$\Delta F=-4(p\to e^+\pi^o)$--Yes	10^{14} GeV
via SO(10) or otherwise	ΔF=0,-2,-6 No	
$<\phi^{\{AB\}}_{\{CD\}}> >> <255>$	$\Delta B=2(n\leftrightarrow\bar{n}$ oscill.)No	
III SU(16)	$\Delta F=-4(p\to e^+\pi^o)$--Yes	10^{14} GeV
→SU(12)$_q$xSU(4)$_\ell$xU(1)	$\Delta B=2(n\leftrightarrow\bar{n}$ oscill.)Yes	10^4-10^5 GeV
→SU(2)xSU(2)xSU(3)xU(1)	ΔF=0,-2,-6 No	

*For the first route of spontaneous descent, there exist subpossibilities under which ΔF=0 and ΔF=-2-proton decay modes can supercede the ΔF=-4-decay mode or vice versa. For this and for the case permitting ΔF=-6 proton decay (i.e. p → 3$\bar{\ell}$+mesons), see Ref. 15.

We see that some of the alternative routes of spontaneous descent require the existence of an intermediate mass-scale as low as about 10^4 GeV with or without a higher intermediate scale in the range of 10^8-10^{11} GeV. It has been shown elsewhere[8] that (consistent with the renormalisation group equations for the running coupling constants and the observed values of $\sin^2\theta_W$ and α_s, such intermediate mass-scales are indeed permissible, if we allow basic family distinctions for quarks and leptons and gauge their maximal symmetries such as $[SU(16)]^3$ or SU(48) as mentioned in the introduction.

It can be argued[15] that the co-existence of the alternative proton-decay modes ($\Delta F=0,-2,-4$ and -6) with or without n-$\bar{n}$ oscillations, symbolising different characteristic mass-scales does not pose any conflict with the cosmological generation of baryon excess for a universe, assumed initially to be matter-antimatter symmetric. The spontaneous rather than explicit[24] nature of B, L, F violations plays a crucial role in avoiding such a conflict.

Finally, it may be observed that in the event SU(16) descends spontaneously through a single mass-scale to the low energy symmetry $SU(2)_L \times U(1) \times SU(3)^c$,

(i) there would arise a long extending "desert" between 10^2-10^{15} GeV.[5]

(ii) one would obtain[5] $\sin^2\theta_W \approx 0.21$.

(iii) the proton would decay predominantly via the $\Delta F=-4, \Delta(B-L)=0$ mode (i.e. via $e^+\pi^0$, $\bar{\nu}\pi^+$-modes, etc.). All other decay modes of the proton as well as n-$\bar{n}$ oscillations would be strongly suppressed, and

(iv) the "standard" calculations carried out in the context of the minimal SU(5)-model for proton-decay rate together with radiative corrections for $\sin^2\theta_W$ would apply, barring perhaps small corrections owing to the presence of mirror fermions in SU(16). Thus proton would be expected to have a lifetime $\tau_p \approx 10^{30.5\pm2}$ years.

In short, the physics of SU(16) descending spontaneously via a single stage hierarchy would coincide essentially with that of SU(5) at low energies. The extra righthanded neutrino would have a super-heavy Majorana mass[14] as also the gauge particles coupling to V+A flavor currents. They would be unobservable. Proton-decay would perhaps be the only observable striking consequence of grand unification. This situation, while logically permissible within SU(16), is no doubt rather dull compared to what it would be if SU(16) or rather SU(48) descends hierarchically in multiplet stages permitting intermediate mass-scales. It would be exciting to see traces of exotic leptoquark-interactions and of left-right symmetry through manifest V+A-interactions at an energy-scale of order 10 TeV, which might be achievable in the near future. But then, who knows what Nature has indeed chosen for the hierarchy of spontaneous descent?

What one has argued[2,3,6,8] is that exciting new physics at low (1 GeV) and moderate energies ($\lesssim$ 100 TeV) can legitimately be expected, consistent with presently known phenomena, if we believe that the underlying unity of quarks and leptons is based upon a maximal symmetry. Ongoing searches for proton-decay and n-$\bar{n}$ oscillations can help us judge better in the near future as to whether some of these expectations are likely to be borne out. In particular, the observation of the $\Delta F=0$, or -6 mode (i.e. $p \to 3\ell$ or $3\bar{\ell}$ + mesons) and/or n-$\bar{n}$ oscillation at any level within conceivable future will strongly suggest the existence of new physics at 10 to 100 TeV region and thereby motivate building high energy machines in this range. For this reason, second and third generation experiments for proton decay and n-$\bar{n}$ oscillation must be planned to look for all modes listed here as possible rare processes, in case they are not found in the first generation experiments. As we now sit at low energies, this might be the best way to probe into the design of grand unification.

It is a pleasure to thank Abdus Salam and John Strathdee for stimulating discussions. The research was supported in part by a grant from the U. S. National Science Foundation.

REFERENCES AND FOOTNOTES

1. J. C. Pati and Abdus Salam, "Lepton hadron unification" (unpublished), reported by J. D. Bjorken in the Proceedings of the 15th High Energy Physics Conference held at Batavia, Vol. 2, p. 304, September (1972); J. C. Pati and Abdus Salam, Phys. Rev. D8, 1240 (1973).
2. J. C. Pati and Abdus Salam, Phys. Rev. Letters 31, 661 (1973); Phys. Rev. D10, 275 (1974); Phys. Letters 58B, 333 (1975).
3. J. C. Pati, Abdus Salam and J. Strathdee, Il Nuovo Cimento 26A, 77 (1975); J. C. Pati, Proceedings of the Second Orbis Scientiae, Coral Gables, Florida, p. 253, January (1975); J. C. Pati, S. Sakakibara and Abdus Salam, ICTP, Trieste, preprint IC/75/93 (1975), unpublished.
4. H. Georgi and S. L. Glashow, Phys. Rev. Letters 32 438 (1974).
5. H. Georgi, H. Quinn and S. Weinberg, Phys. Rev. Lett. 33, 451 (1974).
6. V. Elias, J. C. Pati and Abdus Salam, Phys. Rev. Lett. 40, 920 (1978); V. Elias and S. Rajpoot, Phys. Lett. (1979).
7. H. Fritzsch and P. Minkowski, Ann. Phys. (NY) 93, 193 (1975); H. Georgi, Proc. AIP Conf., Williamsburg (1974).

8. B. Deo, J. C. Pati, S. Rajpoot and Abdus Salam, "Multiplet stage descent of maximal symmetries," unpublished 1979; J. C. Pati and Abdus Salam; Proc. Grand Unification Workshop held at Durham, New Hampshire (1980).

9. J. C. Pati and Abdus Salam, Phys. Rev. D10, 275 (1974); Proceedings of the EPS International Conference on High Energy Physics, Palermo June (1975), p. 171 (Ed. A. Zichichi); J. C. Pati, Abdus Salam and J. Strathdee, Phys. Lett. 59B, 265 (1975); J. C. Pati "Magnetism as the origin of preon binding," University of Maryland Technical Report 80-095 (April 1980); Phys. Lett., Jan 1 (1981); J. C. Pati, Abdus Salam and J. Strathdee, Trieste Preprint, IC/80/180 to appear in Nuclear Physics.

10. Several other authors have also worked on composite models of quarks and leptons from different aspects. An incomplete listing is as follows: K. Matumoto. Progr. Theoret. Phys. 52, 1973 (1974); O. W. Greenberg, Phys. Rev. Lett., 35, 1120 (1975); H. J. Lipkin, Proceedings of the EPS International Conference on High Energy Physics, Palermo, June (1975), p. 609 (Ed. A. Zichichi); J. D. Bjorken and C. H. Woo (unpublished); W. Krolikowski, ICTP, Trieste, preprints IC/79/144 and IC/80/1; E. Nowak, J. Sucher and C. H. Woo, Phys. Rev D16, 2874 (1977); H. Terezawa, Phys. Rev. D22, 184 (1980); H. Terezawa, Y. Chikashige and K. Akama, Phys. Rev. D15, 480 (1977) H. Harari, Phys. Letters 86B 83 (1979); M. A. Shupe, Phys. Letters 86B, 87 (1979); G. 't Hooft, Cargèse Lectures (1979); R. Casalbuoni and R. Gatto, Geneva preprint UGVA-DPT 1980/02-235; R. Barbieri, L. Maiani and R. Petronzio, CERN preprint TH-2900 (1980); R. Chanda and P. Roy, CERN preprint (1980); F. Mansouri, Yale preprint (1980) H. Harari and M. Seiburg preprint, (1980); D. D. Wu, Harvard preprint (1981); M. Yasue, Phys. Lett. 91B, 85 (1980).

11. J. Pati, Invited talk, 1980-Neutrino Conf. held at Erice (June, 80) To appear in the Proceedings; J. Pati, A review to be published in the Proceedings of "Unified Field Theories and Beyond," Workshop held at Johns Hopkins University (May, 1981). H. Harari, Review at SLAC Summer Workshop, SLAC preprint (1980).

12. S. Weinberg, Phys. Rev. D19, 1277 (1979); L. Susskind, Phys. Rev. D20, 2619 (1979); For other relevant references see review on "Technicolour" by E. Farhi and L. Susskind, CERN preprint Ref. TH2975

13. J. C. Pati and Abdus Salam, Phys. Rev. D10, 275 (1974); R. N. Mohapatra and J. C. Pati, Phys. Rev. D11, 566, 2558 (1975); G. Senjanović and R. N. Mohapatra, Phys. Rev. D12, 1502 (1975).

14. M. Gell-Mann, P. Ramond and R. Slansky, Rev. Mod. Phys. 50, 721 (1978); T. Rizzo and G. Senjanović (Phys. Rev. Lett. 46, 1315 (1981) Suggest a scale for W_R-mass lighter than 1 TeV.

14a. A. DeRujula, CERN preprint (1980).

15. J. C. Pati, Abdus Salam and J. Strathdee, Trieste Preprint "Probings through proton-decay and n-n̄ oscillations," IC/80/183; Nuclear Physics (to appear).

16. J. C. Pati and Abdus Salam, Phys. Lett. 58B, 333 (1975). J. C. Pati, "An introduction to unification" in Topics in Quantum Field Theory and Gauge Theories, Proc. Salamanca 1977, page 262; ed. by J. A. de Azcarraga, published by Springer-Verlag.

17. Such a difference may have to traced to a preonic origin of quarks and leptons.

18. J. C. Pati, Proc. Scottish Univ. Summer School Lecture notes, held at St. Andrews (1976). V. Elias and S. Rajpoot, Phys. Lett. (1979). Note that a maximal extension of $[SU(6)]^4$ would be SU(72), not SU(48).

19. One may group heavy leptons with (u,d)-quarks and/or choose to introduce auxiliary global symmetries beyond the local symmetry so as to make the proton stable. This possibility has been considered by a number of authors. See for example, M. Gell-Mann, P. Ramond and R. Slansky (Ref. 14); P. Langacker, G. Segre and H. Weldon, Phys. Lett. 73B, 87 (1978); N. Deshpande and P. Mannheim (preprint, 1979). But the simplest and the most natural consequence of the hypothesis of quark-lepton unification is that proton is unstable.

20. B. Deo, J. C. Pati, S. Rajpoot and Abdus Salam, "On Proton Decay" (unpublished), 1979. This work is reported briefly in J. C. Pati and A. Salam, Proc. Grand Unification Workshop, held at Durham, New Hampshire (April, 1980).

21. J. C. Pati, "Grand unification and proton-stability," Proc. of Workshop held at Madison, Wisconsin (Dec. 1978).

22. R. N. Mohapatra and R. E. Marshak, Phys. Rev. Lett. 44, 1316 (1980).

23. See however, remarks by L. Chang and N. P. Chang, CCNY-HEP-80/5 (1980).

24. S. Weinberg, Phys. Rev. D22, 1694, (1980).

25. A. J. Buras, J. Ellis, M. Gaillard and D. V. Nanopoulos, Nucl. Phys. B135, 66 (1978); T. J. Goldman and D. A. Ross, Cal. Tech. Preprint 68-759 (1980); J. Ellis, M. K. Gaillard, D. V. Nanopoulos and S. Rudaz, CERN-Annecy preprint, LAPP-TH-14, CERN TH-2833; W. Marciano and A. Sirlin, Phys. Rev. Lett. 1981. Other references may be found here.

GRAND UNIFICATION AND THE FUNDAMENTAL PROBLEMS OF CLASSICAL COSMOLOGY

Michael S. Turner*

Institute for Theoretical Physics
University of California
Santa Barbara, California 93106

Abstract

The accomplishments of classical cosmology are reviewed. In particular, the hot big bang model provides a reliable framework for understanding the evolution of the universe back to times at least as early as ~ 0.01 s after the big bang. At present there are (at least) six fundamental problems which have not yet been (completely) resolved. They are: (1) the origin of the baryon number-to-entropy ratio, (2) the origin of the isotropy, (3) the origin of the homogeneity and inhomogeneity, (4) the origin of the flatness, (5) the cosmological constant, and (6) the monopole problem. The role that grand unification has played, and may play in the resolution of these puzzles is discussed. Guth's inflationary universe, which addresses five of these six problems, is reviewed.

1. Introduction

Before I begin discussing the six challenging puzzles which I believe remain for classical cosmology to resolve, and the role that grand unification may play in resolving (or exacerbating) them, I will briefly review our present understanding of the evolution of the universe. There is an impressive body of evidence which supports the hot big bang model (the so-called standard model; for a review of the model see ref. [1]). This body of evidence includes: (i) the general expansion of the universe – the 'Hubble Flow' of galaxies, (ii) the 3K microwave background, (iii) the large abundance of ^{4}He (~ 25% by mass) and D(~ 3×10^{-5} by mass), and (iv) on the theoretical side, the singularity theorems of General Relativity [2]. The universal expansion indicates that the universe began from a big bang, and the blackbody spectrum of the microwave background implies that it was a hot big bang. Specifically, the existence of the 3K background provides direct evidence that the temperature of the universe was once at least $\sim$ 4000K, since the only viable means of producing such a very nearly blackbody photon spectrum is by a highly ionized gas. The present photon and matter densities indicate that this ionized gas phase must have occurred when the universe was $\sim$ 1500 times smaller, when the present 3K background was $\sim$ 4000K. The abundance of ^{4}He and D is good evidence that the temperature of the universe was once $\gtrsim 10^{10}$K. Standard stellar models can account for the abundances of the elements heavier than ^{4}He, but not the abundance of ^{4}He and D. There are no straightforward astrophysical explanations for the large abundance of ^{4}He and D. However, their abundances are rather naturally explained by primordial nucleosynthesis. If one assumes that the universe was once hotter than $\sim 10^{10}$K so that n, p, $e^{\pm}$, $\nu\ \bar{\nu}$, and γ were all in thermal equilibrium, then the standard model supplemented by known nuclear reaction rates "predicts" the observed abundances of ^{4}He and D [3]. Detection of the relic cosmic neutrino backgrounds with temperature $T_\nu = (4/11)^{1/3} T_\gamma$ would provide direct evidence that the temperature was once $\gtrsim 10^{10}$K, since for $T \lesssim 10^{10}$K neutrinos are effectively noninteracting and therefore could not have been produced and thermalized.

The excellent concordance between the predictions of big bang nucleosynthesis and the observed abundances of ^{4}He and D also indicate that during the epoch of nucleosynthesis ($T \sim 10^{10}$K, $t \sim 1$s) the universe was well described by the Friedmann-Robertson-Walker (FRW) model with spherically symmetric line element

$$ds^2 = -dt^2 + R(t)^2 \{dr^2/(1-kr^2) + r^2 d\Omega\} , \qquad (1)$$

and scale factor R(t) governed by

$$H^2 \equiv (\dot{R}/R)^2 = \frac{8\pi G\rho}{3} + \frac{\Lambda}{3} - \frac{k}{R^2}, \qquad (2)$$

where R(t) has been scaled so that k = +1 (closed), -1 (open), or 0 (flat), and Λ is a possible cosmological term. Large deviations from the FRW model during primordial nucleosynthesis would upset the concordance. Today the curvature term and possible cosmological term are not significant, and since $\rho \propto R^{-3}$ (matter-dominated), R^{-4} (radiation-dominated) we can neglect them at earlier times. During the radiation-dominated epoch ($T \gtrsim 10^5\ \Omega h^2$ K),

$$H^2 = \frac{4\pi^3}{45} G\, g_*(T) T^4 \quad , \tag{3}$$

$$t = 2.41\ \mu\text{sec}\ g_*^{-\frac{1}{2}}\ T_{GeV}^{-2} \tag{4}$$

where $g_*(T)$ counts the total number of degrees of freedom of all the relativistic species ($\equiv \Sigma\, g_{bose} + (7/8) g_{fermi}$), and throughout units are used such that $\hbar = k_B = c = 1$. Ω is the ratio of the present density to the critical density and the Hubble parameter $H_0 = 100$ h kms^{-1} Mpc^{-1}. For $t \gtrsim t_{pl} \equiv (G\hbar/c^5)^{\frac{1}{2}} \sim 10^{-43}$s, quantum corrections to classical gravitation theory should be small, which corresponds to temperatures as high as $g_*^{-\frac{1}{4}}$ (10^{19} GeV) × 10^{19} GeV. It is clear why the early universe has often been called the ultimate high energy physics laboratory!

The six problems which face classical cosmology are: (1) The baryon number-to-specific entropy ratio, $kn_B/s \cong (4\text{-}7) \times 10^{-11}$. Although today the universe appears to contain negligible quantities of antimatter, the value of this ratio indicates that for $T \gtrsim 1$ GeV ($t \lesssim 10^{-6}$ s) the fractional matter-antimatter asymmetry was $\sim 10^{-10}$. As all of you are certainly aware, GUTs have provided at the very least a framework for understanding this ratio. (2) The isotropy. The universe is remarkably isotropic, despite the fact that isotropy is an unstable property of the initial data. (3) The homogeneity/inhomogeneity. On the largest scales ($>> 10h^{-1}$ Mpc) the universe is very smooth, while on small scales ($\lesssim 10h^{-1}$ Mpc) the universe is quite lumpy (stars, galaxies, clusters, superclusters). (4) Flatness. We live in a very old and large universe. The only natural timescale is $t_{pl} \sim 10^{-43}$ s; our universe has survived to the ripe-old age of $\sim 10^{60}$ t_{pl} without recontracting or becoming dominated by negative curvature (Milne phase). (5) Cosmological constant. In quantum field theory the vacuum expectation value of $T_{\mu\nu}$ is infinite and corresponds to an infinite cosmological constant, whereas observations indicate that $(\Lambda/8\pi G) \lesssim 2 \times 10^{-29}$ g $cm^{-3} \sim 10^{-46}$ GeV^4. The problem is even more apparent in spontaneously broken gauge theories since the vacuum expectation value changes during symmetry breaking by $O(T_c^4)$. (6)

Monopoles. In the most straight-forward picture of the GUT epoch, relic monopoles from the GUT phase transition would today close the universe by a factor of ~ 10^{10} ($\Omega_M \sim 10^{10}$).

A few comments are appropriate at this point. Puzzle (1) may indeed be near resolution; however, it was included since it had remained unexplained until the advent of GUTs, and is certainly the most significant contribution of GUTs to cosmology. Problems (5) and (6) are not purely cosmological since they result from considering quantum field theory in a cosmological context. Nonetheless, they have been included in my list since their resolution seems non-trivial, and hence will probably have great significance both for cosmology and quantum field theory. Although these problems have been enumerated separately, they (or some subset) are probably connected at a deeper level. Finally, no mention of quantum gravity has been made – how it might help to resolve these conundrums, or the additional conundrums which it might add to my list. Because of the uncertainties associated with and the scope of the subject I have chosen to restrict myself to classical cosmology. It may very well be that the resolution of some of these questions will involve quantum gravity.

In the remainder of this talk I will review in detail problems (1) - (6) and the possible role of GUTs in solving (or exacerbating) them. I will then discuss one scenario, Guth's inflationary universe, which addresses and to some extent resolves five of the six puzzles – unfortunately, it is not without its own difficulties. Finally, I will summarize by presenting a "report card" for GUTs with respect to their role in resolving these six fundamental problems.

2. The Baryon Number-to-Specific Entropy Ratio

Although the laws of physics are very nearly matter-antimatter symmetrical (the only known instance of CP violation being the K^o - $\bar{K}^o$ system [4]), there is strong evidence that universe contains negligible amounts of antimatter ($n_{\bar{b}} \ll n_b$) [5]. In addition baryons are relatively rare, their ratio to photons (essentially all of which are in the 3K background) being $n_b/n_\gamma \cong (3\text{-}5) \times 10^{-10}$ [6]. Most of the entropy in the universe is in the photon and neutrino cosmic backgrounds (about equal amounts in each), so that it is simple to convert

n_b/n_γ to the baryon number ($n_B \equiv n_b - n_{\bar{b}} \cong n_b$)-to-specific entropy ratio

$$kn_B/s \cong (1/7)n_b/n_\gamma \cong (4\text{-}7) \times 10^{-11}. \tag{5}$$

One can look at this ratio in two different ways. First, as $s/kn_B \cong (1.4\text{-}2.5) \times 10^{10}$, which indicates that we live in a universe with a large entropy per baryon (for comparison, the entropy per baryon during core collapse and explosive nucleosyntehsis is only 1-10). There have been many attempts to explain this ratio by assuming that the universe began with cold matter ('cold big bang'), and that this large entropy was generated by dissipating anisotropy, etc. [7]. These attempts have been largely unsuccessful; conventional processes (e.g., neutrino viscosity) tend to produce too little entropy, while dissipation by quantum gravitational processes tend to severely overproduce entropy [8,9]. Crudely speaking, the amount of 'gravitational entropy' available to be converted into thermodynamic entropy is enormous, $\gtrsim 10^{40}$ [9].

If one adopts the view (which I shall) that the universe has been hot all the way back to t_{pl}, then the ratio kn_B/s has a different significance. For $T \gtrsim 1$ GeV ($t \lesssim 1$ μsec) baryons and antibaryons were roughly as abundant as photons, so that this ratio implies the matter-antimatter fractional asymmetry was very small, $(n_b - n_{\bar{b}})/n_b \cong 10^{-10}$. The very early universe was nearly baryon symmetric. Before the recent ideas of baryosynthesis, one was forced to set $kn_B/s \cong (4\text{-}7) \times 10^{-11}$ as an initial condition – a rather peculiar one at that!

As early as a decade ago A. Sakharov and others [10] suggested that this quantity might have evolved dynamically, thus providing the glimmer of hope that this mysterious initial condition was unnecessary, in the same way that big bang nucleosyntehsis eliminated the need for specifying a primordial ^{4}He mass fraction ($\cong$ 0.25). For this to happen there are clearly three necessary ingredients: (1) baryon number must be violated; (2) C and CP must be violated – an arrow is needed to specify that matter is preferentially produced when B is violated; and (3) a departure from thermal equilibrium is required – in equilibrium $n_b = \exp(-m_b/T) \equiv n_{\bar{b}} = \exp(-m_{\bar{b}}/T)$ since $m_b \equiv m_{\bar{b}}$ [11]. GUTs very naturally provide (1), and C, CP violations are known to exist, any viable GUT must also have the possibility to provide (2).

The expansion of the universe provides (3); when $H = -(\dot{T}/T) = (\dot{R}/R)$ exceeds the rate of thermalizing interactions, thermal distributions (at least for massive particles) cannot be maintained. The so-called out-of-equilibrium decay scenario incorporates (1), (2), and (3), and detailed numerical calculations indicate that this mechanism works and that the asymmetry which evolves is reasonably well fit by [13]

$$kn_B/s \cong \frac{\varepsilon}{g_*} \frac{1}{1+K^{1.3}}, \tag{6}$$

where $K = (\Gamma_D \sim$ decay rate of the superheavy boson which mediates B nonconservation$)/H\big|_{T=M} \cong 3 \times 10^{17}$ GeV(α/M), ε measures the CP violation in the superheavy boson system ($\equiv$ mean net B number produced by the decay of a superheavy boson-antiboson pair), and $g_* \sim 10^2$ as before counts the number of relativistic degrees of freedom. I might also add that B violating interactions tend to damp any initial baryon asymmetry that might have been present, so that kn_B/s may truly be independent of initial data. [There are, however, modes with net baryon number which cannot be damped.] It is amusing that the same interactions which lead to proton decay and hence the demise of matter in an open universe, are very likely responsible for the existence of matter in the first place.

Big bang baryosynthesis is certainly not on the same firm footing as big bang nucleosynthesis; however, for the first time we have a framework for understanding the origin of kn_B/s. A great number of details remain to be worked out. At present, given a definite GUT, it is possible to write a network code and compute the baryon asymmetry which evolves. However, the GUT of our universe is not known; even worse is our lack of understanding of the CP violation parameter ε. In general $\varepsilon \lesssim O(\alpha) \lesssim 10^{-2}$ [14] - CP is conserved in the Born approximation, i.e., at the tree graph level. In minimal SU(5) (one $\underset{\sim}{5}$ and one $\underset{\sim}{24}$ of Higgs) ε is many orders of magnitude too small, unless one introduces one or two new generations of heavy quarks [15]. One can also obtain ε of the right magnitude by enlarging the Higgs sector with an additional $\underset{\sim}{5}$ [16]. Because the K^0-$\bar{K}^0$ system involves the lightest quarks, and baryon generation seems to involve the heaviest quarks (their Higgs couplings are the largest), it is extremely unlikely that ε can be directly related to the K^0-$\bar{K}^0$ system, although the same underlying mechanism might be able to explain both. It might be that ε can be related to the neutron electric dipole moment; recently, Ellis et al [17] pointed out that in a large class of GUTs, diagrams which contribute to ε also contribute to the renormalization of θ_{QCD}, and hence to d_n. They have used kn_B/s to set a lower bound on d_n,

$$d_n \gtrsim 3 \times 10^{-18}(n_b/n_\gamma) \text{ e-cm.} \tag{7}$$

By far, the most significant contribution of grand unification to cosmology has been baryosynthesis, and because of GUTs we have made great progress toward understanding the origin of kn_B/s.

3. The Isotropy Problem

The universe is to a high degree of precision isotropic. The evidence for the isotropy comes from: (1) the 3K background, which on all angular scales $\gtrsim$ 10' exhibits $\Delta T/T \lesssim 10^{-3}$; (2) the X- and γ- ray backgrounds; and (3) the Hubble flow of galaxies [18]. Attempts to explain the isotropy by classical dissipative processes have been largely unsuccessful [19]. In addition, isotropy is an unstable property of the initial data. Specifically, Collins and Hawking [20] have shown that the set of initial data which evolve to an open universe which is asymptotically isotropic is of measure zero in the space of initial data.

If one examines perturbations of the homogeneous and isotropic FRW model, one finds that there are both growing and adiabatically decaying modes associated with anisotropy. In some sense the question of isotropy is equivalent to asking why there were only decaying modes of anisotropy (or none at all) initially. The issue is even more complex. Recently, Vishniac [21] and Bardeen [22] have shown that the decomposition of perturbations into growing and decaying modes can change if the equation of state of the stress-energy changes, e.g., the equation of state changes when particles like neutrinos decouple and become effectively collisionless. The implications are that a universe which initially is FRW with only decaying modes of anisotropy, may when the equation of state changes, then have an anisotropic perturbation which is a growing mode.

There are indications that particle production by quantum gravitational processes can damp initial anisotropies; however, some of these anisotropies will grow up again [23]. Penrose has suggested that there may be restrictions as to the nature of the initial singularity, which require isotropy [9]. At present, the high degree of observed isotropy remains a mystery, and to date grand unification has shed no light on the possible origin of the isotropy.

4. The Homogeneity Problem

4.1 The Universe is Smooth

On the largest scales ($\gg 10h^{-1}$ Mpc) the universe is very homogeneous as evidenced by the 3K microwave background ($\Delta T/T \lesssim 10^{-4}$ for $\theta \gtrsim 7^{o}$), and by counts of radio sources [24]. This observation is perplexing because of the existence of particle horizons in known solutions to Einstein's equations [25]. The distance over which a light signal can travel from t_i to t is given by (for $k = 0$)

$$d_H = R(t)\int_{t_i}^{t} \frac{dt'}{R(t')} \quad ; \tag{8}$$

d_H is the maximum size of a causally coherent region. For $R(t) \sim t^n$, $1 > n > 0$, d_H is finite as $t_i \to 0$ and $= (1-n)^{-1}\, t$; hence there exist particle horizons. During the radiation-dominated epoch ($t \lesssim 2.5 \times 10^{10}$s $\Omega^{-2}h^{-4}$) the baryonic mass within the horizon is

$$M_H(t) \cong 10 M_\Theta(\Omega h^2)(t/1s)^{3/2} \quad . \tag{9}$$

At decoupling ($T \sim 4000$K, $t \sim 3.5 \times 10^{12}(\Omega h^2)^{-\frac{1}{2}}$s), when the 3K photons were in some sense emitted, the horizon encompassed $\sim 10^{18}\ M_\Theta(\Omega h^2)^{-\frac{1}{2}}$. The horizon size then, today corresponds to $\sim 3^o$ on the sky, so that two points on the sky, a and b, separated by more than $\sim 3^o$ were causally distinct, yet today photons from these two directions have the same temperature within a part in 10^4. In Newtonian terms, the photons from a and b have been redshifted by almost precisely the same amount, so that $(\phi_a-\phi_b)/\phi \sim (\rho_a-\rho_b)/\rho \lesssim 10^{-4}$. This constitutes the large-scale homogeneity problem.

Note that if $R(t)$ increases more rapidly than t, then d_H diverges as $t_i \to 0$. It has been suggested that quantum gravitational processes "soften" gravity so that $R(t) \sim t$, removing the particle horizons [26]. Removing the particle horizons alone, is not sufficient to guarantee homogeneity; one must also have thermalizing processes which are occurring rapidly on the expansion timescale $H \sim t^{-1}$.

4.2 The Universe is Lumpy

On scales smaller than $\sim 10h^{-1}$ Mpc the universe is very inhomogeneous, being comprised of nuclei, atoms, molecules, creatures, planets, stars, galaxies, clusters, and superclusters. To describe this quantitatively, consider the density contrast, $\delta\rho/\rho$, which is a function of scale and is defined as the r.m.s. deviation of ρ divided by $\bar{\rho}$,

$$\left.\frac{\delta\rho}{\rho}\right|_{\text{scale}} = \frac{\left.\langle(\rho-\bar{\rho})^2\rangle^{\frac{1}{2}}\right|_{\text{scale}}}{\bar{\rho}} . \tag{10}$$

The scale could be a distance, but because of the universal expansion, it is usually more convenient to use the average baryonic mass contained within the comoving volume corresponding to that scale. On the scale of a galactic mass $\delta\rho/\rho \sim 10^6$ today. Both parts of the homogeneity problem can be summarized (today) by

$$\begin{aligned} \frac{\delta\rho}{\rho} &\gtrsim 1 \quad d \lesssim 10h^{-1}\text{Mpc}(\lesssim 10^{15}M_\odot\Omega h^{-1}) , \\ \frac{\delta\rho}{\rho} &\lesssim 1 \quad d \gtrsim 10h^{-1}\text{Mpc}(\gtrsim 10^{15}M_\odot\Omega h^{-1}) . \end{aligned} \tag{11}$$

Because the universe becomes smoother on scales approaching the present horizon ($\sim 10^{22}M_\odot$), the FRW model supplemented by small perturbations should provide an adequate global description of the universe. Peebles has argued that this must always have been the case [18], otherwise, we would observe a profusion of black holes. [If $\delta\rho/\rho > 1$ when a given scale enters the horizon, black hole formation is inevitible.] Let us assume that we can describe the universe by the FRW model with small density perturbations. When a given mass scale comes into the horizon (cf. equation [9]), density fluctuations on that scale can grow due to their self-gravity (the usual Jeans instability), if that mass is greater than the Jeans mass, M_J. As long as the matter and radiation are strongly coupled ($T \gtrsim 4000$K, $t \lesssim 3.5 \times 10^{12}$s $(\Omega h^2)^{-\frac{1}{2}}$), the radiation makes the fluid very stiff (v_s =

$c/\sqrt{3}$) and $M_J \sim M_H$. When matter and radiation decouple, M_J drops far below M_H, and perturbations on scales $\gtrsim 10^5 M_\Theta$ ($\sim M_J$) can grow, and $\delta\rho/\rho \sim t^{2/3}$.

During the radiation-dominated epoch density perturbations can be divided into two types: adiabatic and isothermal. Adiabatic fluctuations are characterized by $\delta\rho/\rho \sim \delta T/T \neq 0$ and $\delta(kn_B/s) = 0$, while isothermal perturbations are characterized by $\delta T(=\delta\rho) = 0$ and $\delta(kn_B/s) \neq 0$. Isothermal fluctuations correspond to a constant temperature and density universe, where baryon number has been non-uniformly distributed. The process of galaxy formation is very different for isothermal and adiabatic initial perturbations [27].

During decoupling the photon mean free path becomes very large, and due to photon diffusion adiabatic perturbations on scales $\lesssim M_s$ (Silk mass),

$$M_s \cong 10^{12} M_\Theta (\Omega h^2)^{-2}/(1 + 0.04/(\Omega h^2))^3 , \tag{12}$$

are damped, so that only perturbations on scales larger than M_s survive. After decoupling these perturbations grow and the first structures to form are $O(10^{14}\ M_\Theta)$ objects, which Zel'dovich and his co-workers [28] argue should be non-spherical, pancake-shaped objects. These pancakes correspond to superclusters, and must hydrodynamically fragment to form galaxies. This picture predicts fluctuations in the microwave background on angular scales corresponding to superclusters ($\theta \gtrsim 10''$) since $\delta T/T \sim \frac{1}{3}\ \delta\rho/\rho$. If the pancakes are to form, go non-linear ($\delta\rho/\rho > 1$), and shatter before the present epoch, $(\delta\rho/\rho)_{dec}$ must have been $\gtrsim O(10^{-3})$ since the maximum growth factor $(t/t_{dec})^{2/3} \sim$ 1500. At present, this prediction $\delta T/T \sim 10^{-3}$ is in contradiction with the current upper limit on this scale, $\delta T/T \lesssim 10^{-4}$ [29]. The adiabatic picture predicts a preferred scale, $M_s \sim 10^{14} M_\Theta$; recently, deep sky surveys [30] have found large ($\sim$ 10Mpc) holes in space which are devoid of galaxies. Such holes are expected due to the formation of pancakes [28].

During decoupling only isothermal perturbations on scales $\lesssim 10^5$ M_Θ are damped by photon diffusion. If one assumes that the initial spectrum $(\delta\rho/\rho)_{dec} \sim \delta(kn_B/s)/(kn_B/s) \sim M^{-n}$ (a featureless power law), then perturbations on the smallest scales go non-linear first, and structure is built up by successive clustering of larger and larger units (hierarchical clustering), with no preferred mass scale. Numerical simulations seem to indicate that $(\delta\rho/\rho)_{dec} \sim 1$ on a mass scale of $\sim 10^8 M_\Theta$ and $-1 \leq n \leq 0$ are required to reproduce the clustering presently observed in the universe [18]. Much smaller fluctuations in the 3K background are expected, since simply put $\delta T = 0$ in the isothermal picture. Small fluctuations are produced due to the lumpy distribution of matter which creates fluctuations in the gravitational potential ϕ, so that $\delta T/T \sim \delta\phi/\phi$ [31]. However, in the isothermal picture it is much more difficult to account for the "holes" in space which have recently been discovered.

4.3 The Origin of the Perturbations

During the radiation epoch isothermal perturbations are characterized by $\delta\rho = 0$ and do not grow. However, adiabatic perturbations are real density fluctuations and while a given mass scale is outside the horizon perturbations on that scale do grow, $\delta\rho/\rho \sim t$ [22,23]. The growth factor for an adiabatic fluctuation until it enters the horizon is

$$(\delta\rho/\rho)_H = (\delta\rho/\rho)_i (3 \times 10^{12} s/t_i)(M/10^{18} M_\Theta)^{2/3} (\Omega h^2)^{-2/3} \quad . \tag{13}$$

Note, that if one chooses $t_i \sim 10^{-43}$ s, the growth factor is $\sim 10^{55}$! Therein lies the good news and the bad news. Let us suppose that there were Poisson fluctuations of O(1) on the horizon scale at the Planck time, then $(\delta\rho/\rho)_i \cong (M/10^{-10}\ \text{baryons})^{-1/2} \cong 10^{-34}(M/M_\Theta)^{-1/2}$; recall that at t_{pl} there is ~ 1 photon in the horizon, and $\sim 10^{-10}$ baryons since $n_b/n_\gamma \sim 10^{-10}$. When a galactic mass comes into the horizon, $\delta\rho/\rho \sim 10^{-34} \times (10^{12} M_\Theta/M_\Theta)^{-1/2} \times 10^{55} \times (10^{12} M_\Theta/10^{18} M_\Theta)^{2/3} \sim 10^{11}$. Unless $(\delta\rho/\rho)_i \sim M^{-2/3}$ or is not a featureless power law, on some scale perturbations will grow too large before they enter the horizon.

There is another important point to be made with regard to spontaneously producing a spectrum of adiabatic density fluctuations. It has been shown that if one creates fluctuations using processes which locally conserve energy and momentum, then n must be $\gtrsim 7/6$ (i.e., E-p conservation restricts the allowed spectrum of correlations) [33]. Zel'dovich and Harrison have suggested that an n=2/3 spectrum with normalization 10^{-4} might be a relic of quantum gravitational processes [34]. This spectrum results in $(\delta\rho/\rho)_H \sim 10^{-4}$ on all scales; these are often called constant curvature fluctuations since they result in $\delta k \sim (\delta\phi/\phi) \sim 10^{-4}$ independent of mass scale.

The horizon problem also enters when discussing the production of perturbations. If we suppose that the fluctuations are to be produced by causal processes, then we can expect perturbations of at most order unity on the horizon scale. For isothermal perturbations produced by a random process

$$\delta_I \sim \delta(kn_B/s)/(kn_B/s) \sim (M/M_H)^{-\frac{1}{2}}$$
$$\delta_I \sim (M/M_\Theta)^{-\frac{1}{2}}(t/1s)^{3/4} \quad , \tag{14}$$

so that in order to obtain $\delta_I \sim 10^{-3}$ for $M \sim 10^{12}M_\Theta$, the fluctuations must be created when $t \sim 10^4$ sec. [Note, this argument also applies to n=7/6 adiabatic perturbations; when the growth factor (13) is taken in account they are equivalent to n=1/2 isothermal perturbations.]

4.4 The Role of GUTs in Galaxy Formation

To date, grand unification has had two important impacts on galaxy formation: (1) baryosynthesis, and (2) massive neutrinos (at present the sign of these effects is not obvious). Clearly baryosynthesis and the origin of isothermal perturbations are intimately connected. It has been shown that isothermal perturbations cannot be created during baryosynthesis in a nearly FRW universe [35]. Very simply, this is because $kn_B/s = f(\varepsilon,g_*,K)$ (cf. equation [6]), and ε,g_*,K only depend upon particle physics parameters. However, in a universe with large-scale inhomogeneous shear $H^2 \cong T^4/m_{pl}{}^2 + \Sigma(\underset{\sim}{x})/R^6$ and $K = (\Gamma_D/H)\big|_{T=M}$ is a function of $\underset{\sim}{x}$($m_{pl} = G^{-\frac{1}{2}} \sim 10^{19}$GeV, and $\Sigma(\underset{\sim}{x})$ parameters the shear "energy density"). Since the universe is not presently dominated by shear, we must demand that the shear perturbation be in a decaying mode (such models exist). The spectrum of isothermal perturbations mirrors the initial spectrum of shear (which itself decays away adiabatically) [36]. In some sense, it can be said that GUTs favor adiabatic perturbations.

If light stable, massive neutrinos exist and the sum of their masses exceeds 3.5 eV, then the universe is neutrino-dominated, and the usual picture of galaxy formation changes significantly. First, because neutrinos are non-interacting, density fluctuations in the neutrinos on scales $\lesssim$ the distance neutrinos can traverse in an expansion time damp due to neutrino free-streaming [37]; as long as neutrinos are relativistic $M_{D\nu} \sim M_H$. As the neutrinos become non-rela–

tivistic ($T \sim m_\nu$) $M_{D\nu}$ ceases to increase and perturbations on scales $\gtrsim$ the maximum value of $M_{D\nu}$,

$$M_{D\nu}(\max) \sim 4 \times 10^{15} M_\Theta (m_\nu/30\text{eV})^{-2} \quad , \tag{15}$$

survive. One is naturally led to the pancake (or fragmentation) theory. There is one very important difference; since neutrinos decouple at $T \sim 1$ MeV, as soon as they go non-relativistic, perturbations in the neutrino sea can begin to grow (recall decoupling occurs at $T \sim 1$ eV), allowing an additional growth factor of $O(m_\nu/1 \text{ eV})$. Smaller initial perturbations are needed, thus easing the situation with respect to small scale fluctuations in the 3K background. When matter and radiation decouple, the baryons immediately fall into the potential wells already created by the neutrinos. The baryons can dissipate their binding energy, fragment and form galaxies; however, the neutrinos cannot dissipate, and only relax via violent relaxation, and hence remain in an extended halo around the supercluster (and not in galactic halos [38]).

The isothermal picture is adversely affected in a neutrino-dominated universe. When the baryons decouple, on scales $\lesssim M_{D\nu}(\max)$ $\delta\rho/\rho \cong \delta\rho_b/\rho \cong (\delta\rho_b/\rho_b) \times (\rho_b/\rho_\nu)$, i.e., their initial amplitude is reduced by $\sim (\rho_b/\rho_\nu)$. In addition, because the neutrino sea provides a smooth background on scales $\lesssim M_{D\nu}(\max)$ the growth rate of isothermal perturbations is $\sim t^p$ rather than $\sim t^{2/3}$, where $p=[(1+24(\rho_b/\rho))^{\frac{1}{2}}-1]/6 \leq 2/3$. With very large initial isothermal fluctuations the isothermal picture could proceed, and even trap neutrinos in galactic halos (if they are more massive than ~ 20 eV [39]). It is particularly desireable to get neutrinos into galactic halos since studies of galactic rotation curves indicate that ratio of dark (form as of yet unknown)-to-luminous matter is $\gtrsim 10$.

Thus far GUTs have not led to significant progress in the theory of galaxy formation. Both baryosynthesis and massive neutrinos tend to make galaxy formation more difficult in the isothermal picture. In some sense baryosynthesis favors adiabatic fluctuations (although permits isothermal); massive neutrinos seem to be just what is needed to revitalize the "pancake" picture, although it appears difficult to get them into galactic halos where they could also explain the dark matter in galaxies. However, grand unification appears to offer no explanation for the origin of the fluctuations which are needed.

"Laboratory" experiments (proton decay, neutrino masses, etc.) will have important implications for the theory of galaxy formation. More accurate observations of the structure of the 3K background (quadrupole, octupole, ...) should help discriminate between different theories of galaxy formation in the near future, and provide us with a direct probe of the spectrum of perturbations at decoupling on very large-scales ($\gtrsim 10^{18}\ M_\odot$). Already, the recently reported quadrupole anisotropy [40] has been used to constrain both the isothermal and adiabatic pictures [31,41].

5. The Flatness Problem [42]

The different FRW models are specified by $k = \{0, \pm 1\}$ and the infinity of possible absolute scalings of $R(t)$. In Newtonian language the $-k/R^2$ term is the kinetic plus potential energy of a unit mass, so that as the scale of $R(t)$ varies and $k = +1, 0, -1$ all possible initial energies (from tightly bound to completely unbound) are spanned. Since the curvature term (cf. equation [2]) $\alpha\ R^{-2}$ and the energy density $\rho\ \alpha\ R^{-3}$ or R^{-4}, it becomes more important as the universe evolves. If one were to start the universe at t_{pl} with energy density and curvature terms of roughly the same magnitude (in Newtonian language: kinetic, potential, and total energy all of comparable size), then in a few Planck times the universe would recollapse or become negative-curvature dominated (Milne phase). The universe has not recollapsed yet, nor is it strongly negative-curvature dominated ($k/R^2 \lesssim 8\pi G\rho/3$). Stated in this context, the flatness problem is, why do we live in such an old universe ($t \sim 10^{60}\ t_{pl}$)?; or in Newtonian language, how was it that the initial kinetic and potential terms were so accurately balanced (to a part in $\sim 10^{56}$) so that the universe is still expanding without having become curvature-dominated?

To be quantitative, for $T \gtrsim 1\text{eV}$ the universe was radiation dominated, and during that epoch

$$H^2 = \frac{4\pi^3}{45}\, g_* \frac{T^4}{m_{pl}^2} - \frac{k\ T^2}{(RT)^2} \tag{16}$$

(ignoring a possible cosmological term for now). If the expansion is adiabatic, then $(RT) \cong$ constant; more precisely, entropy conservation implies $g_*(T)^{1/3}(RT)$ = constant. At $T \sim 1$ eV the curvature term was less than the radiation energy density term, so that for earlier times it was smaller by a factor of $\sim (T/1\ \mathrm{eV})^2$. For $T_i \sim 10^{19}$ GeV, the initial ratio of the curvature term to the radiation term was 10^{-56}.

If the expansion has been adiabatic one can express (RT) in terms of the number of photons within the horizon volume today. The horizon volume

$$V_H(t) = 4\pi\, R(t)^3\, \theta_H(t)\ , \tag{17a}$$

$$\theta_H(t) = \begin{cases} k=1 & \frac{1}{2}(\phi(t) - \sin\phi(t)\cos\phi(t)) \\ k=0 & \frac{1}{3}\,\phi(t)^3 \\ k=-1 & \frac{1}{2}(\sinh\phi(t)\cosh\phi(t) - \phi(t)) \end{cases} \tag{17b}$$

$$\phi(t) = \int_0^t \frac{dt'}{R(t')} \tag{17c}$$

Physically, $\phi(t)$ is the coordinate distance to the present horizon ($R(t)\ \phi(t)$ is the physical distance), and $\theta_H(t)$ is the coordinate volume contained within the present horizon. Since we do not detect any strong affects of curvature today $\phi(t)$ must be $\lesssim 1$ (cf. the line element given by equation [1]), so $\theta_H(t) \cong (1/3)\phi(t)^3 \lesssim 1$ for $k = 0$, ± 1. The number of photons within the horizon is

$$\begin{aligned} N_\gamma &= (2\zeta(3)/\pi^2)T^3\ V_H(t) \\ &= (RT)^3(8\zeta(3)/\pi)\theta_H(t)\ , \end{aligned} \tag{18}$$

so that $(RT)^{-2} \lesssim N_\gamma^{-2/3}$. The number of photons within the horizon is just, $N_\gamma \gtrsim (400\ \mathrm{cm}^{-3}) \times 4\pi/3 \times (10^{28}\mathrm{cm})^3 \sim 10^{87}$, and $N_\gamma^{-2/3} \sim 10^{-58}$. The ratio of the curvature term, $kT^2/(RT)^2$, to the radiation term at the Planck time is just $\sim N_\gamma^{-2/3} \sim 10^{-58}$. In this context the flatness problem is equivalent to asking why the entropy of the universe is so large? As we shall see, relaxing the assumption of adiabaticity (RT = constant) allows a possible natural explanation for the near flatness of the universe, if one can find a mechanism for increasing $(RT)^3$ by a factor of $\sim 10^{87}$. Finally I mention a possible connection between the flatness and isotropy problems; Collins and Hawking [20] also proved that the set of initial data which evolve to an asymptotically isotropic and flat universe is of finite measure.

6. The Cosmological Constant [43]

The most general form of the Einstein equations is

$$G_{\mu\nu} = 8\pi G T_{\mu\nu} + \Lambda g_{\mu\nu} , \tag{19}$$

where Λ is a constant, i.e., $\Lambda,_{\nu} \equiv 0$. Historically, Einstein introduced such a term in order to obtain a static cosmological model. $\Lambda/8\pi G$ corresponds to an ever constant, uniform energy density. The equation for the expansion can be written as

$$H^2 = \frac{8\pi G}{3}(\rho + \rho_0) - \frac{k}{R^2} , \tag{20}$$

where $\rho_0 = \Lambda/8\pi G$. Today the universe is not strongly dominated by a Λ term, so that $\rho_0 \lesssim 2 \times 10^{-29}$ gcm^{-3} $\sim 10^{-46}$ GeV^4 (in more convenient units).

As long as the energy density associated with the vacuum remains constant, it has the same effect as a cosmological term,

$$\rho_V g_{\mu\nu} = \langle 0|T_{\mu\nu}|0\rangle . \tag{21}$$

As is well known, for an unbroken U(1) theory (Q.E.D.) ρ_V diverges, corresponding to an infinite cosmological term. Computationally, this can be circumvented by normal ordering; however, the Casimir effect demonstrates the physical reality of the zero point energy ρ_V.

In a broken gauge theory, ρ_V changes during spontaneous symmetry breaking (SSB) by $O(T)_c^4$, where T_c is the transition temperature. Even if one knew how to handle the infinite part of ρ_V, there are now finite parts which vary with temperature; recall, $\rho_0 \lesssim 10^{-46}$ GeV^4, which corresponds to $T_c \lesssim 3 \times 10^{-3}$ eV $\sim$ 30K.

One ad hoc patch is to introduce an initial, compensatory "true" cosmological term ρ_{oi} to cancel the induced cosmological term associated with the quantum field theory. Ignoring the infinite part of ρ_V, ρ_{oi} must be $O(T_c^4)$, where T_c is now the highest transition temperature of the various stages of SSB ($T_c = 10^{15}$ GeV?, T_c = 300 GeV?). In order to be consistent with the present observations, ρ_{oi} must be specified with a precision, $\Delta\rho_{oi} \lesssim 10^{-46}$ GeV^4, so that

$$\frac{\Delta\rho_{oi}}{\rho_{oi}} < \frac{10^{-46}\ \mathrm{GeV}^4}{\rho_{oi}} = \begin{cases} 10^{-106} & T_c = 10^{15}\mathrm{GeV} \\ 10^{-56} & T_c = 300\ \mathrm{GeV} \\ 10^{-44} & T_c = 1\ \mathrm{GeV} \end{cases}$$

at a time (t_{pl}) when the symmetry is still unbroken. This certainly qualifies as a fine tuning problem!

Supersymmetry offers a glimmer of hope (but also dashes it!). The presence of an unbroken global supersymmetry guarantees that the fermionic and bosonic contributions to ρ_V exactly cancel. However, when the supersymmetry is broken a finite $\rho_V \sim O(T_c^4)$ is predicted by the theory, thus requiring supersymmetry to be valid down to $T \sim 3 \times 10^{-3}$ eV if it is to provide the solution.

Why the cosmological term is nearly zero is troublesome even for unbroken gauge theories, and the problem is even more apparent for spontaneously broken theories. The resolution of this puzzle more than likely will have to come at a very fundamental level of both quantum field theory and gravitational theory. Since we do not presently understand why $\Lambda \cong 0$, perhaps we should take the approach that for reasons yet unknown it is and has always been exactly zero. However, as we shall see shortly, if we take its value seriously it can have some rather interesting consequences.

7. The Monopole Problem [44]

Polyakov and 't Hooft have shown that when a simple group G breaks to a group G_1 which contains a U(1) factor, there exist classical solutions which correspond to finite energy U(1) monopoles. If the breaking occurs at a scale M, then the energy associated with the monopole is $\sim M/\alpha$. These monopoles are topological entities, corresponding in some sense to knots in the Higgs field. Because of causality it is difficult to see how, during the phase transition $G \to G_1$, less than O(1) monopole per horizon volume could be produced since the correlation length of the Higgs field must be $\lesssim d_H$. Their subsequent annihilation, even in clustered systems such as galaxies, etc. is negligible [45]. The number of baryons in the horizon is

$$N_B \cong 10^3\ \Omega h^2 (10^{15}\ \mathrm{GeV}/T)^3\ , \tag{22}$$

so that n_M/n_B should be $\gtrsim 10^{-3}(\Omega h^2)^{-1}$ $(T/10^{15}\mathrm{GeV})^3$. Their present contribution to the energy density of the universe is then

$$\Omega_M \gtrsim h^{-2}10^{-3}(T/10^{15}\mathrm{GeV})^3(M_{\mathrm{GeV}}/\alpha) \ , \tag{23}$$

where T is the temperature at which the transition $G \to G_1$ is completed. For canonical SU(5) numbers, $T \sim 10^{14}$ GeV and $M/\alpha \sim 10^{16}$ GeV, $\Omega_M \gtrsim 10^{10}$, whereas we know $\Omega \lesssim 2$! Lowering the scale M to $\gtrsim 10^{11}$ GeV results in $\Omega_M \lesssim 1$, however, the proton lifetime is also lowered to $\lesssim 10^{18}$ yrs.

A variety of solutions have been suggested; however, none of them is completely satisfying. From (23) it is clear that by delaying the completion of the transition to $T \lesssim 10^{10}$ GeV, they can be suppressed to an acceptable level. There are a variety of schemes for doing this: first order phase transition [46], symmetry breaking schemes where the U(1) factor does not appear until very low energies [47], and a universe which undergoes an inflationary phase (to be discussed in the next section) [48]. Other solutions involve the initial suppression of their production, taking advantage of the non-Abelian magnetic charges they may also carry [49], or pushing the transition scale M close to m_{pl}, so that the monopoles are more massive than m_{pl} and then hoping that quantum gravitational effects (Hawking evaporation, etc.) may dispose of them [50].

Grand unification has given us baryosynthesis, but it has also presented us with the as of yet unresolved problem of the monopoles. In many of the proposed solutions monopoles would still be present in non-negligible numbers, and thus their detection might not be out of the question [51] (note, all previous searches have been for light, $\ll 10^{16}$GeV, monopoles). Of course, it goes without saying that their detection would have great implications both for quantum field theory, grand unification, and cosmology.

8. The Born Again Universe

Recently, Guth has suggested a scenario ('the inflationary universe') which addresses to some degree five of the six fundamental problems which I have been discussing [48]. I will now briefly review it. His picture is based on the underlying the assumption that the induced cosmological terms associated with SSB are to be taken seriously. To wit, he puts in by hand an initial compensatory, true cosmological term, $\rho_o \gtrsim 0$, which after SSB takes place will be cancelled, leaving a net zero cosmological term. His $\rho_o \sim O(T_c{}^4)$ and $T_c \sim 10^{15}$GeV is the temperature associated with the GUT phase transition. The equation for the expansion rate is given by

$$H^2 = \frac{4\pi^3}{45} Gg_* T^4 + \frac{8\pi G}{3}\rho_o - \frac{kT^2}{(RT)^2}; \tag{24}$$

when T falls to $\sim T_c$, $\rho_{rad} \sim T_c{}^4$ is $\sim O(\rho_o)$, and if the transition occurs slowly (i.e., nucleation rate of bubbles of true vacuum is slow), then the universe will remain in the symmetric phase where ρ_o is uncancelled and ρ_o will dominate the expansion dynamics. If this occurs, the universe undergoes a de Sitter expansion with $R \propto \exp(\chi t)$, $T \propto R^{-1} \propto \exp(-\chi t)$, and supercools (i.e., remains in the false, symmetric vacuum until a temperature $T_s \ll T_c$). Here $\chi^2 = 8\pi G\rho_o/3 \sim t_{GUT}{}^{-2}$, and $t_{GUT} \sim 10^{-35}$s. When the transition is finally complete, $t = t_* \cong \eta t_{GUT}$, the compensatory term ρ_o is cancelled, the universe is reheated to $\sim T_c$ by the latent heat release, and (hopefully!) FRW behaviour resumes (see fig. 1a). If the GUT transition is strongly first order (e.g., Coleman-Weinberg symmetry breaking), such a scenario is not only plausible, but seems to be required.

What is the effect of the de Sitter phase? Before the de Sitter phase: (1) $d_H \sim t_{GUT}$ ($\sim 10^3$ baryons), and (2) the curvature term might well have been as large as the radiation term, i.e, $RT \sim O(1)$. After the transition is complete $R \sim R(t_{GUT}) \exp(\eta)$, $d_H \sim t_{GUT} \exp(\eta)$, and $(RT)^{-2} \sim O(1) \exp(-2\eta)$ [since $T(t_*) \sim T_c$ and $R^2 \sim \exp(2\eta)$]. Assuming that microscopic processes had smoothed the universe on scales of $d_H \sim t_{GUT}$ before the transition, it is now smooth on scales $\sim t_{GUT} \exp(\eta)$ (the initially smooth patch has been inflated). The importance of the curvature term has been reduced by $\sim \exp(-2\eta)$. As Guth points out for $\eta \cong 65$, $(RT)^2$ is inflated by $\sim 10^{56}$ (reducing the curvature term by 10^{-56}), and the horizon has been inflated to include $\sim (10^{28})^3 \times 10^3$ baryons $\sim 10^{30} M_\odot$ – thus providing a natural explanation for the flatness and large-scale homogeneity problems. Any fossils from the pre-transition epoch are effectively eliminated, e.g., if a relic species had a ratio to photons of ζ prior to t_{GUT}, after inflation it would have a ratio to photons of $\zeta \exp(-3\eta)$ (assuming that it was not produced during the latent heat release). Any initial baryon or lepton number would be detected; but, since the universe is reheated to $\sim T_c$, baryosynthesis can proceed after the transition [52]. Any initial spectrum of adiabatic perturbations (which are equivalent to spatial variations in k) are reduced by $\exp(-2\eta)$. In essence the universe is born again.

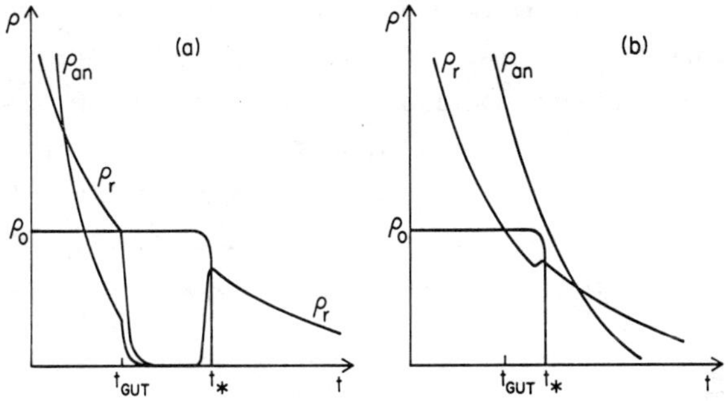

Figure 1 - The contributions to the righthand side of (26) are shown as a function of time. (a) If $\rho_{an} \lesssim \rho_0$ at t_{GUT} the inflationary phase will ensue and both ρ_r and ρ_{an} decrease exponentially. When the transition is complete, $t = t_*$, the latent heat release increases ρ_r to $\sim T_c^4$. (b) If $\rho_{an} \gtrsim \eta^{\frac{1}{2}} \rho_0$ at t_{GUT}, the inflationary phase does not occur, and the universe continues to be anisotropic. When the transition is complete, there is a latent heat release increasing ρ_r slightly. The anisotropy is always decaying adiabatically, $\rho_r/\rho_{an} \sim R^2$.

The enormous horizon size [$\sim 10^3\exp(3\eta)$ baryons] means that one monopole per horizon volume is no problem whatsoever. Having erased any pre-transition fluctuations, a new spectrum of perturbations must be produced for galaxy formation. The large horizon size also helps in this regard. The release of latent heat in bubble wall collisions may give rise to large-scale inhomogeneous shear, which as discussed earlier can produce isothermal perturbations during baryosynthesis. It is also possible that causal processes could produce a viable spectrum of fluctuations. For example, suppose the CP violating parameter ε of baryosynthesis = $\varepsilon_H + \varepsilon_s$, where ε_H is due to intrinsic CP violation, and ε_s is due to spontaneous CP violation and could, therefore, have a different sign in different vacuum domains. Causality implies that these domains must be smaller than the horizon size. Such a model would lead to

$$\delta(kn_B/s)/(kn_B/s) \sim \frac{\varepsilon_s}{\varepsilon_s+\varepsilon_H}\left(\frac{M}{M_D}\right)^{-\frac{1}{2}} , \tag{25}$$

where M_D is the baryonic mass in a domain and is $\lesssim M_H$ which might be $10^{20}M_\odot$ or more [53]. Local fluctuations in the equation of state due to the entropy release could give rise to a similar spectrum (25).

The Guth scenario is not without its difficulties. First, as originally proposed it ignores the isotropy problem. If the early universe were anisotropic, then a very good description of its expansion dynamics are given by a generalized Friedmann equation,

$$H^2 = \frac{4\pi^3}{45} Gg_* T^4 + \frac{8\pi G}{3}\rho_o - \frac{kT^2}{(RT)^2} + \frac{\Sigma^2 T^6}{(RT)^6} , \tag{26}$$

where $\rho_{an} = \Sigma^2/R^6$ is the anisotropy "energy density" which decays adiabatically $\alpha\ R^{-6}$. If $\rho_{an} \gtrsim \eta^{\frac{1}{2}}\rho_o \sim 10\ \rho_o$ at t_{GUT}, then ρ_{an} will dominate the expansion dynamics during the entire phase transition, thus preventing the inflationary phase [54] (see fig. 1b). Of course, if $\rho_{an} \lesssim \eta^{\frac{1}{2}}\rho_o$, then the inflationary phase would proceed, and in addition would reduce any anisotropies (both growing and decaying modes) exponentially.

A very serious difficulty pointed out by Guth is the problem of gracefully returning to FRW behaviour. Usually a phase transition is completed by percolation (bubbles expanding and overlapping), but during the de Sitter phase bubbles do not expand when viewed in comoving coordinates (crudely, because both the bubbles and the universe are expanding at v = c). Consequently, the transition must be completed by every point being nucleated. The universe probably never percolates, and bubble wall collisions which are the mechanism for entropy release are rare.

A possible resolution of this is the cooling of the universe below T_1, the metastability point of the false vacuum ($T > T_1$ $\phi = 0$ is a local minimum, $T < T_1$ $\phi = 0$ is no longer a local minimum). For $T < T_1$ the transition must occur. However, without a fine tuning of the parameters $T_1 \sim O(T_c)$ (in Coleman-Weinberg $T_1 = 0$), and since the inflation factor $\exp(\eta) \sim T_c/T_1$, there would be little inflation. Press and Eardley [55] have explored the possility the Hawking temperature associated with the event horizon of the de Sitter geometry might provide a temperature which as $R \propto \exp(\chi t)$ does not drop exponentially. When the Hawking temperature has fallen to T_1, R might be ~ $\exp(\eta)$, thus saving the beneficial features of the Guth scenario.

The Guth scenario has some very attractive features, and possibly provides natural explanations (involving GUTs) for five of the six puzzles, and relies upon a particular interpretation of the sixth, the question of the cosmological term. However, the inflationary phase might be prevented by large initial anisotropies, or it may be that there is no way to smoothly resume FRW behaviour. It presupposes that at some epoch the cosmological term was very important, although it is insignificant today.

9. Report Card for GUTs

Puzzle	Comments	Grade
Baryon to photon ratio	Provides framework for understanding its origin	++
Isotropy	Sheds no new understanding;	0
	Possible stumbling block for Guth's scenario	+/-
Homogeneity		
(i) Large-scale	Guth scenario	+
(ii) Small-scale	Favors adiabatic fluctuations (but can accommodate isotherm-als); Massive neutrinos may	+
	save the pancake picture;	+
	Guth scenario	+
Flatness	Guth scenario	+
Cosmological term	Exacerbated by SSB;	-
	Supersymmetry	+/--

Monopoles	Problem created by GUTs;	--
	Guth scenario	+
		++++

* Permanent Address: Astronomy and Astrophysics Center, The University of Chicago, Chicago, IL 60637.

Acknowledgements

I am grateful to all the members of the Early Universe Program at the ITP (UCSB), especially R. Kolb, C. Hogan, and G. Steigman, for helping me to clarify the ideas discussed here. This work was supported in part by NSF PHY77-27084 (at UCSB) and DOE AC02-80ER10773 (at Chicago).

References

[1] S. Weinberg, Gravitation and Cosmology (Wiley, NY, 1972), Chap. 15; or D.N. Schramm and R.V. Wagoner, Ann. Rev. Nucl. Sci. 27, 37 (1977).

[2] S.W. Hawking and G.F.R. Ellis, The Large Scale Structure of Space-Time (Cambridge Univ. Press, Cambridge, 1973).

[3] P.J.E. Peebles, Astrophys. J. 146, 542 (1966); R.V. Wagoner, W.A. Fowler, and F. Hoyle, Astrophys. J. 148, 3 (1967); for an updated review of the consistency of the standard model, J. Yang, M.S. Turner, G. Steigman, D.N. Schramm, and K.A. Olive, in preparation (1981).

[4] J.H. Christenson, J.W. Cronin, V.L. Fitch, and R. Turlay, Phys. Rev. Lett. 13, 138 (1965); Phys. Rev. B140, 74 (1965).

[5] G. Steigman, Ann. Rev. Astron. Astrophys. 14, 339 (1976).

[6] J. Yang, M.S. Turner, G. Steigman, D.N. Schramm, and K.A. Olive, in preparation (1981).

[7] C.W. Misner, Astrophys. J. 151, 431 (1968).

[8] Ya.B. Zel'dovich and A.A. Starobinsky, JETP 34, 1159 (1972); B.L. Hu and L. Parker, Phys. Rev. D17, 933 (1978); J.B. Hartle and B.L. Hu, Phys. Rev. D20, 1772 (1979); D21, 2756 (1980).

[9] R. Penrose, in General Relativity: An Einstein Centenary Survey, eds. S.W. Hawking and W. Israel (Cambridge Univ. Press, Cambridge, 1979).

[10] A. Sakharov, JETP Lett. 5, 24 (1967); L. Parker, in Asymptotic Structure of Space-Time, eds. F.P. Esposito and L. Witten (Plenum, NY, 1977); S. Weinberg, in Lectures on Particles and Field Theory, eds. S. Deser and K. Ford (Prentice-Hall, NJ, 1964); M. Yoshimura, Phys. Rev. Lett. 41, 281 (1978).

[11] A. Sakharov, JETP Lett. 5, 24 (1967); S. Dimopoulos and L. Susskind, Phys. Rev. D18, 4500 (1979); D. Toussaint, S.B. Treiman, F. Wilczek, and A. Zee, Phys. Rev. D19, 1036 (1979).

[12] S. Weinberg, Phys. Rev. Lett. 42, 850 (1979); A. Ignatiev, N. Krasnikov, V. Kuzmin, and A. Taukhelidze, Phys. Lett. 76B, 436 (1978); J. Ellis, M.K. Gaillard, and D.V. Nanopoulos, Phys. Lett. 80B, 360 (1978).

[13] E.W. Kolb and S. Wolfram, Phys. Lett. 91B, 217 (1980), Nucl. Phys. B172, 224 (1980); J.N. Fry, K.A. Olive, and M.S. Turner, Phys. Rev. D22, 2953 (1980); D22, 2977 (1980); Phys Rev. Lett. 45, 2074 (1980); for a recent review of baryosynthesis, M.S. Turner, "Big Bang Baryosynthesis and Grand Unification" to appear in Proceedings of the VPI Workshop on Weak Interactions as Probes of Unification (1981).

[14] D.V. Nanopoulos and S. Weinberg, Phys. Rev. D20, 2484 (1979).

[15] S. Barr, G. Segrè, and H.A. Weldon, Phys. Rev. D20, 2494 (1979); G. Segrè and M.S. Turner, Phys. Lett. 99B, 399 (1981).

[16] A. Yildiz and P.H. Cox, Phys. Rev. D21, 906 (1980).

[17] J. Ellis, M.K. Gaillard, D.V. Nanopoulos, and S. Rudaz, Phys. Lett. 99B, 101 (1981).

[18] P.J.E. Peebles, The Large-Scale Structure of the Universe, (Princeton Univ. Press, NJ, 1980).

[19] J.D. Barrow and R.A. Matzner, Mon. Not. R. Astr. Soc. 181, 719 (1977); J.M. Stewart, Mon. Not. R. Astr. Soc., 145, 347 (1969).

[20] C.B. Collins and S.W. Hawking, Astrophys. J. 180, 317 (1973).

[21] E.T. Vishniac, Astrophys. J., in press (1981).

[22] J.M. Bardeen, Phys. Rev. D22, 1882 (1980).

[23] see, e.g., B.L. Hu, "Effect of finite temperature quantum fields on the early universe," Phys. Lett. B, in press (1981).

[24] R. Kron, Ph.D. thesis, Univ. of Calif. Berkeley (1978).

[25] W. Rindler, Mon. Not. R. Astr. Soc. 116, 663 (1956).

[26] L. Parker and S.A. Fulling, Phys. Rev. D7, 2357 (1973); M.V. Fischetti, J.B. Hartle, and B.L. Hu, Phys. Rev. D20, 1757 (1979); A.A. Starobinsky, Phys. Lett. 91B, 99 (1980); A. Zee, Phys. Rev. Lett. 44, 703 (1980).

[27] Most of what is discussed in this section can be found in greater detail and with a more complete list of references to the original literature in ref. [18].

[28] Ya. B. Zel'dovich, Astron. Astrophys. 5, 84 (1970); IAU Symposium No. 79, 409 (1978).

[29] M.L. Wilson and J. Silk, Astrophys. J. 243, 14 (1981).

[30] M. Davis and W.H. Press, in preparation (1981); R.P. Kirshner, A.G. Oemler, and P.L. Schechter, Astron. J. 84, 951 (1979).

[31] P.J.E. Peebles, Astrophys. J. Lett. 243, L119 (1981).

[32] W.H. Press and E.T. Vishniac, Astrophys. J. 239, 1 (1980).

[33] Ya. B. Zel'dovich, Adv. Astron. Astrophys. 3, 241 (1965).

[34] E.R. Harrison, Phys. Rev. D1, 2726 (1970); Ya. B. Zel'dovich, JETP 37, 33 (1973).

[35] M.S. Turner and D.N. Schramm, Nature 279, 303 (1979); S. Weinberg, Phys. Scripta. 21 (1980); D. Lindley, Nature 291, 133 (1981).

[36] J.D. Barrow and M.S. Turner, Nature, in press (1981); J.R. Bond, E.W. Kolb, and J. Silk, Univ. of Calif. Berkeley preprint (1981).

[37] I. Wasserman, Astrophys. J., in press (1981); J.R. Bond. G. Efstathiou, and J. Silk, Phys. Rev. Lett. 45, 1980 (1980); F.R. Klinkhamer and C.A. Norman, Astrophys. J. Lett. 243, L1 (1981); H. Sato and F. Takahara, Kyoto Univ. preprint RIFP-400 (1980); M. Davis, M. Lecar, C. Pryor, and E. Witten, Astrophys. J., in press (1981).

[38] J.R. Bond and A.S. Szalay, in preparation (1981); A.G. Doroshkevich, M. Yu. Khlopov, R.A. Sunyaev, A.S. Szalay, and Ya. B. Zel'dovich, Proceedings of the Xth Texas Symposium on Relativistic Astrophysics (1981).

[39] S. Tremaine and J.E. Gunn, Phys. Rev. Lett. 42, 407 (1979).

[40] R. Fabbri, I. Guidi, F. Melchiorri, and V. Natale, Phys. Rev. Lett. 44, 1563 (1980); S.P. Bough, E.S. Cheng, and D.T. Wilkinson, Astrophys. J. Lett. 243, L113 (1981).

[41] J. Silk and M.L. Wilson, Astrophys. J. Lett. 244, L37 (1981).

[42] This problem has been stressed by R.H. Dicke and P.J.E. Peebles, General Relativity: An Einstein Centenary Survey, eds. S.W. Hawking and W. Israel (Cambridge Univ. Press, Cambridge, 1979), and A. Guth, Phys. Rev. D23, 347 (1981).

[43] S.A. Bludman and M.A. Ruderman, Phys. Rev. Lett. 38, 255 (1977).

[44] This section is meant to be a brief overview of the monopole problem and not a thorough review. The first discussions of this problem were by Ya. B. Zel'dovich and M. Yu. Khlopov, Phys. Lett. 79B, 239 (1978); J. Preskill, Phys. Rev. Lett. 43, 1365 (1979). The existence of monopoles was discussed by G. 't Hooft, Nucl. Phys. B79, 276 (1974); A. Polyakov, JETP Lett 20, 194 (1974).

[45] J.N. Fry, Astrophys. J. Lett., in press (1981); T. Goldman, E.W. Kolb, D. Toussaint, Phys. Rev. D23, 867 (1981).

[46] A. Guth and S.-H. Tye, Phys. Rev. Lett. 44, 631 (1980); M.B. Einhorn and K. Sato, Nucl. Phys. B, in press (1981).

[47] P. Langacker and S.-Y. Pi, Phys. Rev. Lett. 45, 1 (1980); J. Pati (in this volume).

[48] A. Guth, Phys. Rev. D23, 347 (1981).

[49] F.A. Bais and S. Rudaz, Nucl. Phys. B170, 507 (1980); A.D. Linde, Phys. Lett. 96B, 293 (1980); G. Lazarides and Q. Shafi, Phys. Lett. 94B 149 (1980); 97B, 87 (1980).

[50] S. Dimopoulos (in this volume).

[51] G. Lazarides, Q. Shafi, and T.F. Walsh, Phys. Lett. 100B, 21 (1981); D. Cline, talk at ITP Workshop on the Interaction of Particle Physics and Astrophysics (UCSB, May 1981).

[52] Baryosynthesis after a first-order phase transition when particle distributions are likely to be non-thermal has recently been discussed by M.S. Turner, submitted to Phys. Rev. D (1981); M.S. Turner and J.N. Fry, submitted to Phys. Rev. Lett. (1981).

[53] J.D. Barrow and M.S. Turner, Nature, in press (1981).

[54] J.D. Barrow and M.S. Turner, Nature, in press (1981).

[55] D. Eardley and W.H. Press, presented at Moriond Workshop on Astrophysics (1981).

CONSTRAINTS ON GUTS FROM COSMOLOGICAL BARYON ASYMMETRY

M. Yoshimura

National Laboratory for High Energy Physics (KEK)
Oho-machi, Tsukuba-gun
Ibaraki-Ken 300-32, Japan

The theory of cosmological baryosynthesis has presumably a long history[1] but it is only a couple of years ago[2] that the subject became well motivated in the context of modern gauge theories. Since then, many papers have appeared and the basic mechanism of baryon production is well understood by now. One can even discuss at a semi-quantitative level what kinds of constraints are imposed on grand unified theories by the cosmological baryon production. I shall discuss in this talk those aspects most relevant to models of grand unification. With a specific model given, one can derive mass bounds for superheavy particles, constrain the way CP violation is incorporated etc. One can even exclude some "minimal" models. It would be fun to see how the single event of matter creation in the distant past and the single number[3], the baryon to photon ratio $N_B/N_\gamma \sim 10^{-9}$, left over from this event, can tell us something about microscopic physics.

Let me first remind you of basic ideas behind the theory of baryon production. The whole story starts at a cosmological time, 10^{-36} seconds after the initial bang. The corresponding temperature was high enough to produce many X bosons and their antibosons as massive as 10^{15} GeV. The most important role of X bosons is to mediate baryon nonconserving processes. As the universe expands and the temperature drops, production rate of baryons decreases because of lower densities and lower energies. In order to produce a net baryon number starting from an initial configuration which is symmetric between baryons and antibaryons, three conditions must be satisfied; (1) Baryon nonconservation, (2) C-, and CP-nonconservation or equivalently time reversal noninvariance, (3) Departure from thermal equilibrium. The first two conditions are necessary to give the asymmetry and an arrow of direction at the microscopic level, and the third condition is necessary to give them at the macroscopic level. To appreciate a subtlety of the last condition, consider the energy spectrum of microwave background in the present universe. It is of a black body form despite that these photons went out of thermal equilibrium a long time ago. This is because one can prove[4] from a Boltzmann-like equation in an expanding universe that the distribution function of massless particles keeps an equilibrium form having a temperature simply red-shifted according to the expansion law once the massless particles were previously in equilibrium. Reflecting this theorem, reactions among light (light with respect to 10^{15} GeV/c^2) particles can hardly produce a net baryon number. The scenario calls for processes with a threshold around the mass of X boson m_X. An obvious candidate [4,5] is then a decay of heavy ($\sim m_X$) particle. For instance, consider a pair decay of X boson and its antiparticle $\overline{X}$;

$$X \to ql \text{ with a rate } \gamma_1, \qquad \to \overline{qq} \text{ with } \gamma_q,$$

$$\overline{X} \to \overline{ql} \text{ with } \overline{\gamma}_1, \qquad \to qq \text{ with } \overline{\gamma}_q.$$

A net baryon number produced by this pair decay is

$$\varepsilon = \frac{\frac{1}{3}\gamma_1 - \frac{2}{3}\gamma_q}{\gamma_1 + \gamma_q} + \frac{-\frac{1}{3}\overline{\gamma}_1 + \frac{2}{3}\overline{\gamma}_q}{\overline{\gamma}_1 + \overline{\gamma}_q} = \frac{\overline{\gamma}_q - \gamma_q}{\gamma_1 + \gamma_q},$$

because of the CPT theorem $\gamma_1 + \gamma_q = \overline{\gamma}_1 + \overline{\gamma}_q$. Obviously, C- and CP-noninvariance are necessary for a nonvanishing ε.

Before going into discussions of the baryon to photon ratio, let me elaborate a little more on the decay mechanism, assuming that the pair asymmetry ε does not vanish. From this argument a mass bound for the superheavy parent particle will be derived. One crucial question about this decay mechanism is how many X bosons considered as a parent are available to produce the baryon asymmetry. As the universe expands and cools down, X bosons and their antibosons that were once abundant decouple, namely, their number in the comoving volume is frozen. Suppose that at some time later than the decoupling time these bosons start to decay freely without appreciable compensation by the inverse decay. This scenario is called delayed-decay mechanism[4,5], and it demands that when the decay lifetime becomes comparable to the age of the universe, the temperature is low enough for the inverse decay not to take place appreciably, because of the Boltzmann factor $\exp(-m_x/T)$. For a two-body decay of the gauge boson X_V this condition gives

$$T_d \lesssim m_X \text{ with } \alpha_G m_X \sim \frac{T_d^2}{m_*}, \quad m_* \simeq 10^{19} \text{ GeV}.$$

T_d is a temperature below which the decay proceeds faster than the expansion, and m_* is the Planck mass. This consideration leads[5] to a very crude bound on m_X,

$$m_X \gtrsim \alpha_G m_* .$$

A more precise computation in the minimal SU(5) model gives a substantial reduction[6,7] by a factor of ∿100 on the right hand side, but it seems still higher for the mass of gauge boson which is estimated[8] to be (4-6) 10^{14} GeV from evolution of the low energy coupling constants. On the other hand, since the Higgs mass is largely undetermined and the Higgs coupling α_H may be smaller than the gauge coupling α_G, a Higgs boson X_S, a relative of gauge X boson, has a better chance of producing the baryon asymmetry. Detailed studies[7] indicate that the decay mechanism of this Higgs X_S is viable if

$$m_H \gtrsim \alpha_H \cdot 10^{17} \text{ GeV}.$$

Another candidate that has drawn[9-11] much attention recently is a heavy neutral lepton F contemplated in certain models like SO(10) and E_6. Depending on parameters of the model, main decay modes of baryon production may be

$$F \to qqq \text{ and } \overline{qqq} .$$

If this happens, the delayed-decay mechanism demands that

$$T_d \lesssim m_F \text{ with } \alpha_G^2 \frac{m_F^5}{m_X^4} \sim \frac{T_d^2}{m_*} .$$

Note that the decay rate depends very sensitively on the mass m_F because of the effective four Fermi interaction. For simplicity, we assumed that the Fermi interaction is mediated by a gauge boson.

From this one gets an upper bound

$$m_F \lesssim (\alpha_G^{-2} m_X^4/m_*)^{1/3} .$$

Thus, one sees that an effective nonrenormalizable decay vertex such as the Fermi type is crucial for a relatively light particle ($\ll 10^{15}$ GeV) to produce the baryon asymmetry. This is important particularly in a partial unification scheme in which an intermediate energy scale of 10^6 GeV or above gives baryon nonconservation with the selection rule $\Delta B = 2$[12]. Whatever the agent of baryon production may be, the baryon number in the comoving volume is frozen shortly after the epoch of baryon production and a typical baryon to photon number in the present universe is estimated to be[5-7]

$$\frac{N_B}{N_\gamma} = O(10^{-2}) \cdot \varepsilon,$$

unless a substantial generation of entropy occurs between the time of baryon production and the present.

Let me now proceed to computation of the asymmetry ε in elementary processes. Suppose, for illustration, a partial decay rate of the form,

$$\gamma = |g_1 f_1 + g_2 f_2|^2.$$

g_i lumps together coupling factors and f_i is a genuine dynamical amplitude. For the corresponding rate of its antiparticle one finds that

$$\bar{\gamma} = |g_1^* f_1 + g_2^* f_2|^2.$$

Note that only coupling factors get complex conjugated. Thus,

$$\gamma - \bar{\gamma} = -4\mathrm{Im}(g_1 g_2^*)\mathrm{Im}(f_1 f_2^*) .$$

Besides a complex coupling due to CP violation, one needs an interplay of different dynamical phases to get a finite ε. An example of such a dynamical phase is a rescattering phase due to a final state interaction, which arises from interference of a Born term and a one-loop diagram. Necessity of the interference guarantees that the asymmetry is bounded from above,

$$|\varepsilon| \lesssim O(\alpha) .$$

Usually, the coupling α involves a Higgs boson because CP violation is needed, and tends to be smaller than the gauge coupling constant. It is important to recognize that there is a natural mechanism for suppressing the magnitude of the baryon asymmetry.

At which order of perturbative expansion a finite asymmetry occurs depends on details of a specific model, most importantly, how CP violation is incorporated. The CP violation in question takes place in the baryon nonconserving sector, hence, in principle, it may have no direct connection to the observed CP violation in K-$\overline{K}$ complex. In what follows I shall mention a sample of specific results for the asymmetry ε in some popular models. It is intended to illustrate a delicate constraint on model building. First of all, the minimal SU(5) model[13] which is the simplest GUT has a problem. This is because in the minimal SU(5) there is essentially one source of CP violation which shows up in a complex Higgs fermion coupling matrix h. A nonvanishing asymmetry occurs[14] first in the eighth order

$$\varepsilon \sim \text{Imtr}\ (f^2 h^\dagger f^2 h h^\dagger h)/[\text{tr}(f^2)+\text{tr}(hh^\dagger)].$$

Only if a fourth generation of fermions heavier than 30 GeV exists, may this asymmetry become marginally consistent[15] with the observed baryon to photon ratio. Another problem of simple SU(5) models is that B-L is exactly conserved so that cosmological B-L must be given by an initial condition, which is not attractive. In more expanded schemes, especially in models with many Higgs multiplets, the situation gets less tight, and the asymmetry occurs[16] in lower order of perturbation. An interesting case is a class of SO(10) models[17] in which a generation of fermions is arranged in the spinorial (16) representation,

$$\begin{pmatrix} u_1 & u_2 & u_3 & \nu \\ d_1 & d_2 & d_3 & e \end{pmatrix}_L + \begin{pmatrix} \overline{u}_1 & \overline{u}_2 & \overline{u}_3 & \overline{N} \\ \overline{d}_1 & \overline{d}_2 & \overline{d}_3 & e^+ \end{pmatrix}_L .$$

We allow a possibility of N_R, the chiral partner of left-handed neutrino ν_L, having its own, independent Majorana mass. One-loop diagrams may give[18] the asymmetry ε to be of order α_H for the Higgs decay mode $X_S \to \overline{ud}$, but two contributions, one containing ν_L in the intermediate state and the other containing N_R, tend to cancel[18] each other. The degree of cancellation depends on a mass ratio $m(N_R)/m(X_S)$, and the smaller the mass ratio is, the more complete the cancellation. The reason for this cancellation is that by combining the parity operation and an SO(10) transformation, conjugate couplings are related.

$$\overline{\psi_L}\,\psi'_R \xrightarrow{P} \overline{\psi_R}\,\psi'_L \xrightarrow{SO(10)} \overline{\psi'_R}\,\psi_L = (\overline{\psi_L}\,\psi'_R)^\dagger .$$

Thus, two couplings related by the parity operation are a pair of a complex number and its conjugate so that their sum is real. In order to have a sizable asymmetry proportional to the imaginary part of these couplings, phase spaces of the two contributions must be quite different, namely, masses of the two neutral leptons must be enormously different.

From a more detailed analysis[18,19] of this kind one can obtain[19] an interesting bound for a combination of the heaviest quark mass and the heaviest neutrino mass. Specifically, if one adopts the Higgs decay just mentioned, the magnitude of the asymmetry is given by

$$\varepsilon \simeq \frac{1}{8\pi}\,\frac{\mathrm{Im}[\mathrm{tr}(f'^\dagger f\, f'^\dagger f\, m_N^\dagger m_N)]}{(m_X^2 + m_{X'}^2)\ \mathrm{tr}(f^\dagger f)}$$

where f and f' are Yukawa couplings of Higgs bosons, X_S and X'_S. Existence of at least two Higgs bosons are demanded to lift the degeneracy, $m_u = m_d$. Assuming three generations and a minimal set of Higgs multiplets in the SO(10), one finds

$$\varepsilon \sim \frac{\sqrt{2}}{4\pi}\, G_F\, \frac{1}{9}\, \frac{m_t^2 m_N^2}{m_X^2} \times O(1)$$

$$\sim \frac{\sqrt{2}}{4\pi}\, G_F\, \frac{1}{9^3}\, \frac{m_t^6}{m_X^2 m_\nu^2}$$

by keeping terms which contain the largest coupling. Here I used the relation between the Higgs and gauge coupling constants, $f_{max} \sim g m_t/m_W$, and the relation,

$$m_\nu^{max} = m_t^2/m_N,$$

which holds in this simple SO(10) model. A factor of 1/9 was put in by taking into account of the renormalization effects. A maximal CP violation is assumed, and the asymmetry given here is independent of the usual mixing matrix. There are both upper and lower bounds for the Higgs mass m_X. The upper bound,

$$m_X < (4\text{-}6)\cdot 10^{14}\ \text{GeV},$$

is taken here because one does not want the produced baryon number to be damped by the decay and inverse decay of the gauge boson X_v. The lower bound is derived from the out-of-equilibrium condition,

$$m_X \gtrsim \alpha_H \cdot 10^{17}\ \text{GeV} \sim \left(\frac{m_t}{\text{GeV}}\right)^2 \cdot 10^{11}\ \text{GeV}$$

Thus, there is an allowed region in (m_t, m_ν) plane and from the present lower bound of m_t ($\sim$18 GeV) one gets

$$10^{-2}\ \text{eV} \lesssim m_\nu^{max} \lesssim 100\ \text{KeV}$$

This bound is subject to many uncertainties. On a theoretical side it rests with the simplest choice of the Higgs fermion coupling. Independent of this detailed analysis, what is interesting in a more general framework[18,20] is that existence of a basic left-right symmetry and its apparent, large spontaneous breakdown can explain both the smallness of neutrino mass and a sizable baryon asymmetry. This idea is nicely realized in a simple SO(10) model as discussed here, but it seems more general than this example.

A few comments are now in order;

(1) Throughout the discussion here, a second order or weakly first order phase transition of GUT has been assumed. Actually, even in a strongly first order phase transition most of our scenario will follow if the universe is reheated to a temperature around m_X due to the latent heat released after the transition. The first order phase transition may be favored as a mechanism of suppressing monopole production.[21]

(2) We did not discuss here the initial baryon number, rather we assumed a symmetric initial state. Several people[6,7,22,23] have made detailed investigations on the problem of initial condition. To prepare an initial baryon-neutral state, one needs to damp an arbitrary baryon number (perhaps comparable to the photon number) given from a previous epoch. Damping is a first order process unlike the production we have so far discussed, and one needs not worry about CP-noninvariance. It is only necessary to show that thermalization of baryon number takes place prior to the delayed-decay process. A possible scenario, worked out by a detailed numerical analysis of Chicago group[24], is first to erase an arbitrary baryon number by decay and inverse decay of the X gauge boson X_V, and then to produce a net baryon number by nonequilibrium decay of the Higgs boson X_S. This is an economical scheme because the gauge boson tends to wipe out the baryon number more efficiently due to a larger coupling, while the Higgs boson produces a net baryon number more easily due to a complex CP-violating coupling. For this scheme to work, the Higgs X_S must be less massive than the gauge X_V.

(3) Recently, several people[25] proposed a mechanism of producing the baryon asymmetry at a much lower temperature $\sim 10^6$ GeV/k_B. Dynamical processes responsible for this baryon production are related to $\Delta B=2$ processes such as the n-$\bar{n}$ oscillation, which may occur if an intermediate unification scale exists between 100 GeV and 10^{15} GeV. Unfortunately, the model contains more arbitrary parameters than a simple GUT, and much more work seems necessary to construct a detailed scenario. Even if this type of low energy process is effective for the baryon production, one has to check[26] whether this does not wipe out the baryon number produced at the GUT epoch.

In conclusion, without more handles on physics beyond the standard gauge theory it is difficult to pinpoint a specific dynamical process of baryon production. But, from recent studies it is already clear that the cosmological baryon excess gives an important clue to the structure of unification beyond what we believe we know.

References

[1] For a partial list of earlier suggestions, see S. Weinberg, in Lectures on Particles and Field Theory, ed. by S. Deser and K. Ford (Prentice Hall, Englewood Cliffs, NJ 1964) p. 482; A.D. Sakharov, Zh. Eksp. Theor. Fiz. Pis'ma 5 (1967) 32; L. Parker, in Asymptotic Structure of Space-Time, ed. by F.P. Esposito and L. Witten (Plenum, New York, 1977) pp. 118, 160.

[2] M. Yoshimura, Phys. Rev. Lett. 41 (1978) 281; 42 (1979) 746 (E); A. Yu Ignatiev, N.V. Krasnikov, V.A. Kuzmin and A.N. Tavkhelidze, Phys. Lett. 76B (1978) 436.

[3] G. Steigman, Ann. Rev. Astr. Astrophys. 14 (1976) 339.

[4] D. Toussaint, S.B. Treiman, F. Wilczek and A. Zee, Phys. Rev. D19 (1979) 1036.

[5] S. Weinberg, Phys. Rev. Lett. 42 (1979) 850.

[6] M. Yoshimura, Phys. Lett. 88B (1979) 294.

[7] J.N. Fry, K.A. Olive and M.S. Turner, Phys. Rev. D22 (1980), 2953; 2977.

[8] T.J. Goldman and D.A. Ross, Phys. Lett. 84B (1979) 208; W.J. Marciano, Phys. Rev. D20 (1979) 274; J. Ellis, M.K. Gaillard, D.V. Nanopoulos and S. Rudaz, Nucl. Phys. B176 (1980) 61.

[9] T. Yanagida and M. Yoshimura, Phys. Rev. Lett. 45 (1980) 71.

[10] J.A. Harvey, E.W. Kolb, D.B. Reiss and S. Wolfram, Nucl. Phys. B177 (1981) 456.

[11] R. Barbieri, D.V. Nanopoulos and A. Masiero, Phys. Lett. 98B (1981) 191.

[12] R.N. Mohapatra and R.E. Marshak, Phys. Rev. Lett. 44 (1980) 1316.

[13] H. Georgi and S.L. Glashow, Phys. Rev. Lett. 32 (1974) 438.

[14] J. Ellis, M.K. Gaillard and D.V. Nanopoulos, Phys. Lett. 80B (1979) 360; 82B (1979) 464 (E).

[15] G. Segre and M.S. Turner, Phys. Lett. 99B (1981) 399.

[16] S.M. Barr, G. Segre and H.A. Weldon, Phys. Rev. D20 (1979) 2494; A. Yildiz and P. Cox, Phys. Rev. D21 (1980) 906; D.V. Nanopoulos and S. Weinberg, Phys. Rev. D20 (1979) 2454; T. Yanagida and M. Yoshimura, Nucl. Phys. B168 (1980) 534.

[17] H. Georgi, in Particles and Fields, ed. by C.E. Carlson (AIP Conf. No. 23, New York, 1975); H. Fritzsch and P. Minkowski, Ann Phys. (N.Y.) 93 (1975) 193; M.S. Chanowitz, J. Ellis and M.K. Gaillard, Nucl. Phys. B128 (1977) 506; H. Georgi and D.V. Nanopoulos, Nucl. Phys. B155 (1979) 52.

[18] T. Yanagida and M. Yoshimura, Phys. Rev. D. in press.

[19] M. Fukugita, T. Yanagida and M. Yoshimura, in preparation.

[20] V.A. Kuzmin and M.E. Shaposhnikov, Phys. Lett. 92B (1980) 115.

[21] J.P. Preskill, Phys. Rev. Lett. 43 (1979) 1365; A.H. Guth and S.H. Tye, Phys. Rev. Lett. 44 (1980) 631; M.B. Einhorn, D. Stein and D. Toussaint, Phys. Rev. D21 (1980) 3295.

[22] M. Honda and M. Yoshimura, Progr. Thoer. Phys. 62 (1979) 1704.

[23] E.W. Kolb and S. Wolfram, Nucl. Phys. B172 (1980) 224.

[24] J.N. Fry, K.A. Olive and M.S. Turner, Phys. Rev. Lett. 45 (1980) 2074. See also Ref. (7).

[25] A. Masiero and R.N. Mohapatra, Max Plank preprint (1980).

[26] S. Weinberg, Phys. Rev. D22 (1980) 1694.

FIXED POINTS: FERMION MASS PREDICTIONS

Christopher T. Hill

Fermi National Accelerator Laboratory
P.O. Box 500, Batavia, Illinois 60510

Familiar theoretical relationships exist amongst the masses of the light fermions in SU(5):[1]

$$\frac{m_d}{m_e} \approx \frac{m_s}{m_\mu} \simeq 4 \pm \tfrac{1}{2} \qquad\qquad \frac{m_b}{m_\tau} \simeq 2.75 \pm .25 \quad . \tag{1}$$

These result from essentially three ingredients: (a) the Higgs mechanism which relates the fermion mass to the Higgs-Yukawa (HY) coupling constant by $m_f = g_f\langle\phi^o\rangle/\sqrt{2} \approx g_f \cdot (175 \text{ GeV})$; (b) the renormalization group (RG) equations which describe the evolution of the (HY) couplings down from M_X to low energies and (c) the boundary conditions from SU(5), assuming a $\underset{\sim}{5}$ and $\underset{\sim}{24}$ of Higgs, which relate $g_{-1/3 \text{ quark}}(M_X) = g_{-1 \text{ lepton}}(M_X)$ within a given generation. Thus one obtains:

$$\frac{m_{-1/3}}{m_{-1}} = \frac{g_{-1/3}(\mu)}{g_{-1}(\mu)} \approx \left(\frac{\alpha_3(\mu)}{\alpha_3(M_X)}\right)^{8/2b_o} + (\text{electroweak})\Big|_{\mu \simeq m_{quark}} \tag{2}$$

and the numerical results of eq. (1) follow to one loop accuracy with $\Lambda_s \sim .4$ GeV. We note that ingredients (b) and (c) do not apply to models with composite Higgs bosons (e.g. ETC) or composite fermions on a scale of $M' < M_X$ since the RG equations demand pointlike particles over the entire range of the desert. Though the results for m_d/m_e and m_s/m_μ are questionable we note that they are qualitatively correct. We take the "prediction" for m_b/m_τ to be a successful result of SU(5) and we seek to extend the above analysis to much heavier objects; either a very heavy top-quark or a heavy fourth generation of quarks and leptons.

For very heavy quarks and leptons the RG equations of the HY coupling constants, which lead to eq. (2), must be modified to include the effects of the (now large) HY couplings themselves. The equations thus become nonlinear in g_f when $g_f \sim (g_2$ or $g_3)$ and will thus fix the absolute scale of g_f. Hence, the relevant fermion mass scale for which these effects become important is expected to be, assuming a single $\underset{\sim}{5}$ of Higgs, $m_{quark} \sim g_3(200)\cdot 175 \sim 240$ GeV (with several isodoublet Higgs having VEVs this scale decreases).

Furthermore, as first emphasized by Frogatt and Nielsen[2] and more recently by Pendleton and Ross(PR),[3] there will be "fixed points" which determine $g_f(\mu)$ independent of $g_f(M_X)$ (though in general dependent upon μ, M_X, g_3, g_2, g_1, etc.). Thus for sufficiently large $g_f(M_X)$ (typically $g_f(M_X) \gtrsim 1$) we can hope to make predictions for heavy fermion masses from the fixed point structure of the RG equations without knowing any details of the initial conditions at M_X! However, the nature of these fixed points is slightly subtle,[4] as we now show.

Pendleton and Ross considered the possibility of a heavy t-quark (we will focus upon a fourth generation below, which we expect to be a more realistic possibility; the t-quark will serve as a paradigm for the mathematics) for which the RG equations of g_t and g_3 ($\equiv g_{QCD}$) become:

$$\left.\begin{aligned} 16\pi^2 \frac{d}{dt} \ln g_t &= \left(\frac{9}{2} g_t^{\,2} - 8g_3^2 \right) \\ 16\pi^2 \frac{d}{dt} \ln g_3 &= -b_0 g_3^2 \quad ; \quad t = \ln \mu,\ b_0 = 7 \end{aligned}\right\} \quad . \tag{3}$$

Combining:

$$16\pi^2 \frac{d}{dt} \ln (g_t/g_3) = \frac{9}{2} g_t^{\,2} - (8 - b_0) g_3^2 = \frac{9}{2} g_t^{\,2} - g_3^2 \tag{4}$$

and hence, the Pendleton-Ross "quasi-stable fixed point" is the vanishing of the rhs of eq. (4):

$$g_t(\mu) = \sqrt{\frac{2}{9}}\, g_3(\mu) \quad . \tag{5}$$

If $g_t(\mu)$ is ever near the value given by eq. (5) in terms of $g_3(\mu)$, it will remain "locked in" to this relationship for all subsequent, decreasing μ. Taking $\mu \sim 100$ GeV one has $m_{top} = g_t \cdot 175 = 110$ GeV, which becomes 135 GeV upon including electroweak corrections.

Physically we must ask, however, given an arbitrary initial $g_t(M_X)$ what is the most probable final result for $g_t(\mu)$? In fact, eq. (4) is just the "Bernoulli equation" and can be solved analytically:

$$g_t^{\,2}(\mu) = \frac{g_t^{\,2}(M_X)\left(g_3^{\,2}(\mu)/g_3^{\,2}(M_X) \right)^{8/b_0}}{\left\{ 1 + \dfrac{9g_t^{\,2}(M_X)}{2g_3^{\,2}(M_X)} \left(\left(\dfrac{g_3^{\,2}(\mu)}{g_3^{\,2}(M_X)} \right)^{1/b_0} - 1 \right) \right\}} \quad . \tag{6}$$

In the limit $\left(g_3^{\,2}(\mu)/g_3^{\,2}(M_X)\right)^{1/b_0} >> 1$ we reach the PR fixed point:

$$g_t^{\,2}(\mu) \rightarrow \frac{2}{9}\left(\frac{g_3^{\,2}(\mu)}{g_3^{\,2}(M_X)}\right)^{7/b_0} g_3^{\,2}(M_X) = \frac{2}{9}\, g_3^{\,2}(\mu)\Big|_{b_0=7} \quad . \quad (7)$$

However, to be at a fixed point in the sense that $g_t(M_X)$ no longer influences $g_t(\mu)$ it is sufficient that

$$\frac{9}{2}\,\frac{g_t^{\,2}(M_X)}{g_3^{\,2}(M_X)}\left(\left(\frac{g_3^{\,2}(\mu)}{g_3^{\,2}(M_X)}\right)^{1/b_0} - 1\right) >> 1 \quad (8)$$

which can easily occur for $\left(g_3^{\,2}(\mu)/g_3^{\,2}(M_X)\right)^{1/b_0} \backsim 1$, long before the limit leading to the PR fixed point. Defining $R = g_3^{\,2}(\mu)/g_3^{\,2}(M_X)$ we now assume $1/b_0 \ln R \backsim 0$. Expanding eq. (6) one finds:

$$g_t^{*2}(\mu) \equiv g_t^{\,2}(\mu) = \frac{2b_0}{9}\,\frac{g_3^{\,2}(\mu)}{\ln R}\left\{1 + \frac{1}{2b_0}\ln R + \frac{7}{12b_0^{\,2}}(\ln R)^2 + \ldots\right\} \quad (9)$$

Equation (9) defines a moving fixed point in μ and M_X which we refer to as the "intermediate fixed point." It is the physically interesting asymptotic behavior for $g_t(\mu)$ as $\mu/M_X \rightarrow 0$ and it sets in before the decoupling limit, $\mu \gtrsim m_t$.

In Figure 1 we illustrate how an arbitrary initial $g_t(M_X)$ tends to be swept toward $g_t^*(\mu)$ (we also plot the decoupling limit, $\mu = 175\ g_t(\mu)$) provided $g_t(M_X) \gtrsim 1$. This is a valid perturbative estimate provided $g_t^{\,2}(\mu)/16\pi^2 \lesssim 1$ or $g_t \lesssim 4\pi$. Hence, there is a large perturbative domain of attraction corresponding to $g_t^*(\mu) \backsim 1.3$ for $\mu = 200$ GeV. The resulting "prediction" for our hypothetical single heavy t-quark is 240 GeV (± 10%) including full electroweak corrections.[4] This is also equivalent to the absolute upper bound of Cabibbo et al.[5] Is this a reasonable value for the mass of the physical t-quark? Recently Buras[6] has obtained a limit of $m_t \lesssim 33$ GeV and we remark that a t-quark heavier than $\backsim$ 200 GeV will destroy the quantitatively successful m_b/m_τ relationship.[7] Otherwise, this is consistent with all bounds.

Probably more interesting and relevant to the real world are the consequences for a fourth SU(5) generation. Here we must numerically integrate the equations for g_E, g_T, g_B (where T, B, E refer to +2/3,

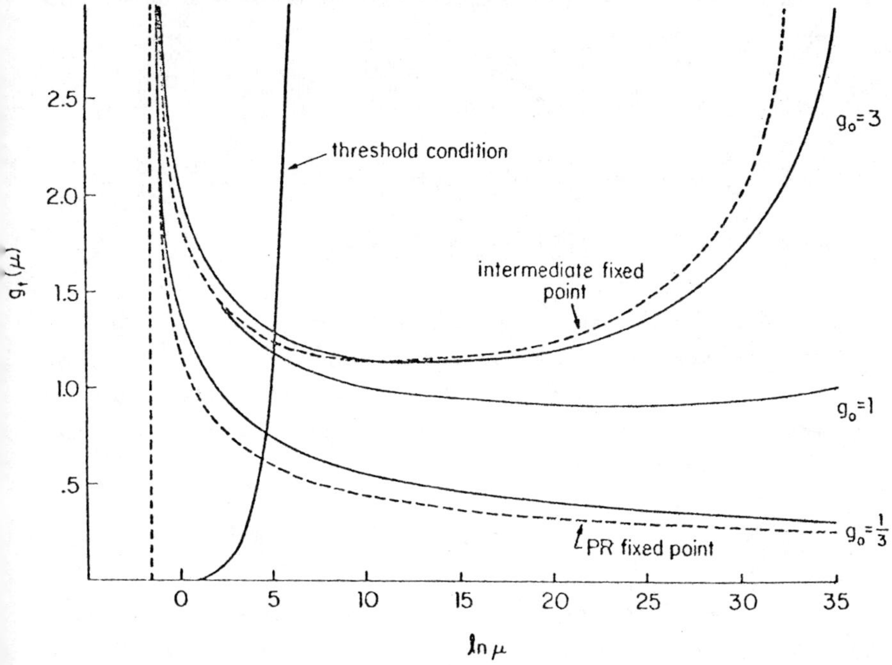

Figure 1. The evolution of $g_t(\mu)$ for three initial values, $g_t(M_X)$. Only the portion to the right of the "threshold condition" is physical. The intermediate fixed point of eq. (9) is the dotted line.

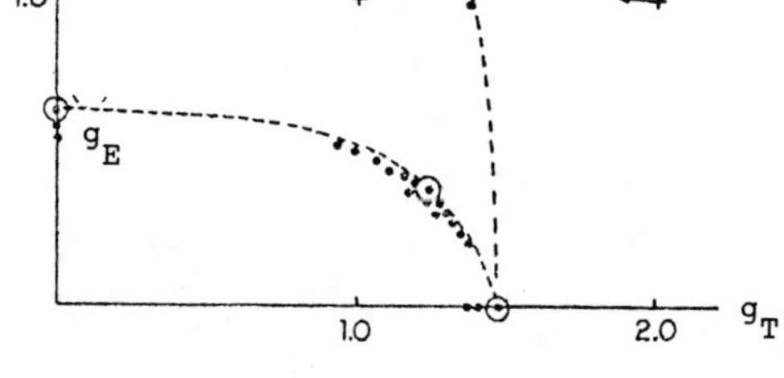

Figure 2. The results of the numerical integration of eq. (10) for a 5 × 5 integer array of $(g_T(M_X), g_B(M_X) = g_E(M_X))$. Points cluster about the fixed points and boundary curves.

−1/3 quarks and the -1 lepton respectively; we assume a light neutrino here):

$$16\pi^2 \frac{d}{dt} \ln g_T = \frac{9}{2} g_T{}^2 + \frac{3}{2} g_B{}^2 + g_E{}^2 - 8g_3{}^2 - \frac{9}{4} g_2{}^2 - \frac{17}{12} (g')^2$$

$$16\pi^2 \frac{d}{dt} \ln g_B = \frac{9}{2} g_B{}^2 + \frac{3}{2} g_T{}^2 + g_E{}^2 - 8g_3{}^2 - \frac{9}{4} g_2{}^2 - \frac{5}{12} (g')^2$$

$$16\pi^2 \frac{d}{dt} \ln g_E = \frac{5}{2} g_E{}^2 + 3g_T{}^2 + 3g_B{}^2 - \frac{9}{4} g_2{}^2 - \frac{15}{4} (g')^2 \quad . \quad (10)$$

The results of this analysis have been detailed in ref. (4). For lack of space we quote only the essential results.

(I) There is a nontrivial fixed point for these equations with all g_i nonvanishing analogous to the intermediate fixed point discussed above. It corresponds to:

$$m_T = 220 \text{ GeV} \qquad m_B = 215 \text{ GeV} \qquad m_E = 60 \text{ GeV} \qquad (11)$$

to within 10% uncertainty at one loop. We have also assumed no Cabibbo mixing to lighter generations (the results are quite insensitive to this; rotating say the B quark maximally (90°) out of the weak current for this generation displaces the fixed point 27%).

(II) The bound of Cabibbo et al.[5] generalizes to an allowed region in the space of g_T, g_B and g_E. Only points within this region are physical (have finite values of g_T, g_B, g_E over the entire desert). Moreover, the boundary of this region acts somewhat like a "generalized fixed point" since arbitrary, large initial points $g_T(M_X)$, $g_B(M_X)$ and $g_E(M_X)$ are mapped preferentially to the fixed point or the boundary, the fixed point lying on the boundary. Figure 2 illustrates the distribution of values of g_B(200 GeV) and g_E(200 GeV) vs. g_T(200 GeV) resulting from the numerical integration of eq. (10) for an initial 5×5 array of points $(g_B(M_X), g_T(M_X))$ and $(g_E(M_X) \equiv g_B(M_X), g_T(M_X))$. We see that the points cluster near the fixed point along the boundary curve (the fixed point corresponds to the mass values of eq. (11)). Hence, in addition to a relationship between m_B and m_E, we also obtain a further relationship between all three masses m_B, m_E and m_T for a sufficiently heavy fourth generation! These results can be generalized to many succeeding heavy generations.

(III) These heavy masses are fully consistent with known bounds on fermion masses, e.g. the ρ-parameter,[8] unitarity[9] and the stability of the Higgs potential.[10]

Recently we have considered the effects upon the standard evolution of g_1, g_2 and g_3 in SU(5) from large g_f, at the two loop level, and thus the effects upon M_X and $\sin^2\theta_W$.[11] These are found to be miniscule with:

$$\frac{\Delta M_X}{M_X} \lesssim 2.5\% \text{ per heavy fermion} \tag{12a}$$

$$0 > \frac{\Delta \sin^2\theta_W}{\sin^2\theta_W} \gtrsim -.15\% \text{ per heavy fermion} \tag{12b}$$

where eq. (12a) is the change in M_X from the large HY coupling effects but does not include the change from a new flavor threshold. Surprisingly, the change in M_X coming from the addition of a fourth generation neglecting HY effects, but with m_T, $m_B \sim 200$ GeV is found to be a factor of ~ 1.25, much less than the 1.8 quoted earlier[12] but consistent with recent estimates of Marciano.[13] The effects of a fourth generation on m_b/m_τ are not known at present.

Searches for heavy quarks in the 200 to 240 GeV region and leptons in the vicinity of 60 GeV may make interesting grist for the Tevatron and collider mills of the future.

REFERENCES

1. A. Buras, J. Ellis, M.K. Gaillard and D.V. Nanopoulos, Nucl. Phys. B155, 189 (1979).
2. C.D. Frogatt and H.B. Nielsen, Nucl. Phys. B136, 115 (1978).
3. B. Pendleton and G.G. Ross, Phys. Letters 98B, 291 (1981).
4. C.T. Hill, Fermilab-Pub-80/97-THY, Dec. 1980, to be published in Phys. Rev. D.
5. N. Cabibbo, L. Maiani, G. Parisi, R. Petronzio, Nucl. Phys. B158, 295 (1979).
6. A.J. Buras, Fermilab-Pub-81/27-THY, March 1981.
7. M. Machacek, M. Vaughn, Northeastern Univ. Preprint, NUB-2493.
8. M. Veltman, Nucl. Phys. B123, 89 (1979).
9. M. Chanowitz, M. Furman, I. Hinchcliffe, Nucl. Phys. B153 402, (1979).
10. P.Q. Hung, Phys. Rev. Lett. 42, 873 (1979); H.D. Politzer and S. Wolfram, Phys. Lett. 82B, 242 (1979).
11. M. Fischler and C.T. Hill, Fermilab-Pub-81/43-THY, May 1981.
12. T. Goldman, D. Ross, Nucl. Phys. B171, 273 (1980); J. Ellis, M. Gaillard, D. Nanopoulos, S. Rudaz, Nucl. Phys. B176, 61 (1980).
13. W. Marciano, Talk presented at 2nd Workshop on Grand Unification and private communication.

FERMION MASSES: ANOTHER LOOK*

Marie Machacek**

Department of Physics
Northeastern University
Boston, Massachusetts 02115

Grand unified theories, despite their obvious aesthetic appeal, leave few experimentally accessible remnants at present energies. It is important that each of these testable predictions of unified theories be reviewed periodically in light of recent advances in the measurement of low energy parameters pertinent to these predictions and advances in the calculational techniques necessary to extrapolate them to unification energies. Let us, therefore, take another look at one of the famous, although perhaps least understood, predictions of unified theories, the prediction of fermion masses and mass ratios.

Assume for this discussion that fermion masses arise from Yukawa couplings of fermions to fundamental Higgs scalars in the Lagrangian symmetric under the unification group. This unification symmetry imposes a technically natural zeroth order "mass relation" [1] upon the Yukawa coupling constants in the theory, which is determined in tree level at superheavy unification energies M. For example, in SU(5) the down quark and charged lepton masses arise from a single SU(5) $5\text{-}5\text{-}\overline{10}$ Yukawa coupling F. This requires that at $M \sim 10^{15}$ GeV $F_D(M) = F_L(M) = F(M)$ where F_D (F_L) are the down quark (charged lepton) Yukawa couplings in the $SU(3) \times SU(2) \times U(1)$ effective Lagrangian. Radiative corrections to the zeroth order relation are finite and calculable. The coupling constants in the effective theory are extrapolated from M to the scale M', where the effective symmetry is again broken, by the use of renormalization group (RG) equations appropriate to the operators which appear in the effective

*This work is supported in part by the National Science Foundation under Grant No. PHY79-05922.

**On leave at Lyman Laboratory of Physics, Harvard University, Cambridge, Massachusetts 02138.

Lagrangian [2,3,4]. In particular, for theories where $SU(3) \times SU(2) \times U(1)$ is the effective symmetry between M and $M' \sim M_W$, the mass operator $m_F \bar{\psi}\psi$ is forbidden by chiral symmetry. The operators of interest in determining the RG equations for "mass" relations in this energy regime are the $SU(3) \times SU(2) \times U(1)$ Yukawa interaction terms $f_{ij} \bar{\psi}_i \psi_j \phi$ between fermions ψ and Higgs scalars ϕ. The fermion mass $m_F(Q)$ for $M_W \leq Q \leq M$ is defined as the product of the Yukawa coupling f and vacuum expectation value v, where each has been extrapolated to the energy Q, i.e.,

$$m_F(Q) = f(Q)\, v(Q). \tag{1}$$

(Initially v must be determined at energies $M' \sim v$, where perturbation theory for the effective potential is valid, and then must be rescaled upward to Q using the anomalous dimension of the corresponding scalar field.) [4] For any theory where a single vacuum expectation value or combination of vacuum expectation values enters into the determination of the fermion masses m_1 and m_2, the ratio of fermion masses is precisely the ratio of corresponding Yukawa couplings,

$$\frac{m_1(Q)}{m_2(Q)} = \frac{f_1(Q)\, v(Q)}{f_2(Q)\, v(Q)} = \frac{f_1(Q)}{f_2(Q)}. \tag{2}$$

Below $M' \sim M_W$ the new effective theory is $SU(3) \times U(1)$ symmetric. The mass operator $m_F \bar{\psi}\psi$ exists in the effective Lagrangian so that the RG equations used to extrapolate masses below M_W are the familiar ones for the mass operator itself.

Let us now demonstrate this procedure for theories in which the zeroth order relation $F_D(M) = F_L(M)$ holds at $M \sim 10^{15}$ GeV. In most past extrapolations of this relation to ordinary energies [2,5,6], the fermions are all assumed light compared to M_W. Thus terms cubic in the Yukawa couplings, i.e. $\sim O(f^3)$, were neglected in the one-loop $SU(3) \times SU(2) \times U(1)$ RG equations relative to $f g_j^2$ terms, where g_j, $j = 1,2,3$ are the gauge couplings for SU(j), respectively. For a sufficiently heavy t quark (experimentally we know $m_t \gtrsim 18$ GeV!), or for any new heavy generation of fermions which couple to Higgs particles in the ordinary way, this assumption will no longer be true. Instead $f \sim O(g_j)$ such that f^3 terms must appear in the RG equations at the one-loop level. The full set of these equations is listed in Ref. [7], and their consequences have been studied recently by various authors [7,8,9].

The equations for the Yukawa couplings are coupled matrix

equations of the form

$$\frac{d(\ln F_i)}{dt} = C_i(F_k, k = U,D,L; g_j, j = 1,2,3) \qquad (3)$$

where F_i, $i = U,D,L$ is the matrix in generation space of Yukawa couplings and C_i is the corresponding matrix of coefficients in the RG equations. The equation for the extrapolation of mass ratios becomes simply

$$\frac{d}{dt}\left[\ln\left(\frac{F_i}{F_j}\right)\right] = C_i - C_j. \qquad (4)$$

Thus any diagram which contributes equally to the RG equation for F_i and F_j cancels in the mass ratio. In particular, any strictly SU(2) contribution to the extrapolation of F_D and F_L cancels in the ratio F_D/F_L since L and D have the same properties under SU(2). Also, any contribution to the RG equations for F_i from the anomalous dimensions of the field ϕ are exactly cancelled by the simultaneous extrapolation of v to energy scale Q when defining the fermion mass at Q [4].

If we further neglect mixing between the third and lighter generations, we find the equation [7]

$$16\pi^2 \frac{d}{dt}\left[\ln\left(\frac{f_b}{f_\tau}\right)\right] = -8g_3^2 + 2g_1^2 + \frac{3}{2}(f_b^2 - f_\tau^2 - f_t^2) + \cdots \qquad (5)$$

Thus m_b/m_τ depends sensitively, through f_t, on the mass of the t quark! Terms not explicitly specified by Eq. (5) are all suppressed by mixing angles. This is true even if yet heavier generations exist. The major effect of additional heavy generations on m_b/m_τ comes through the usual dependence of the gauge coupling β functions on the number of generations.

In order to complete the numerical calculation of m_b/m_τ at ordinary energies, we must specify the values of effective coupling constants as we cross particle thresholds and the boundaries of different symmetry regions. A consistent calculation requires that these boundary conditions be known to one less order than that of the RG equations. Thus for one-loop RG equations the zeroth order relations are themselves sufficient so that $g_1(M) = g_2(M) = g_3(M) = g(M)$, the gauge coupling of the grand unified theory, and $F_D(M) = F_L(M) = F(M)$. Similarly the initial conditions for the new RG equations as particles decouple below their thresholds are found by smoothly matching all couplings at the particle threshold.

In Figure 1 the result of the one-loop RG analysis of

m_b/m_τ as a function of $\underline{t}$ quark mass are presented for three (solid curves) and four (dashed curves) generations of fermions. It is clear that m_b/m_τ begins to increase sharply for $m_t \gtrsim 100$ GeV. The masses of the fourth generation of fermions are chosen for definiteness to correspond to extrapolated solutions of the SU(5) fixed point equations [7,10]

$$\frac{d}{dt}\left(\frac{F}{g}\right) = 0 \quad \text{and} \quad \frac{d}{dt}\left(\frac{H}{g}\right) = 0 \tag{6}$$

where g is the SU(5) gauge coupling and F(H) is the 5-5-$\overline{10}$ (5-10-10) matrix of Yukawa couplings, respectively. This gives $M_U \simeq 165$-185 GeV, $m_D \simeq 140$-165 GeV, and $m_L \simeq 60$ GeV for the fourth generation. Although speculative, such a procedure could be interesting for heavy generations since the RG equations tend to draw initially large Yukawa couplings rapidly toward such values. Other amusing solutions to Eq. (6) are also discussed in Refs. [7 and 10].

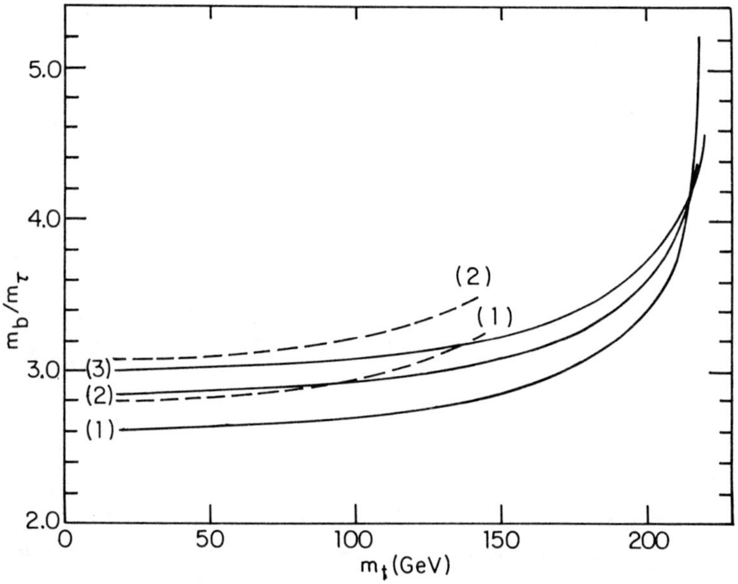

Figure 1: The mass ratio m_b/m_τ as a function of t quark mass m_t with three (solid curve) and four (dashed curve) fermion generations for three values of the unification mass M; (1) M = 5.0×10^{14} GeV, (2) M = 1.07×10^{15} GeV, (3) M = 1.65×10^{15} GeV.

Over much of the allowable range for m_t ($m_t \lesssim 100$ GeV) the f^3 terms in the RG equations produce $\lesssim 10\%$ effect on m_b/m_τ. This effect is of the same order as one might expect from other two-loop corrections in the RG equations. Let us close with a few remarks concerning the two-loop extrapolation of mass ratios.

First, the use of two-loop RG equations (using $\overline{MS}$ subtraction) requires that the radiative corrections to the boundary conditions be calculated to one-loop. Thus at M the zeroth order relations

$$g_1(M) = g_2(M) = g_3(M) \tag{6}$$

are modified by finite, calculable amounts

$$\frac{1}{g_j^2(\mu)} = \frac{1}{g^2(\mu)} = \lambda_j(\mu), \quad j = 1,2,3 \tag{7}$$

by integrating out the contributions of superheavy particles at $\mu \sim M$ [3,11,12]. Similarly the zeroth order relation between Yukawa couplings

$$F_D(M) = F_L(M) = F(M)$$

is modified to

$$F_D(\mu) = F(\mu)[1 + \delta_D(\mu)]$$

$$F_L(\mu) = F(\mu)[1 + \delta_L(\mu)] \tag{8}$$

through contributions of all one-loop graphs involving the exchange of superheavy particles. In the minimal SU(5) model with a 5 and 24 of Higgs particles, F_D and F_L are split at the boundary by

$$F_D - F_L = -\frac{1}{(4\pi)^2}\left\{\frac{5}{2}g^2 + g^2\left[1 - \frac{M_S^2}{(M_V^2 - M_S^2)}\ \ell n\left(\frac{M_V^2}{M_S^2}\right)\right]\right.$$
$$\left. + \frac{7}{8}H^+H + \frac{3}{4}H^+H\ \ell n\left(\frac{M_V^2}{M_S^2}\right)\right\}F \tag{9}$$

where g,F,H are the SU(5) gauge, 5-5-$\overline{10}$, and 5-10-10 couplings, respectively, M_V is the mass of the (3,2) component of the 24 of vector bosons, M_S is the mass of the (3,1) component of the 5 of scalar bosons, and all couplings are evaluated at the scale $\mu = M_V$. Thus this boundary condition lowers m_b/m_τ, but, with even the most optimistic values of m_t and M_S, by at most a few percent.

Second, the full two-loop RG equations for the extrapolation of fermion masses and mass ratios should include all terms of order

f^5, fg^4, f^3g^2, $f^3\lambda$ (λ is the scalar quartic coupling), ··· as well as those considered previously, particularly when the possibility of yet heavier generations is contemplated. This formidable calculation is being attempted by several groups [7,9].

As a first step let us consider the calculation in Feynman gauge of the "light" fermion case ($f \sim g^2$) keeping only f^3 and fg^4 pieces to this order. The two-loop contributions to the anomalous dimensions of the fermion fields are well-known [13]. The two-loop contributions to the light scalar field cancel, as in the one-loop case, against the renormalization of the vacuum expectation value $v(\mu)$ when defining $m_F(\mu)$. The remaining 1PI (one particle irreducible) contributions to the Yukawa vertex renormalization are of two types. The first involves graphs similar in form to those which renormalize the mass operator $m\bar{\psi}\psi$ below m_W, i.e., no gauge particle couples to the light scalar leg. Recently these graphs have been recalculated by two groups independently [14] for SU(3), and, after a few minor algebraic and group theoretic corrections to the previous result [15], these three groups agree. The second 1PI contribution to the vertex involves the exchange of at least one SU(2) or U(1) gauge boson from the light scalar leg to the fermion legs. The most important contributions to m_b/m_τ from graphs of this type involve, in addition, the emission and reabsorption of a gluon on the fermion lines, i.e. are of the form $f g_2^2 g_3^2$ and $f g_1^2 g_3^2$. The results of this calculation are forthcoming [16].

In conclusion, fermion mass ratios such as m_b/m_τ are highly model dependent. Within the context of such a model, they can, however, set useful limits on quark masses, such as m_t, and (given the unification mass M) on the number of heavy fermion generations which couple in the standard way. The present calculations indicate that a fourth generation of fermions is not excluded in the minimal SU(5) model provided $m_t < 100$ GeV and $M \sim 5 \times 10^{14}$ GeV. The definitive determination of the number of generations allowed by m_b/m_τ requires, however, a careful two-loop RG equation analysis of the problem. This calculation is currently in progress.

References

[1] T. Hagiwara and B.W. Lee, Phys. Rev. D 7 (1973) 459;
S. Weinberg, Phys. Rev. Lett. 29 (1973) 388;
B.W. Lee, Proc. 16th Int. Conf. on High-Energy Physics, Chicago-

Batavia, Illinois, 1972, eds. J.D. Jackson and A. Roberts, vol. 4, p. 249, (NAL, Batavia, Illinois, 1973);

H. Georgi and S.L. Glashow, Phys. Rev. D 6 (1972) 2977; D 7 (1973) 2457;

H. Georgi and A. Pais, Phys. Rev. D 10 (1974) 539.

[2] H. Georgi, H. Quinn, and S. Weinberg, Phys. Rev. Lett. 33 (1974) 451.

[3] S. Weinberg, Phys. Lett. 91B (1980) 51.

[4] H. Georgi and M. Machacek, Nucl. Phys. B173 (1980) 32.

[5] A. Buras, J. Ellis, M.K. Gaillard, and D.V. Nanopoulos, Nucl. Physics B135 (1978) 66;

H. Georgi and D.V. Nanopoulos, Nucl. Phys. B159 (1979) 16;

S. Dawson and H. Georgi, Nucl. Phys. B179 (1981) 477.

[6] D.V. Nanopoulos and D. Ross, Nucl. Phys. B157 (1979) 273.

[7] M. Machacek and M.T. Vaughn, Northeastern University preprint #2493, to appear in Phys. Lett. (1981).

[8] N. Cabibbo, L. Maiani, G. Parisi and R. Petronzio, Nucl. Phys. B158 (1979) 295;

B. Pendleton and G.G. Ross, Phys. Lett. 98B (1981) 291.

[9] C.T. Hill, FERMILAB-PUB-80/97-THY (1980);

C.T. Hill in Proceedings of the Second Workshop on Grand Unification (1981).

[10] M.T. Vaughn, Northeastern University preprint, 1981.

[11] L. Hall, Nucl. Phys. B178 (1981) 75.

[12] W. Marciano in Proceedings of the Second Workshop on Grand Unification, 1981, and references therein.

[13] E.G. Floratos, D. Ross, and C.T. Sachrajda, Nucl. Phys. B129 (1977) 66.

[14] R. Tarrach, Marseille preprint, 1980; O. Nachtmann and W. Wetzel, Heidelberg preprint no. HD-THEP-81-1.

[15] D.V. Nanopoulos and D.A. Ross, Nucl. Phys. B157 (1979) 273.

[16] M. Machacek and M.T. Vaughn, Northeastern University preprint, in preparation.

PETITE UNIFICATION: AN ALTERNATIVE VIEWPOINT

P.Q. Hung

Lawrence Berkeley Laboratory, University of California
Berkeley, California 94720

Grand unification ideas[2] were proposed long before the experimental confirmation of the standard $SU(2)_L \times U(1)_Y$ model[1] as the most viable candidate describing "low-energy" electroweak phenomena. The most popular and, certainly, most economical grand unified theories are SU(5)[3] and SO(10).[4] There are good reasons, such as the approximate agreement of the measured value of $\sin^2\theta_W$ with the theoretical expectations and the economical assignment of known fermions to SU(5) representations, to take the grand-unification idea very seriously. However, there are also well-known difficulties, in particular the large number of arbitrary parameters and especially the lack of a credible scenario of spontaneous symmetry breaking. Furthermore, it requires a rather bold extrapolation of thirteen orders of magnitude, from $M_W \sim 80$ GeV to $\sim 10^{15}$ GeV with relatively little new phenomena in between.

Given this situation, it may be of importance to carefully examine less ambitious alternatives. I shall describe here a recent work done in collaboration with J.D. Bjorken and A. Buras[5] related to such alternatives. We assume that at some distance scale, not too many orders of magnitude less than the Compton wavelength of intermediate bosons $W^{\pm}$ and Z°, the $SU(3)_c \times SU(2)_L \times U(1)_Y$ gauge theory, characterized by three coupling constants, becomes embedded in a gauge theory $G_S \times G_W$ characterized by only two coupling constants, g_S and g_W. The strong group G_S and weak group G_W are assumed each to be either simple or pseudo-simple i.e. a direct product of simple groups with identical coupling strengths. We call such a possibility petite unification. Any subsequent unification of the strong force with the weak at still shorter distances we shall leave unconsidered.

We adopt here a "building-up" procedure, that is to say we use the available inputs from the "low-energy" theory

$SU(3)_c \times SU(2)_L \times U(1)_Y$ to restrict the choices of G_S and G_W. The inputs we use are the experimental value of $\sin^2\theta_W$ and the known fermion representations. We found that the choices of G_W are quite restricted. The smallest acceptable G_W turns out to be $[SU(2)]^4$. We also found that the most efficient choice of a strong group is SU(4) built à la Pati and Salam,[6] which is the simplest case for which the electroweak $U(1)_Y$ generator is a linear combination of both G_S and G_W generators. Furthermore, leptons provide the fourth color degree of freedom achieving thus an early quark-lepton unification.

We then examine in some details the phenomenology of the minimal petite unification model $SU(4) \times [SU(2)]^4$. In particular we look into rare transitions such as $K_L \to \mu e$ which are induced by the $SU(4)/SU(3)_c \times U(1)_S$ leptoquark gauge bosons ($U(1)_S$ is a subgroup of SU(4)). Interesting bounds (upper and lower) on leptoquark masses are found by combining rare decays with $\sin^2\theta_W$. The model also predicts the existence of heavy mirror fermions which are singlets under the first two SU(2)'s identified as $SU(2)_L \times SU(2)_R$ for ordinary fermions.

1. Choices of G_W

In this section, we derive a formula for $\sin^2\theta_W(M_W^2)_{Theo.}$ in the general framework of $G_S(g_S) \times G_W(g_W)$ and compare it to the experimental value $\sin^2\theta_W(M_W^2)_{exp.}$. This comparison serves to restrict the choices of possible weak groups G_W.

We assume the following symmetry breaking pattern: $G_S(g_S) \times G_W(g_W) \xrightarrow{M} SU(3)_c(g_3) \times \widetilde{G}_S(\widetilde{g}_S) \times G_W(g_W) \xrightarrow{\widetilde{M}} SU(3)_c(g_S) \times SU(2)_L(g_2) \times U(1)_Y(g') \xrightarrow{M_W} SU(3)_c \times U(1)_{e.m.}^{(e)}$. To simplify the discussion, we specialize to the so-called "unlocked" case where the electroweak $SU(2)_L$ is one of the unbroken subgroups of G_W. The generators of $U(1)_{e.m.}$ and $U(1)_Y$, Q and T_o respectively, are given by

$$Q = T_{3L} + T_o \quad , \tag{1.a}$$

$$T_o = \sum_\alpha C_{\alpha W} T^\circ_{\alpha W} + \sum_i C_{iS} \tilde{T}^\circ_{iS} \quad , \tag{1.b}$$

where $T^\circ_{\alpha W} \in G_W$, $\tilde{T}^\circ_{iS} \in \tilde{G}_S$. For an adjoint representation of G_W (the gauge bosons of G_W) or any representation which is singlet under G_S, the second term on the right-hand side of Eq. 1.b is inoperative. It is easy to obtain

$$[1 + \sum_\alpha C^2_{\alpha W}]^{-1} = [TrT^2_{3L}/TrQ^2]_{adjoint} \quad . \tag{2}$$

Equation 2 is used below in the search for the appropriate G_W groups.

Using the definition $\sin^2\theta_W(M^2_W) = e^2(M^2_W)/g^2_2(M^2_W)$ and the evolution equations for g_2, g', g_3 and $\tilde{g}_S$, we can derive

$$\sin^2\theta_W(M^2_W) = \sin^2\theta^\circ_W \left\{ 1 - C^2_S \frac{\alpha(M^2_W)}{\alpha_S(M^2_W)} - 8\pi\alpha(M^2_W) \left[K \ln \frac{\tilde{M}}{M_W} + K' \ln \frac{M}{\tilde{M}}\right] \right\} \tag{3}$$

where $\sin^2\theta^\circ_W = [1 + \sum_\alpha C^2_{\alpha W}]^{-1}$, $C^2_S \equiv \sum_i C^2_{iS}$, $\alpha(M^2_W) \equiv e^2(M^2_W)/4\pi$, $\alpha_S(M^2_W) \equiv g^2_3(M^2_W)/4\pi$, $K = b_1 - (\sum_\alpha C^2_{\alpha W})b_2 - C^2_S b_3$, $K' = C^2_S(\tilde{b} - b_3)$, with b_i and $\tilde{b}$ being coefficients of g^3_i in the relevant renormalization group β_i function. Equation 3 is very reminescent of the Georgi-Quinn-Weinberg[7] formula with two diferences. The first one is the fact that $\sin^2\theta^\circ_W \equiv TrT^2_{3L}/TrQ^2$ only applies to representations which are singlets under G_S. The second one is the presence of the term $C^2_S\alpha(M^2_W)/\alpha_S(M^2_W)$ which represents the possibility that the electroweak $U(1)_Y$ resides partly in the strong group G_S.

Since we are interested in the mass scale range $M \approx 10^{5\pm1}$ GeV and $\tilde{M} \approx 300$ GeV - a few TeV, the quantity inside the curly brackets of Eq. 3 is more sensitive to the term $C_S\alpha(M^2_W)/\alpha_S(M^2_W)$. Denoting that quantity by R (i.e. $\sin^2\theta_W(M^2_W) = R\sin^2\theta^\circ_W$), we found

$R \approx 0.95$, 0.85, 0.65 for $C_S^2 = 1/6$, $2/3$, $8/3$ respectively. We notice in passing that $C_S^2 = 1/6$, $8/3$ corresponds to the case in which there are higher charged fermions while $C_S^2 = 2/3$ accomodates conventionally charged quarks and leptons. Experimentally, $\sin^2\theta_W(M_W^2) \approx 0.22 \pm 0.014$. Consequently, only gauge groups which satisfy

$$0.23 < \sin^2\theta_W^\circ < 0.30 \qquad \text{for } C_S^2 = 1/6,\ 2/3 \quad , \tag{4.a}$$

$$0.30 < \sin^2\theta_W^\circ < 0.4 \qquad \text{for } C_S^2 = 8/3 \quad , \tag{4.b}$$

will be considered.

We now derive $\sin^2\theta_W^\circ \equiv (\mathrm{Tr}T_{3L}^2/\mathrm{Tr}Q^2)_{\mathrm{adj.}}$ within the framework of $G_W = \left[SU(N)\right]^k$. We obtain the following results.

i.) Gauge bosons with charges ± 3 or higher are not allowed. Indeed, for $k = 1$, $|Q_{\mathrm{adj}}^{\mathrm{max}}| = 3$, $N \geqslant 4$, $\sin^2\theta_W^\circ \leqslant [12 - (8/N)]^{-1} \leqslant 1/10$. $Q_{\mathrm{adj}}^{\mathrm{max}}$ is the highest charge of the gauge bosons.

ii.) There can only be at most <u>two</u> doubly charged gauge bosons. In particular, for $k = 1$, $|Q_{\mathrm{adj}}^{\mathrm{max}}| = 2$, $N \geqslant 3$, $\sin^2\theta_W^\circ = 0.25$. With four or more doubly charged gauge bosons, $\sin^2\theta_W^\circ < 0.20$.

iii.) For $k = 1$, $|Q_{\mathrm{adj}}^{\mathrm{max}}| = 1$, $N \geqslant 2$, we derive

$$2/N \leqslant \sin^2\theta_W^\circ \leqslant [2(1 - (1/N)]^{-1} \qquad N \text{ even} \quad , \tag{5.a}$$

$$2N/(N^2 - 1) \leqslant \sin^2\theta_W^\circ \leqslant [2(1 - (1/N)]^{-1} \qquad N \text{ odd} \quad . \tag{5.b}$$

Notice that for $k = 1$, $N = 2$, i.e. $G_W = SU(2)$, $\sin^2\theta_W^\circ = 1$. Also for $G_W = [SU(N)]^k$ with permutation symmetry assumed, one obtains

$$\sin^2\theta_W^\circ = (1/k)\sin^2\theta_W^\circ\big|_{k=1} \quad . \tag{6}$$

More general cases can be found in Ref. 5. We see that in the case of

SU(2) we need k = 4 i.e. $G_W = [SU(2)]^4$ in order to obtain a reasonable value for $\sin^2\theta_W^\circ$ which is here equal to 1/4. Since the upper bound of $\sin^2\theta_W^\circ$ is always less than or equal to one, we can see that for $G_W = [SU(N)]^k$ with $|Q_{adj}^{max}| = 1$, the maximum number of factors allowed is k = 4. Notice that within our framework of petite unification, it is hard to accomodate the case $G_W = [SU(2)]^2$ since here $\sin^2\theta_W^\circ = 1/2$ and we would need a partial unification mass of the order $\sim 10^{12}$ GeV in order to have the correct value for $\sin^2\theta_W^\circ(M_W^2)$.

The G_W groups which give $\sin^2\theta_W^\circ = 1/4$ are $[SU(2)]^4$, $[SU(3)]^3$, $[SU(4)]^2$ and SU(8) for $N \leqslant 8$. Other groups which satisfy the constraints (4.a,b) can be found in Ref. 5. It turns out that when we include fermions, many of these groups are themselves ruled out. In what follows, we shall discuss in more details the most minimal one (in a sense of gauge groups) among them, namely $G_W = [SU(2)]^4$.

2. Choices of G_S

What could G_S be? If G_S were just $SU(3)_c$, $\sin^2\theta_W^\circ \equiv TrT_{3L}^2/TrQ^2$ would apply to <u>all</u> representations of G_W, fermions <u>and</u> gauge bosons. Since fermions and gauge bosons should give the same $\sin^2\theta_W^\circ$ in this case, we would need extra fermions with or without unconventional charges. To illustrate this point, let us take, as an example, $G_S = SU(3)_c$ and $G_W = SU(3)_W \supset SU(2)_L \times U(1)_Y$. There are two cases.

i.) $|Q_{adj}^{max}| = 1$. In this case, one obtains $\sin^2\theta_W^\circ|_{adj} = 3/4$. Furthermore quarks and leptons belong to $\underline{3}$ and $\underline{8}$ of $SU(3)_W$ respectively. There are <u>extra</u> charge - 1/3 quarks and charge -1 leptons. This case is of no interest since $\sin^2\theta_W^\circ$ is too big.

ii.) $|Q_{adj}^{max}| = 2$. As discussed earlier, this case, containing two doubly charged bosons, gives $\sin^2\theta_W^\circ|_{adj} = 1/4$. Again, quarks $\in \underline{3}$ and leptons $\in \underline{8}$ of $SU(3)_W$. In addition to quarks and leptons with conventional charges, there are charge - 4/3 quarks and charge -2 leptons.

The moral of the above example is that it is nicer to generalize $SU(3)_c$ in order to have quark-lepton unification. Furthermore, our point of view is that, in the absence of strong interactions, quarks and leptons are indistinguishable with respect to the weak group G_W.

It turns out that the simplest extension of $SU(3)_c$ is SU(4) à la Pati and Salam[6] where leptons play the role of the fourth color. Therefore, we take $G_S = SU(4)$ and the petite unification group to be $SU(4) \times [SU(N)]^k$. The strong group SU(4) can be arranged to break down to $SU(3)_c \times U(1)_S$. The generator of $U(1)_S$ which is the 15th generator of SU(4), T_{15}, is proportional to B - L. In fact for a fermion representation which transforms as (4; N, 1,..., 1) under $SU(4) \times [SU(N)]^k$, one has

$$1/2\ (B - L) = \sqrt{2/3}\ T_{15} = \begin{pmatrix} 1/6 & & & 0 \\ & 1/6 & & \\ & & 1/6 & \\ 0 & & & -1/2 \end{pmatrix}, \tag{7}$$

where $\mathrm{Tr}T_{15}^2 = 1/2$. The generator of $U(1)_Y$, T_o, is then given by

$$T_o = \sqrt{2/3}\ T_{15} + \sum_\alpha C_{\alpha W}\ T^\circ_{\alpha W} \quad . \tag{8}$$

The quantity $C_S^2 = 2/3$ for this case. The fermion representation (4; N, $\overline{N}$, 1,..., 1) gives $C_S^2 = 8/3$. It turns out that for weak groups which give $\sin^2\theta^\circ_W = 1/4$, $C_S^2 = 8/3$ is unacceptable because it gives too small a value for $\sin^2\theta_W(M_W^2)$. Furthermore this case gives rises to higher charge fermions. It also turns out that a more detailed study of the constraints on fermions reveals that many weak groups which satisfy the constraints (4.a,b) are ruled out.

3. Minimal Petite Unification Model

The minimal model[5] which we study in more detail is $G_S \times G_W = SU(4) \times SU(2)_L \times SU(2)_R \times SU(2)_{L'} \times SU(2)_{R'}$. This model gives $\sin^2\theta^\circ_W = 1/4$ which upon using Eq. 3 predicts a value of $\sin^2\theta_W(M_W^2)$ consistent with experiment for the petite unification mass $< 10^8$ GeV. Fermions transform according to:

$\Psi = (4; 2, 1; 1, 1)_L$; $\Psi^* = (\bar{4}; 1, 2; 1, 1)_L$; $\Psi' = (4; 1, 1; 2, 1)_L$; $\Psi'^* = (\bar{4}; 1, 1; 1, 2)_L$. Symbolically,

$$\Psi = \underbrace{\begin{pmatrix} u_1 & d_1 \\ u_2 & d_2 \\ u_3 & d_3 \\ \nu & e \end{pmatrix}_L}_{\leftarrow SU(2)_L \rightarrow} \updownarrow SU(4) \quad , \qquad \Psi' = \underbrace{\begin{pmatrix} U_1 & D_1 \\ U_2 & D_2 \\ U_3 & D_3 \\ N & E \end{pmatrix}_L}_{\leftarrow SU(2)_{L'} \rightarrow} \updownarrow SU(4) \quad , \tag{9}$$

and similarly for Ψ^* and Ψ'^* corresponding to $SU(2)_R$ and $SU(2)_{R'}$ respectively. $\Psi^{(*)}$ are ordinary fermions transforming non-trivially under the left-right symmetric $SU(2)_L \times SU(2)_R$. A new feature emerges. It is the existence of a "mirror" left-right symmetric gauge group $SU(2)_{L'} \times SU(2)_{R'}$ and its "mirror" fermions $\Psi'^{(*)}$.

Why do we need the "mirror" fermions $\Psi'^{(*)}$? It is because we want to have a correct $\sin^2\theta_W^\circ$. In fact, from Eqs. (1.a,b) it is easy to derive

$$\sin^2\theta_W^\circ = [1 + \sum_\alpha C_{\alpha W}^2]^{-1} = TrT_{3L}^2/(TrQ^2 - C_S^2 TrT_{15}^2)_{\text{fermions}} . \tag{10}$$

Notice, in passing, that Eq. 10 reduces to Eq. 2 for the weak gauge bosons. Using Eq. 10, it is easy to see that if we had only the ordinary fermions $\Psi^{(*)}$, $\sin^2\theta_W^\circ = 1/2$ in contradiction with the value obtained using the adjoint representation of G_W.

How heavy could the mirror fermions be? They must be heavy enough to escape detection and light enough — i.e. $m_{\Psi'} < \tilde{M}$ with $\tilde{M}$ being the scale of $[SU(2)]^4$ breaking — to ensure the permutation symmetry among the SU(2)'s when $[SU(2)]^4$ is a good symmetry. It means that all the fermions (ordinary and "mirror") would have to contribute to the relevant renormalization group functions for $Q^2 > \tilde{M}^2$. It is then natural to expect the mirror fermions to populate the mass range between 20 GeV to a few TeV's (see below for more details on mass scale range).

Even with this minimal model some new interesting physics can occur. The first one is rare transitions induced by leptoquark gauge bosons which connect quarks to leptons. They are triplets under $SU(3)_c$ and carry charges ±2/3. In the breakdown of SU(4) to $SU(3)_c \times U(1)_S$, they gain masses of the order M. Of particular interest is the muon-number changing effective interactions for $Q^2 \ll M^2$. Making some kind of "kinship" hypothesis whereby one has (d_i, e^-), (s_i, μ^-), ... with i = 1, 2, 3, the effective Lagrangian describing $d\mu \to es$ is given by

$$\mathcal{L}_{eff}(d\mu \to es) = \sqrt{2}\, G_S \sum_{i=1}^{3} (\bar{d}_i \gamma_\mu e \bar{\mu} \gamma^\mu s_i + h.c.) \quad , \tag{11}$$

where $\sqrt{2}\, G_S = g_S^2/2M_G^2$ with g_S and M_G ($\approx$M) being the gauge coupling constant of SU(4) and mass of the leptoquark gauge bosons respectively. A Fierz-Michel rearrangement of Eq. 11 given an effective Lagrangian describing $ds \to \mu e$. We then use it to compute the branching ratio $B(K_L \to \mu e)$.

We compare $B(K_L \to \mu e)$ with $B(K_L \to \mu\bar{\mu})$. Using $B(K_L \to \mu e) < 2 \times 10^{-9}$ and $B(K_L \to \mu\bar{\mu}) = (9.1 \pm 1.8) \times 10^{-9}$, we obtain the following bound

$$\alpha_S^2(M_G)/M_G^4 \lesssim 10^{-24}\ \mathrm{GeV}^{-4} \quad , \tag{12}$$

where $\alpha_S(M_G) = g_S^2(M_G)/4\pi$. According to our "minimal" petite unification scheme, $\alpha_S(M_G) = \alpha_3(M_G)$ where $\alpha_3(M_G)$ is the $SU(3)_c$ coupling constant. Using the evolution equation for α_3 taking into account the effect of "mirror" fermion threshold, the bound (12) is translated into

$$M_G > 300\ \mathrm{TeV} \quad , \tag{13}$$

Not only do we have a lower bound on M_G ($\approx$M) from rare decays, but we also have an upper bound as well which comes from the experimental value of $\sin^2\theta_W(M_W^2)$. For M or M_G too large, $\sin^2\theta_W(M_W^2)_{Theor.}$ will not agree with $\sin^2\theta_W(M_W^2)_{exp.}$. The upper bound on M_G depends on

$\widetilde{M}$ ($\approx$ mass of W_R, $W_{L'}$, $W_{R'}$). In fact,

$$M_G < 10^8 \text{ GeV} \qquad \text{for } \widetilde{M} \approx 300 \text{ GeV} \quad , \tag{14.a}$$

$$M_G < 1000 \text{ TeV} \qquad \text{for } \widetilde{M} \approx 1 \text{ TeV} - 10 \text{ TeV} \quad . \tag{14.b}$$

Any value of M_G exceeding the bounds (14.a,b) is forbidden by $\sin^2\theta_W(M_W^2)_{exp.}$. An interesting feature emerges. The upper bounds on M_G imply lower bounds on $B(K_L \to \mu e)$. The bound (14.b) is of considerable interest since 300 TeV $< M_G <$ 1000 TeV implies

$$10^{-11} < B(K_L \to \mu e) < 10^{-9} \quad . \tag{15}$$

This range of $B(K_L \to \mu e)$ may be feasible experimentally. We make a plea to experimentalists to give some thoughts on how to improve the precision of these experiments.

In our model as it stands, the proton is stable. However by complicating the Higgs system, it is in principle possible to generate proton decay in higher orders.

Finally, I wish to say a few words about the "mirror" fermions. These "mirror" fermions are the exact duplicates of the ordinary fermions — same electric charges, same SU(4) interactions — except that they only have $SU(2)_{L'} \times SU(2)_{R'}$ weak gauge interactions. They do not couple directly to the electroweak $W_L^{\pm}$ bosons. We have mentioned earlier that they may be expected to populate the mass range between 20 GeV to a few TeV's. How do we distinguish the lightest among the charged "mirror" fermions from its ordinary counterparts? Let us take for example $E^{\pm}$ to be the lightest charged "mirror" leptons. They could be produced in reactions like $e^+ + e^- \to E^+ + E^-$. What distinguishes $E^{\pm}$ from just an ordinary heavy sequential lepton is the possibility that $E^{\pm}$ could live longer than one would expect from an ordinary weak decay. This could occur if there is no significant mixing between $W_L^{\pm}$ and $W_{L'}^{\pm}$, which could be induced by a Higgs representation which transforms as a (2,2) under $SU(2)_L \times SU(2)_{L'}$. Other induced decays e.g. through a (2,2) Higgs exchange could be expected to be small if the Yukawa coupling is itself small. It would be interesting to see if the next heavy lepton (if any) is something new or just another sequential lepton. Similar considerations could be applied to "mirror" quarks as well.

In summary, I have described here an attempt to fill up the desert, at least a few orders of magnitude beyond M_W. It remains to be seen if such a step is necessary but it is certainly worthwhile to contemplate effective theories describing various mass scales.

Because of the lack of space I apologize for not being able to give a complete list of references which can be found in Ref. 5.

Acknowledgment

This work was supported by the Director, Office of High Energy and Nuclear Physics, Division of High Energy Physics of the U.S. Department of Energy under Contract No. W-7405-ENG-48.

References

[1] S.L. Glashow, Nucl.Phys. 22 (1961) 579; A. Salam and J.C. Ward, Phys. Lett. 13 (1964) 168; S. Weinberg, Phys. Rev. Lett. 19 (1967) 1264; A. Salam, in Elementary Particle Theory, ed. N. Svartholm (Almquist and Wiksells, Stockholm, 1969) p. 367.

[2] For a complete list of contributors to these ideas, the reader is urged to consult the review paper of P. Langacker, SLAC-PUB-2544 (1980), to be published in Physics Reports.

[3] H. Georgi and S.L. Glashow, Phys. Rev. Lett. 32 (1974) 438.

[4] H. Georgi, Particles and Fields 1974, ed. C.E. Carlson (AIP, NY, 1975) p. 575; H. Fritzsch and P. Minkowski, Ann. Phys. 93 (1975) 193.

[5] P.Q. Hung, A.J. Buras and J.D. Bjorken, FERMILAB-Pub-81/22-THY (1981).

[6] J.C. Pati and A. Salam, Phys. Rev. D10 (1974) 275.

[7] H. Georgi, H.R. Quinn and S. Weinberg, Phys. Rev. Lett. 33 (1974) 451.

SUPERSYMMETRIC TECHNICOLOR

Mark Srednicki

Department of Physics, Princeton University
Princeton, New Jersey 08544

I will discuss some work I recently completed with M. Dine and W. Fischler on supersymmetric technicolor [1]. E. Witten [2] and S. Dimopoulos and S. Raby [3] have considered similar ideas.

Our central idea is to combine supersymmetry and technicolor to produce a "natural" theory which is capable of reproducing all the known phenomenology of particle physics, especially the quark-lepton mass spectrum and the absence of flavor changing neutral currents. Supersymmetry allows us to introduce fundamental scalars which are naturally light. Some of these scalars play the role of Higgs fields, and give mass to quarks and leptons via ordinary Yukawa couplings (which are chosen so that we get the correct masses and mixing angles). The supersymmetric partners of all known particles turn out to be too heavy to have been observed in experiments to date; many of them, however, weigh less than 100 GeV.

We want to generate all low energy symmetry breaking dynamically, in accord with the principle of naturalness. In particular, supersymmetry must break down dynamically. Fortunately, it is quite plausible that a supersymmetric version of QCD would dynamically break supersymmetry. In ordinary QCD we know that $\overline{\psi}\psi$ has a nonzero vacuum expectation value (VEV); in supersymmetric QCD, we can write

$$\{Q, \overline{A}\psi + \overline{\psi}A\} = \overline{\psi}\psi \qquad (1)$$

where A $(\overline{A})$ is the partner of ψ $(\overline{\psi})$, and Q is a supersymmetry generator. If we now assume

$$\langle\overline{\psi}\psi\rangle \neq 0 \quad , \qquad (2)$$

then Eq. (1) implies that

$$Q|0\rangle \neq 0 \quad , \tag{3}$$

i.e., supersymmetry is broken. Furthermore, the composite $\overline{A}\psi + \overline{\psi}A$ is the Goldstone fermion.

In our supersymmetric model, we assume the existence of a strongly interacting gauge sector, supercolor, for which $\langle\overline{\psi}\psi\rangle \simeq (10 \text{ TeV})^3$. This breaks supersymmetry. We let both right and left handed superquarks be doublets of $SU(2)_L$, and carry hypercharge +1. This ensures that the nonzero VEV of $\overline{\psi}\psi$ leaves $SU(2)_L$ and $U(1)_Y$ unbroken, but results in mass for the winos and bino (the fermionic partners of the $SU(2)_L$ and $U(1)_Y$ gauge bosons). These masses can be estimated; they are

$$m_w \simeq (e^2/\sin^2\theta_W)(10 \text{ TeV}) \tag{4}$$

$$\simeq 4 \text{ TeV} \tag{5}$$

$$m_b \simeq (e^2/\cos^2\theta_W)(10 \text{ TeV}) \tag{6}$$

$$\simeq 1.2 \text{ TeV} \quad . \tag{7}$$

Scalar fields will gain mass from one loop diagrams involving the now massive winos and bino. These masses range from 45 GeV to 470 GeV depending on the weak quantum numbers of the scalar. The 45 GeV scalars are the partners of right handed down quarks; each decays to its partner plus a gluino (the fermionic partner of the gluon) in 10^{-23} seconds.

The gluino gains mass via two loop diagrams involving a wino, ordinary quarks, and scalar partners of quarks. A very rough estimate of the gluino mass is 5 GeV. It decays to a Goldstone fermion and a gluon in about 10^{-10} seconds.

In order to break $SU(2)_L \times U(1)_Y$, we use a separate technicolor interaction. The Higgs fields get nonzero VEV's from their couplings to technicolored fields. These couplings also give masses of about 30 GeV to the charged and neutral higginos (the fermionic partners of the Higgs scalars). They decay to quark-antiquark-gluino in about 10^{-20} seconds. The branching ratio is largest for decay to the heaviest available quark-antiquark pair.

Our model does not have the usual strong CP problem. There are enough anomalous U(1) symmetries to rotate away the CP violating instanton angle in each strongly interacting sector (supercolor, technicolor, and color). The U(1) symmetry with a color anomaly is spontaneously broken by the supercolor condensates at 10 TeV. This results in a very weakly coupled axion with a mass of 1 keV. Such an axion is unobservable in the laboratory, but can probably be ruled out via astrophysical arguments. It may be possible to complicate the model and avoid the axion. In any case, the model's solution to the strong CP problem can be realized in other contexts, such as ordinary grand unified theories [4]. In these theories, the axion is very light and completely harmless.

I must mention that our model is actually unnatural as it stands; one mass parameter (associated with $U(1)_Y$) must be arbitrarily set equal to zero. However, Witten [2] has shown that if $U(1)_Y$ is unified into any non-Abelian group at any energy scale, then this mass parameter is automatically zero. Thus, a grand unified version of our model would solve its residual naturalness problem, and could prove to be an appealing theory of particle physics.

I am supported in part by NSF Grant No. PHY80-19754 at Princeton University.

[1] M. Dine, W. Fischler, and M. Srednicki, "Supersymmetric Technicolor," Princeton University preprint, to be published in Nuclear Physics B.

[2] E. Witten, "Dynamical Breaking of Supersymmetry," Princeton University preprint.

[3] S. Dimopoulos and S. Raby, "Supercolor," Stanford University preprint.

[4] M. Dine, W. Fischler, and M. Srednicki, "A Simple Solution to the Strong CP Problem with a Harmless Axion," IAS preprint.

SUPERSYMMETRIC GUTS

Savas Dimopoulos*†

Institute of Theoretical Physics
Stanford University

Institute of Theoretical Physics
University of California, Santa Barbara

University of Michigan, Ann Arbor

and

Howard Georgi**

Lyman Laboratory of Physics
Harvard University
Cambridge, Massachusetts 02138

We construct an explicit realistic SU(5) model in which softly broken supersymmetry is used to protect the Higgs doublets from quadratic mass renormalization. The model requires one natural but incredibly accurate adjustment of parameters. We argue that such an adjustment will be required in any supersymmetric GUT in which baryon number is not conserved.

In any theory in which supersymmetry is relevant down to 1 TeV, the supersymmetric partners of the leptons are spinless bosons with the right $SU(2) \times U(1)$ properties to be Higgs mesons. However, for various reasons, they cannot be. The most serious objection to using the lepton partners as Higgs mesons is that the partners of the charge -1/3 quarks then mediate baryon number changing interactions. To eliminate this difficulty, the Higgs doublets must be part of a dif-

*Address after January 1, 1982: Lyman Laboratory of Physics, Harvard University, Cambridge, MA 02138.

†This research is supported in part by the National Science Foundation under Grant No. PHY77-27084.

**This research is supported in part by the National Science Foundation under Grant No. PHY77-22864.

ferent SU(5) multiplet whose color triplet components are superheavy. To ensure that baryon number is conserved except by the usual SU(5) baryon number changing interactions, we require that the Higgs multiplets couple to pairs of quark and lepton multiplets just as in the usual SU(5) model.

We will call the quark and lepton supermultiplets matter supermultiplets. The above constraint can be restated by saying that the matter supermultiplets always couple in pairs to the Higgs supermultiplets.

In a supersymmetric theory, the supermultiplets carry a handedness, that of the fermions they contain. Only supermultiplets of the same handedness can have Yukawa couplings to one another. Thus the Higgs which give mass to charge 2/3 quarks and those which give mass to the charge -1/3 quarks cannot be charge conjugates of one another--the handedness of associated fermions must be the same. We need two left-handed Higgs doublet supermultiplets, one with average charge 1/2 which gives mass to the charge 2/3 quarks and one with average charge -1/2 which gives mass to the charge -1/3 quarks.

We next review the subject of coupling constant renormalization in a supersymmetric SU(5) theory. The basic difference between supersymmetric SU(5) and the standard SU(5) theory is that the supersymmetric partners of the light gauge bosons are not superheavy. This makes the SU(3), SU(2), and U(1) gauge couplings approach one another more slowly than in the usual theory, so the unification scale is increased. This doesn't affect the prediction of the weak mixing angle. But the presence of extra light Higgs doublets does tend to increase $\sin^2\theta$, as we will see quantitatively below. We will also want to keep track of the number of matter multiplets to check that the gauge couplings do not get large before unification.

Call the number of Higgs doublet supermultiplets h and the number of matter doublet supermultiplets m (there are four doublets for each generation of quarks and leptons, so $m = 12$ for three generations). Then the gauge couplings at M_W satisfy

$$\begin{aligned}
\frac{2\pi}{\alpha_s} &= \frac{2\pi}{\alpha_0} + (9-m/2)\,\ln\frac{M_W}{M_u} \\
\sin^2\theta\,\frac{2\pi}{\alpha} &= \frac{2\pi}{\alpha_0} + (6-h/2-m/2)\,\ln\frac{M_W}{M_u} \qquad (1)\\
\frac{3}{5}\cos^2\theta\,\frac{2\pi}{\alpha} &= \frac{2\pi}{\alpha_0} + (-3h/10-m/2)\,\ln\frac{M_W}{M_u}
\end{aligned}$$

where α_0 is the gauge coupling at unification and M_u is the mass of the leptoquark gauge supermultiplets. Solving (1), we find

$$\frac{1}{\alpha_0} = \frac{(9-m/2)/\alpha - (6-h-4m/3)/\alpha_s}{18+h},$$

$$\ell n \frac{M_u}{M_W} = \frac{2\pi/\alpha - 16\pi/3\alpha_s}{18+h}, \tag{2}$$

$$\sin^2\theta = \frac{(3+h/2) + (10-h/3)\alpha/\alpha_s}{18+h}.$$

Inserting a reasonable value for the ratio α/α_s at M_W, $\alpha/\alpha_s = 5.7 \times 10^{-2}$ (corresponding roughly to $\Lambda = .3$), we find for $h = 2$

$$\sin^2\theta = .226$$

$$M_u \simeq 10^{17} \text{ GeV}. \tag{3}$$

These results are independent of m, as usual in a one loop calculation. But as (2) shows, m is highly constrained by the constraint of perturbative unification, the requirement that α_0 be small. α_0 is of order one if $m = 20$. Thus at most five generations are allowed by perturbative unification.

A general supersymmetric Yang-Mills theory is characterized by specifying the gauge group and gauge coupling constants, the representation under which all the left-handed supermultiplets transform, and a single gauge invariant function v of the left-handed fields. The following notation is very convenient [3]: Let ψ^a (A=1 to N) be the left-handed fermion fields associated with the scalar supermultiplets and ϕ^a be the corresponding complex spinless meson fields. Let A^μ_α be the gauge fields (where μ is a vector index and α labels the generators of the gauge group) and χ_a be their supersymmetric partners which we can take to be left-handed fermion fields. The covariant derivative on the ψ and ϕ fields is

$$D^{\mu a}_b = \partial^\mu \delta^a_b + i\, g_\alpha A^\mu_\alpha T^a_{\alpha b} \tag{4}$$

where T_α is the gauge generator and g_α is the gauge coupling constant. The gauge invariant function is $v(\phi)$. We will also need the derivatives:

$$v_a(\phi) = \frac{\partial}{\partial\phi^a} v(\phi),$$

$$v_{ab}(\phi) = \frac{\partial^2}{\partial\phi^a\partial\phi^b} v(\phi), \tag{5}$$

$$v_{abc}(\phi) = \frac{\partial^3}{\partial\phi^a \partial\phi^b \partial\phi^c} v(\phi).$$

Gauge invariances of v implies

$$v_a(\phi)\, T^a{}_b \phi^b = 0 \tag{6}$$

(summation over repeated indices implies). The penultimate pieces of notation we need is for the complex conjugates of ϕ^a;

$$\bar{\phi}_a = \phi^{a*}$$

$$\bar{v}(\bar{\phi}) = v(\phi)^*$$

$$\bar{v}^a(\phi) = v_a(\phi)^* \tag{7}$$

etc.

Finally, we define the functions

$$K_\alpha = g_\alpha \bar{\phi}_a T^a_{\alpha b} \phi^b \tag{8}$$

(no sum on α) with derivatives defined as in (5) and (7).

To put all this together into the Lagrangian for the supersymmetric theory, it is convenient to work in a Majorana basis for the fermion fields. The kinetic energy terms and the gauge couplings are the usual ones with the covariant derivatives given by (4). The Yukawa couplings and fermion mass terms are

$$-\frac{1}{2}\psi^{aT}\gamma^0 v_{ab}(\phi)\psi^b + \text{h.c.}$$

$$-\chi_\alpha \gamma^0 K_{\alpha a}(\bar{\phi})\,\psi^a + \text{h.c.} \tag{9}$$

The scalar meson potential is

$$v(\phi,\bar{\phi}) = v_a(\phi)\,\bar{v}^a(\bar{\phi}) + \frac{1}{4}K_\alpha K_\alpha. \tag{10}$$

The vacuum expectation value (VEV) of the ϕ's is determined by minimizing (10). In most cases, the minimum will occur when

$$v_a(\phi) = \bar{v}^a(\bar{\phi}) = K_\alpha = 0. \tag{11}$$

If (11) can be satisfied, then although the gauge symmetries may be spontaneously broken, the supersymmetry remains unbroken. It is sometimes possible to arrange $v(\phi)$ so that (11) is not allowed, so that (10) has its minimum for nonzero values $v_a(\phi)$ and K_α. In this case the supersymmetry is broken spontaneously. We will discuss spontaneously broken supersymmetry further in Reference [9]. We will not pursue it further here because, as we shall see, it does not lead to a sensible phenomenology in tree approximation.

From (9-11) it is clear that if the theory is to be renormalizable by conventional power counting, $v(\phi)$ can be at most cubic in the fields. We will assume that this constraint is satisfied.

If this were an ordinary field theory, we would expect quantum loop corrections to renormalize the parameters in $v(\phi)$ and to generate effective (finite) nonrenormalizable terms in v. The weird, astonishing, unique thing about supersymmetric theories is that these things do not happen. The parameters in $v(\phi)$ are modified by radiative corrections only through wave function renormalization [1].

We can now describe a simple SU(5) model which illustrates the first half of our program; the breaking of SU(5) down to SU(3) x SU(2) x U(1) preserving supersymmetry and producing a pair of massless Higgs doublets.

As noted above, we can characterize the theory by giving the SU(5) transformation properties of the fields and specifying the SU(5) invariant function v. The left-handed supermultiplets are the following (all of these correspond to the ϕ^a of section III):

$$H^x,\ H'_x:$$
$$M_j^{xy} = -M_j^{yx},\ M'_{jx};$$
$$\Sigma^x_y. \qquad (12)$$

The x and y are SU(5) indices and j is a flavor index. H^x is a 5 and H'_x is a $\bar{5}$. The SU(2) doublet components of these are the Higgs supermultiplets. M_j^{xy} are 10's and M_{jx} are $\bar{5}$'s. These are the matter supermultiplets. Σ^x_y is a 24 whose VEV will break SU(5).

The function v is given by

$$\lambda_1\left(\frac{1}{3}\,\Sigma^x_y\Sigma^y_z\Sigma^z_x + \frac{m}{2}\,\Sigma^x_y\Sigma^y_x\right)$$
$$+ \lambda_2\, H'_x\left(\Sigma^x_y + 3m'\delta^x_y\right)$$
$$+ f_{jk}\,\varepsilon_{uvwxy}\, H^u\, M_j^{vw} M_k^{xy}$$
$$+ g_{jk}\, H'_x\, M_j^{xy} M'_{ky}. \qquad (13)$$

We can now apply the formalism to (13) with the following results: the potential is minimized for

$$\langle H^x\rangle = \langle H'_x\rangle = \langle M_j^{xy}\rangle = \langle M_{jx}\rangle = 0, \qquad (14)$$

and one of three VEV's for Σ;

$$\langle\Sigma^x_y\rangle = 0, \tag{15a}$$

$$\langle\Sigma^x_y\rangle = \frac{m}{3}\,\delta^x_y - \frac{5m}{3}\,\delta^x_5\,\delta^5_y, \tag{15b}$$

$$\langle\Sigma^x_y\rangle = 2m\,\delta^x_y - 5m\left(\delta^x_4\,\delta^4_y + \delta^x_5\delta^5_y\right) \tag{15c}$$

Supersymmetry is not broken. (15a) does not break SU(5). (15b) breaks it down to SU(4) x U(1). (15c) breaks it down to SU(3) x SU(2) x U(1). At this point, these three physically inequivalent vacua are completely degenerate. This bizarre situation is not uncommon in supersymmetric theories (it may be cosmologically relevant!).

In any event, (15c) is the solution of interest. For this solution, the SU(2) doublet Higgs $H^{4,5}$ and $H'_{4,5}$, have a mass squared which is proportional to $m - m'$. If we set m' exactly equal to m, they are massless, while the color triplet components $H^{1,2,3}$ and $H'_{1,2,3}$ have mass of order m. This is the result we want, so we require in (13)

$$m' = m = O(M_u). \tag{16}$$

It may seem that we have accomplished nothing by introducing supersymmetry. We are still required to make an incredibly accurate adjustment of the parameters in our theory. But there is at least a technical difference between (16) and the adjustments we have to make in a nonsupersymmetric theory. (16) need only be imposed once, at the very beginning, because the ratio m'/m is not renormalized at all. Both m' and m are renormalized like the Σ field. Thus the imposition of (16) is natural in a technical sense [4].

At momenta small compared to m, the theory collapses into an effective SU(3) x SU(2) x U(1) supersymmetric gauge theory, in which the only relevant fields are the matter supermultiplets (the debris of M^{xy}_j and M_{jx}), the SU(3) x SU(2) x U(1) gauge supermultiplets, and the two massless Higgs supermultiplets. The matter fields are massless because they are protected by an unbroken chiral gauge symmetry of the full theory. The gauge symmetry also protects the gauge fields. The Higgs fields are massless because we made them massless. It is true that in the effective low energy theory there is a global chiral symmetry that protects the Higgs fields. But as the above analysis shows, it cannot be a symmetry of the full theory. It is broken by the interactions which give large mass to the color triplet relatives of the Higgs doublets. If we did not have to make the color triplets heavy, we could take $\lambda_2 = 0$. Then both the doublets and the triplets would be massless in the supersymmetric theory. They would be protected by a global chiral symmetry of the full theory. In a standard SU(5) theory, the triplets must be very heavy because they mediate baryon number violating processes. We expect a similar situation in any unified theory in which the Higgs doublets are unified with fields that can produce baryon number changing effects. It is conceivable that in a theory with global baryon number conservation the entire multiplet including the Higgs doublets could be kept massless by a global symmetry, without any adjustments at all.

At this point, except for the multiplicity of Higgs, we have a simple doubling of the degrees of freedom. Degenerate with the twelve SU(3), SU(2), and U(1) gauge bosons are twelve massless gauge fermions. The matter fermions are accompanied by matter bosons with the same quantum numbers, and the Higgs bosons, with Higgs fermions.

One might worry that the matter bosons would have large baryon number violating self interactions because a four-boson term, unlike a four-fermion term, is not necessarily suppressed by powers of M_u^{-1}. In particular, the terms in the potential of the form $K_\alpha K_\alpha$ where α runs over the leptoquark gauge bosons in SU(5) contain baryon number violating four-matter-boson interaction terms. But the contribution of these terms is cancelled by a similar effect from the coupling (again in the $K_\alpha K_\alpha$ term) of the matter bosons to the massive boson in the massive vector supermultiplet with the A_α^μ field. Indeed, the $K_\alpha K_\alpha$ term is essentially just a mass term for this massive boson, and it leaves no trace in the effective low energy theory. This, of course, is what we expected. The supersymmetry of the effective low energy theory should rule out any $K_\alpha K_\alpha$ term except those associated with the unbroken $SU(3) \times SU(2) \times U(1)$ gauge symmetries. But it is nice to see the unwanted terms disappearing explicitly.

In the simple model discussed above, spinless fields do not develop a supersymmetry breaking VEV at any mass scale. However, we could modify the model so that the effective low energy theory would break supersymmetry spontaneously at around a TeV. The question is, do we get a realistic theory with such a strategy? The answer is no, at least not in tree approximation. This can be seen from the following theorem.

Theorem: In an SU(3) x SU(2) x U(1) theory of quarks and leptons with spontaneously broken supersymmetry, the tree approximation mass matrix describes a color triplet spinless meson of one of the following kinds; a charge 2/3 triplet not heavier than the lightest charge 2/3 quark, or a charge -1/3 triplet not heavier than the lightest charge -1/3 quark. We will not prove this here. Loosely speaking, it is true because spontaneous supersymmetry breaking leaves the fermion mass matrix unchanged while it splits the spinless mesons in opposite directions [9].

Thus if we break supersymmetry spontaneously, we must rely on radiative corrections to give a large mass to the supersymmetric partners of the quarks. This is only reasonable if some of the couplings get large, as in a supercolor theory [5].

Here we explore a simpler alternative possibility--that the supersymmetry is broken explicitly, but softly, by terms of dimension less than four in the Lagrangian. We add to the Lagrangian (not to v) the following SU(5) invariant mass terms, all of the order of a TeV:

(1) A positive mass squared term for the matter bosons;

(2) A mass for the Higgs fermions (and their SU(5) partners);

(3) A mass for the Gauge fermions (and their SU(5) partners);

(4) A negative mass squared term for the boson fields in the Σ supermultiplet;

(5) A mixed (with positive and negative eigenvalues) mass squared matrix for the Higgs bosons.

The effect of (1) - (3) is simply to increase the mass of unobserved particles which would otherwise be light or massless. (4) does not do much to the low energy theory, but it does pick out the SU(3) x SU(2) x U(1) symmetry breaking as the lowest energy state of the full theory. (5) produces a nonzero VEV for the Higgs bosons. The form of the Higgs boson mass matrix must be chosen with care to assure that the Hamiltonian is bounded below because the Higgs boson potential in the effective low energy theory is flat in some directions. Also, the supersymmetric low energy Lagrangian has a Peccei-Quinn symmetry which can be broken by the explicit boson mass squared term to eliminate the axion.

We see no essential difference between the soft breaking of supersymmetry and the soft breaking of an ordinary symmetry. The Symanzik theorem ought to apply to either case [6].

Grisaru has given an elegant argument that a scalar meson mass term is a consistent soft breaking by showing that it is equivalent to a coupling to an external vector supermultiplet with a constant auxiliary field component [7]. We know of no similar argument for a fermion mass term, but we think of no reason why it should not make sense.

It is amusing that the two mass scales in this theory arise from completely different physics. The unification scale, M_u, is essentially the mass parameter in the supersymmetric theory. The $SU(2)\times U(1)$ scale, M_W, is determined by the scale of the explicit soft supersymmetry breaking.

The phenomenology of this model is simple. In addition to the usual light matter fermions, gauge bosons and Higgs bosons, we predict heavy matter bosons, gauge fermions and Higgs fermions as supersymmetric partners. We can say little about their masses except that they cannot be very large compared to 1 TeV or the motivation for the model disappears.

The lightest of the supersymmetric partners is stable. The others all decay into it plus ordinary particles. One simple possibility is to imagine that the lightest particle is the U(1) gauge fermion. In this respect, the model differs somewhat from a supercolor model with spontaneously broken supersymmetry in which all the supersymmetric particles can decay into their ordinary partners plus a Goldstone fermion.

One interesting sidelight is the mechanism by which the model avoids large flavor changing effects from loops involving the heavy particles. A potentially dangerous diagram is the $\Delta S = 2$ superbox diagram of Figure (1). So long as we are careful to break the supersymmetry by adding a common mass to all the matter bosons with the same quantum numbers, this diagram is suppressed by a super GIM mechanism. Like the usual box diagram, it is proportional to the square of quark mass differences which are the same as those of their supersymmetric partner matter bosons. In fact, this diagram is completely negligible.

In one respect, the simple model of section III is inadequate to describe the physics we see. As in any SU(5) theory in which the Higgs mesons transform as part of a (5), the charge -1/3 quark and charged lepton masses are equal at the unification scale. In this model, this is bad for at least one and probably two reasons. The most serious is the b/τ mass ratio. Here with so many color states in the low energy theory, the predicted b/τ ratio is about 6 (for three generations),

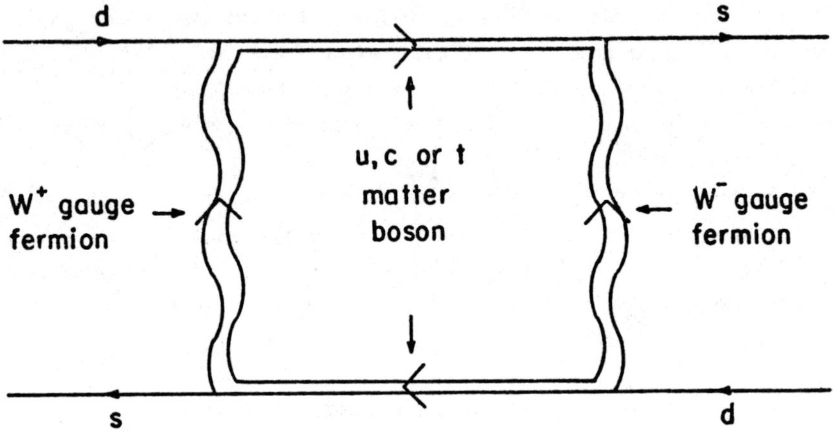

Figure 1

The $\Delta S = 2$ superbox diagram. The virtual particles are the supersymmetric partners of those in the ordinary box diagram.

much larger than it is in the standard model. This is unacceptable. Also, probably unacceptable is the fact that it predicts $b/\tau=s/\mu=d/e$.

The first difficulty can be lessened if the Higgs which gives mass to the charge -1/3 quarks and the charged leptons is part of a $\overline{45}$. This is easy to arrange by a simple generalization of the model of section IV with a 45 and $\overline{45}$ in addition to the 5 and $\overline{5}$. In this generalization the light Higgs doublets come from the 5 and the $\overline{45}$.

If the second difficulty is real, then the light Higgs which gives mass to the charge -1/3 quarks and charged leptons must be a linear combination of the doublets in a $\overline{5}$ and $\overline{45}$. But they probably cannot both be light. That would give an unacceptably large contribution to $\sin^2\theta$. We can, however, construct the theory so that the Higgs doublet which is light is a linear combination of the doublet components of the $\overline{5}$ and $\overline{45}$. The orthogonal component would be superheavy and not get any VEV. Again, this can be arranged. Although the theory is somewhat cumbersome, the idea is exactly the same as in the simple model. The fine adjustments required are all natural.

We have shown that supersymmetry can be incorporated into a unified theory in an interesting way which makes use of the special renormalization properties of supersymmetric theories. We have stressed that while we can use supersymmetry to make the Higgs doublets massless, the masslessness is not automatic or trivial. It requires an exact but natural adjustment of parameters.

In discussing the gauge hierarchy problem, Weinberg said with some justice that the required adjustment of parameters was unnatural but not unattractive [8]. With even more justice, we can say that the adjustment we must make in this theory is unattractive but not unnatural.

In the specific model we discuss, the SU(2) x U(1) breaking scale is introduced through the explicit supersymmetry breaking. It may be possible to introduce this scale dynamically through spontaneous symmetry breaking in a supercolor model or perhaps in some other way [1]. We hope that the explicit breaking scheme we have discussed will be useful, at least as an illustration of how easy it is to use supersymmetry in a realistic model.

The most important moral of the whole business is the comment that the observation of proton decay can demonstrate convincingly that supersymmetry is not relevant at energies anywhere near 1 TeV.

References

[1] For a review of supersymmetry and references to the literature, see E. Witten, "Dynamical Breaking of Supersymmetry", Princeton Preprint (1981).

[2] S. Dimopoulos, S. Raby, and F. Wilczek, to be published.

[3] The notation here is similar in spirit to that of S. Ferrara, L. Girardello and F. Palumbo, Phys. Rev. D 20, 403 (1979).

[4] This is naturalness in the sense of no infinite renormalization, as in H. Georgi and A. Pais, Phys. Rev. D 10, 539 (1974) and references 2-8 therein, rather than naturalness in the sense of G. 't Hooft, lecture at Cargése Summer Institute, 1979, in "Recent Developments in Gauge Theories", ed. by G. 't Hooft *et al*. (Plenum, 1980).

[5] S. Dimopoulos and S. Raby, "Supercolor", Santa Barbara Preprint (1981);

M. Dine, W. Fischler and M. Srednicki, "Supersymmetric Technicolor", Princeton Preprint, 1981.

[6] K. Symanzik, in *Coral Gables Conference on Fundamental Interac-at High Energies II*, edited by A. Perimutter, G.J. Iverson, and R.M. Williams (Gordon and Breach, New York, 1970);

S. Coleman, in Proceedings of the 1971 International School of Physics "Ettore Majorana" (Academic, New York, to be published).

[7] M. Grisaru, private communication.

[8] S. Weinberg, private communication in discussions of his paper in Phys. Letters 82B, 387 (1979).

[9] S. Dimopoulos and H. Georgi, Harvard University preprint HUTP-81/A022.

CONFERENCE SUMMARY

Steven Weinberg

Lyman Laboratory of Physics
Harvard University
Cambridge, Massachusetts 02138

This has been a very stimulating meeting, and I think that I can speak for all the participants in this workshop in thanking the members of the Organizing Committee and also Barbara Nieman for their excellent arrangements. I especially want to express my own agreement with the practice here of interleaving theoretical and experimental talks. Even though experimentalists are not always interested in the haggles over which theories are natural, and the theorists are not always interested in where to put sump pumps, at least the format of this workshop has kept us within each other's fiducial volumes.

I will attempt to summarize this Workshop under three headings. First, theoretical expectations: matters concerning which we are not sure, but about which we feel we have some basis for our guesses. Next, the present and forthcoming experiments on proton decay. Finally, the deep problems that will still be with us whatever these experiments may find.

In going over theoretical expectations, I will start with the points about which I feel we can have the most confidence. Then I will move in stages to matters concerning which in my opinion there is a larger element of uncertainty.

The point which seems to me most plausible is that baryon and lepton number are not conserved. I say this for a number of reasons which have come into prominence at various points of history. First of all, baryon and lepton number are not unbroken gauge symmetries. This we know for sure because we don't see the long-range forces that would be associated with them. At this meeting, Pati has considered the possibility that at least B-L is a local symmetry which is spontaneously broken. Also, Glashow here considered the possibility that B-L is a global symmetry that is spontaneously broken; as he showed, this would have some interesting astrophysical implications. But in any case, if baryon and lepton number are symmetries, they are broken.

Very closely related to the first point is the fact that since baryon and lepton number are not unbroken gauge symmetries, it's possible for anomalies to violate them. In fact we know that triangle anomalies do violate baryon and lepton conservation, though this violation is much too small to be of any practical interest. (At least that's what I always thought. Baluni here in a discussion session raised the possibility that the violation of baryon and lepton conservation due to instantons may actually produce an experimentally observable degree of violation.)

Another reason for believing that baryon and lepton number are not conserved is the fact that we're here, that the universe appears (although it's not certain) to have an excess of baryons over antibaryons. This could be understood if baryon number were violated in the early universe. Details of how this could happen were discussed here by Yoshimura and also by Turner.

Finally, there isn't really any phenomenological need for baryon and lepton conservation as fundamental conservation laws. The situation here is very similar to that of strangeness conservation. You know, strangeness was discovered through the fact that strange particles are only produced in association, with positive strangeness cancelling negative strangeness, and one might have supposed that strangeness was a fundamental symmetry of nature. In fact for a long time one did imagine just that and it was somewhat mysterious why strangeness conservation was violated by weak interactions. Then it was realized when quantum chromodynamics came along that there's simply no way the theory of strong (and electromagnetic) interactions could be complicated enough to violate strangeness. We don't put strangeness into our theory of strong interactions as an input. The theory just can't violate the conservation of strangeness or charm or even for that matter of charge conjugation. So these symmetries are now explained. They are not needed as fundamental symmetries, and that of course makes it immediately understandable why they are violated by other interactions like the weak interactions, because they were never there in the first place in any fundamental sense. I think the same is true of baryon and lepton conservation. Baryon and lepton conservation on the phenomenological level are explained in an $SU(3) \times SU(2) \times U(1)$ gauge theory of strong and electroweak interactions as long as the only fermions which exist are particles of familiar type like quarks and leptons. Or to put it another way, the approximate conservation of baryon and lepton number is explained if the only particles which are not of the familiar type are very heavy. I will call them generically exotic particles, meaning any particles which do not have the usual SU(3) × SU(2) × U(1) quantum numbers. If all exotic particles are heavy then the approximate conservation of baryon and lepton number at low energy is automatic, so we don't need it as a fundamental symmetry, and therefore it is entirely plausible that it's not a fundamental symmetry, and is in fact violated just as strangeness is violated.

In order to explain the observed conservation of baryon and lepton number to a sufficiently high degree of accuracy, you need the exotic particles to be very heavy indeed. There are at least two good reasons I think for suspecting that there is in fact a new scale of energies vastly higher than any we study in our accelerators. One is the phenomenon of gravitation which, as pointed out in 1900, tends to make you suspect that there's a scale in physics of 10^{19} GeV. And the other has to do with the strong and electroweak interactions, which if they are to be unified at all can only be unified at very, very high energies, where the slow decrease of the strong interaction coupling constant brings this coupling to some kind of equality with the electroweak couplings. There is a broad class of theory to which I will come back in which the energy at which the couplings become equal is of the order of 10^{15} GeV. Furthermore, the theories which have been developed to embody a unification of the electroweak and the strong interactions do in fact all contain superheavy exotic particles which have exotic SU(3) × SU(2) × U(1) quantum numbers and therefore naturally violate baryon-lepton conservation.

To some extent my attitude toward baryon conservation is a little different from the one, for example, that Pati expressed here. It's a question of whether or not you think it's the job of physics to explain symmetries or to explain their absence. I personally think that symmetries have to be explained and if there's no good explanation for them you shouldn't believe them. Gauge theories are almost the only kind of sensible dynamics we can imagine for relativistic quantum field theories, and they lead to symmetries -- some exact and some approximate. As far as I'm concerned if the symmetry isn't explained in some way like that we should be very suspicious about it. Pati, on the other hand, takes the point of view, which is interesting and worth pursuing, that you should look for the maximum possible symmetry and then, whatever you don't see in nature, explain how it got broken.

Next, and with somewhat less assurance, I think we can also expect that baryon and lepton conservation are violated predominantly by those operators in an effective Hamiltonian which have minimum dimensionality. The reason for the condition of minimum dimensionality is that an operator of dimensionality $[\text{mass}]^d$ is likely to have a coupling roughly of order M^{4-d}, where M is the very large (10^{15} GeV?) mass of the exotic particles, so the only terms we have a chance of detecting are those for which d is as small as possible. It has been pointed out that for baryon nonconservation the lowest dimensional terms which respect the $SU(3) \times SU(2) \times U(1)$ symmetries of the strong and electroweak interactions all have $d = 6$, and take the form

$$\begin{aligned} \mathcal{H}_{eff} = \; & g_1\, d_R\, u_R (u_L\, e_L - d_L\, \nu_L) + g_2\, d_L\, u_L\, u_R\, e_R \\ & + g_3\, d_L\, u_L (u_L\, e_L - d_L\, \nu_L) + g_4\, d_R\, u_R\, u_R\, e_R \\ & + \mu \text{ and } \tau \text{ lepton terms } + \\ & + \text{strange particle terms} \end{aligned} \tag{1}$$

(In the notation I am using here, d_L (d_R) is the two-component field which destroys a left (right) -handed d quark and creates a right (left) -handed anti-d quark, and likewise for u, e, and ν. Color and spin indices are not explicitly shown, but in fact for each term there is only one rather obvious way that these indices can be contracted: the color indices α, β, γ on the three quark fields must be contracted with the totally antisymmetric SU(3) tensor $\varepsilon_{\alpha\beta\gamma}$, and the two-component spin indices must be contracted with the totally antisymmetric SU(2) tensor ε_{mn}, separately for fields of L type and R type. In the third and fourth terms, where there are four fields of the same type, Fermi statistics still only allows the spin indices to be contracted in just one way.)

Leaving aside strange particles and μ and τ type leptons, the effective Hamiltonian contains just four independent terms. This is I think closely analogous to the sort of $S + V + T + A + P$ effective Hamiltonian that was used for a quarter century in studies of nuclear beta decay. I would like to urge here that experiments on proton decay be phenomenologically analyzed in terms of these four unknown coupling constants $g_1 \ldots g_4$, in much the same way that beta decay was analyzed in terms of $C_S \ldots C_P$ until the form of the effective Hamiltonian was finally worked out in the late 1950's.

Even without knowing the values of the coupling constants g_i, Eq. (1) has a number of important empirical implications. The most striking of these is the conservation of B-L, the fact that proton decay produces antileptons and not leptons. Proton decay also conserves isospin. This is sometimes called the $\Delta I = 1/2$ rule, but if you think of the neutrino-electron doublet as an isospin doublet, and you make a simple correction for the fact that you have right-handed as well as left-handed electrons, then (1) is an isospin-invariant interaction, so you can use isospin selection rules in a very simple way with proton decay.

A still less certain further assumption is that the effective baryon nonconserving interaction is produced by the exchange of exotic vector rather than scalar bosons. Since vector fields couple to bilinears like $\psi_L\psi_R$, vector exchanges can only produce the g_1 and g_2 terms in (1), and not the g_3 and g_4 terms. Then if we do not measure parity-odd effects (like longitudinal polarizations) which can distinguish left- and right-handed leptons, the decay rates for all e^+ and $\bar{\nu}_e$ modes will be independent of the coupling constants g_i, except for common factors $|g_1|^2 + |g_2|^2$ and $|g_1|^2$ in all e^+ and $\bar{\nu}_e$ modes, respectively.

Table 1 gives relative decay rates for some proton and neutron decay modes, calculated from the effective Hamiltonian (1) with only vector exchange (i.e. $g_3 = g_4 = 0$), according to various versions of the quark model. This table was prepared by taking the ratios of the rates for the four modes $p \to \pi^0 e^+$, $p \to \rho^0 e^+$, $p \to \eta e^+$, and $p \to \omega e^+$ (which should be independent of the details of the underlying mechanism for baryon nonconservation, as long as it is based on vector exchange) from the calculations summarized here by Langacker, and then using isospin to determine all other relative decay rates. Within the context of the general assumptions I have outlined here, the preparation of such a table of relative rates is strictly a problem of strong interaction physics -- it does not depend at all on ideas of grand unification, and the branching ratios do not depend on any unknown parameters describing the mechanism for baryon violation, except for the factor $N \equiv |g_1|^2/|g_2|^2 + |g_2|^2$ which gives the ratio of $\bar{\nu}_e$ to e^+ modes. Table 1 shows that the answers that the different authors get do not agree with each other, and I suppose you can take the spread in their answers as some kind of indication of how difficult the strong-interaction part of such calculations really is. Yet even if the strong interaction part of the

Mode	Calculated Relative Rates							
	M	GYOPR	D	G	DGS	KK1	KK2	KK3
$p \to e^+\pi^0$	0.39	0.58	0.10	0.16	0.41	0.55	0.54	0.51
$p \to e^+\rho^0$	0.20	0.03	0.24	0.25	0.28	0.03	0.10	0.15
$p \to e^+\eta$	0.14	0.11	0.03	0.001	0.07	0.11	0.02	0
$p \to e^+\omega$	0.26	0.28	0.63	0.58	0.25	0.32	0.34	0.35
$p \to \bar{\nu}_e\pi^+$	0.78N	1.16N	0.20N	0.32N	0.82N	1.10N	1.08N	1.02N
$p \to \bar{\nu}_e\rho^+$	0.40N	0.06N	0.48N	0.50N	0.56N	0.06N	0.20N	0.30N
$n \to e^+\pi^-$	0.78	1.16	0.20	0.64	0.82	1.10	1.08	1.02
$n \to e^+\rho^-$	0.40	0.06	0.48	0.50	0.56	0.12	0.20	0.30
$n \to \bar{\nu}_e\pi^0$	0.39N	0.58N	0.10N	0.16N	0.41N	0.55N	0.54N	0.51N
$n \to \bar{\nu}_e\rho^0$	0.20N	0.03N	0.24N	0.25N	0.28N	0.03N	0.10N	0.15N
$n \to \bar{\nu}_e\eta$	0.14N	0.11N	0.03N	0.001N	0.07N	0.11N	0.02N	0
$n \to \bar{\nu}_e\omega$	0.26N	0.28N	0.63N	0.58N	0.25N	0.32N	0.34N	0.35N

Table 1. Relative Rates for Nucleon Decay

All rates here are expressed in units of the total rate for proton decay into a positron and any non-strange neutral meson. The first four relative rates for $p \to e^+\pi^0$, $p \to e^+\rho^0$, $p \to e^+\eta$, and $p \to e^+\omega$ are taken from the dynamical calculations of various authors (labelled M, GYOPR, etc.), as summarized here in Langacker's talk. The remaining relative rates are calculated from those using the isospin symmetry of the effective Hamiltonian for nucleon decay. All results presented in this table (including those taken from the calculations summarized by Langacker) do not depend on any hypotheses about the mechanism for nucleon decay, beyond the assumption that the decay is produced by the exchange of very heavy vector bosons. Under this assumption, any dependence on specific models of grand unification can enter only through the value of the parameter N, defined as $|g_1|^2/|g_1|^2 + |g_2|^2$.

calculations could be done with precision, the comparison of the theoretically predicted relative rates of Table 1 with experiment would not tell us very much about the underlying mechanism of baryon violation — it would only serve to determine the single parameter N or $|g_1|^2/|g_2|^2$, and of course to check the guess that baryon decay is due to exchange of vector bosons of very high mass. With regard to the last point, I would like to urge that the calculations quoted here by Langacker, and indeed all calculations regarding proton decay, be redone using the full Hamiltonian (1), so that we can see what sort of experiment might be sensitive to the presence of the scalar-exchange coupling g_3 and g_4. I'll return to this point.

Now I come to strong-electroweak unification. As I already mentioned, this implies a certain idea of the scale of superheavy masses. There is a simple formula for the superheavy mass scale, which was originally derived for grand unified theories of a certain standard class. (A rough version of this formula, useful as a mnemonic, is that the superheavy scale is about $\exp(4\pi^2/11e^2)$ times the QCD scale parameter Λ.) This is the class of theories which, first, have some simple group unifying strong and electroweak interactions, which breaks in one step to $SU(3) \times SU(2) \times U(1)$ and perhaps hypercolor. And second, the $SU(3) \times SU(2) \times U(1)$ generators are normalized by assuming that the simple group has at least one multiplet which contains only fermions of conventional types plus neutrals and maybe a few others. In this class of theories, as was discussed here with all the fine corrections and details by both Langacker and Marciano, the superheavy mass scale M comes out to be about 1.5×10^{15} times the QCD scale Λ, and $\sin^2\theta$ is 0.21 and is very insensitive to anything. As Marciano summarized, the experimental value of $\sin^2\theta$ that is relevant to this comparison is 0.215 ± 0.012. This situation, the comparison between theory and experiment for $\sin^2\theta$, has gone from being embarrassingly bad to being suggestively close with a possible small discrepancy, and now it is almost embarrassing how good the agreement is between theory and experiment for $\sin^2\theta$. (Perhaps the experimental value will continue to go down and we will end up worrying about a discrepancy between a theoretical value of 0.21 and an experimental value of 0.15.) The exotic particle mass M is the chief input required in calculating the proton lifetime; a crude estimate is that the proton lifetime is of order $M^4/\alpha^2 m_p^5$, which for M a little below 10^{15} GeV comes out to be about 10^{31} years. There are more careful calculations of this, but I think that there is an irreducible uncertainty of about an order of magnitude in calculating the proton lifetime — after all it is a hadronic process and we can't take into account all the strong interaction factors, and there's no analog process anywhere in strong interaction physics from which we could take the matrix elements to figure out what is going on here.

The fact that $\sin^2\theta$ comes out right (and in a sense the exotic mass scale M comes out right, because, of course, if it came out much smaller that would be wrong, we would already have seen proton decay) means that this standard class of grand unified theories is a very interesting class of theories on which we might want to concentrate our attention. It is however a large class of theories. It includes theories like SU(5), SO(10), and bigger theories. It also includes theories which are quite different from those. I want to give one example as to the sort of grand unified theory of this class which is going to be necessary to try to distinguish experimentally from other possible theories. The gauge group is one that has been long under study, and was mentioned here by Pati: $SU(4)^4$. Just write out the observed quarks and leptons of the first two generations in a 4×4 array:

$$\begin{bmatrix} u_R & u_W & u_B & \nu_e \\ d_R & d_W & d_B & e \\ c_R & c_W & c_B & \nu_\mu \\ s_R & s_W & s_B & \mu \end{bmatrix} \qquad (2)$$

Then you have four SU(4)'s

$$SU(4)_{L\updownarrow} \times SU(4)_{R\updownarrow} \times SU(4)_{L\leftrightarrow} \times SU(4)_{R\leftrightarrow}$$

the arrows indicating whether the SU(4)'s act horizontally or vertically. If you adjoin a permutation symmetry which says all SU(4)'s are the same, this is a simple group. Also, the fermions are conventional. That is, the fermions in this model are just two of the standard generations, plus right-handed neutrinos, which since they are neutral under $SU(3) \times SU(2) \times U(1)$ don't affect the argument at all. (Other fermions have to be added to cancel anomalies, but this can be done so as also not to affect the results.) And therefore if this group breaks down in one step to $SU(3) \times SU(2) \times U(1)$, then you get the standard formula for $\sin^2\theta$, giving a value again of 0.21. You also get the standard formula for the superheavy mass scale, giving a value 10^{15} GeV or a little less.

There has been some confusion about this and I should explain these remarks. The early papers that dealt with this group tended to emphasize versions of it in which $SU(4)^4$ breaks first into other groups and then only at lower energies into $SU(3) \times SU(2) \times U(1)$. That is the difference between their results and what I am quoting here although I believe you can find this also mentioned as one of the alternative possibilities. I'm just making the conventional assumption here that you have breaking in one step; then this theory behaves in a conventional way.

The one-step breaking of $SU(4)^4$ to $SU(3) \times SU(2) \times U(1)$ can be done with a single irreducible multiplet of Higgs scalars, transforming under the representation $(4,\bar{4},15,1)$ + permutations. In addition, one needs to give large Majorana masses to the right-handed neutrinos, which suggests the introduction of scalar fields transforming according to irreducible representations like (1,6,1,6) + permutations. These scalars couple both to quark-lepton and antiquark-antiquark channels, so they are just of the type that would produce a proton decay of the conventional type, into antilepton plus mesons. All of the exotic scalars can be expected to have masses of the order of the same superheavy mass scale, around 10^{15} GeV, which yields a proton lifetime very roughly of order 10^{31} years. Thus, even if $\sin^2\theta$ is definitely measured to be 0.21 and a proton decay is discovered with a lifetime of order 10^{31} years, and the decay is as expected into a positron and meson, there will still be a number of viable candidates for a grand unified theory, including this one based on $SU(4)^4$. The special thing about this model which might allow it to be distinguished experimentally from other models in the standard class is that proton decay in this $SU(4)^4$ model is due entirely to scalar boson exchange. It is partly for this reason that I earlier emphasized the importance of calculating proton decay branching ratios, polarizations, etc., using the full Hamiltonian (1), including the g_3 and g_4 terms which can only be produced by scalar exchange.

I am now finished with my review of theoretical expectations, and will go on to discuss current experiments on proton decay.

Table 2 gives a list of experiments that are either in progress, being readied, or under consideration. The table is taken from one to be published in Scientific American, but I have added the names of those who have reported here on the various experiments. I have also added a listing for the deep ocean experiment projected here by Learned. (One of the advantages of such experiments is that it is easy to figure out

LOCATION	DATE OF OPERATION	DEPTH (EQUIVALENT METERS OF WATER)	MONITORED MATERIAL	DETECTORS	INSTITUTIONS	REPORTER HERE
HOMESTAKE GOLD MINE SOUTH DAKOTA	NOW	4,400	150 TONS OF WATER (TO BE INCREASED TO 900 TONS)	144 PHOTOMULTIPLIER TUBES IN WATER	UNIVERSITY OF PENNSYLVANIA	STEINBERG
KOLAR GOLD FIELD INDIA	NOW	7,600	150 TONS OF IRON (1/2-INCH PLATES)	GAS PROPORTIONAL COUNTERS BETWEEN PLATES	TATA INSTITUTE OF FUNDAMENTAL RESEARCH, BOMBAY UNIVERSITY OF OSAKA UNIVERSITY OF TOKYO	MIYAKE
BAKSAN VALLEY U.S.S.R.	NOW	850	80 TONS OF LIQUID SCINTILLATOR	1,200 PHOTOTUBES	INSTITUTE FOR NUCLEAR RESEARCH, MOSCOW	--
SOUDAN MINE MINNESOTA	NOW	1,800	30 TONS OF TACONITE CONCRETE AND IRON (TO BE INCREASED TO 1,000 TONS)	3,456 GAS PROPORTIONAL COUNTERS IN CONCRETE	UNIVERSITY OF MINNESOTA ARGONNE NATIONAL LABORATORY	SHUPE
MONT BLANC TUNNEL BETWEEN FRANCE AND ITALY	NOW	4,270	30 TONS OF IRON AND LIQUID SCINTILLATOR (TO BE INCREASED TO 200 TONS)	PHOTOTUBES AND STREAMER TUBES	INSTITUTE FOR NUCLEAR RESEARCH, MOSCOW UNIVERSITY OF TURIN	--
MORTON SALT MINE OHIO	1981 ESTIMATED	1,670	10,000 TONS OF WATER	2,400 PHOTOMULTIPLIER TUBES IN WATER	UNIVERSITY OF CALIF. AT IRVINE UNIVERSITY OF MICHIGAN BROOKHAVEN NATIONAL LABORATORY	SINCLAIR
SILVER KING MINE UTAH	1981 ESTIMATED	1,700	1,000 TONS OF WATER	800 PHOTOMULTIPLIER TUBES IN WATER AND MIRRORED WALLS	HARVARD UNIVERSITY PURDUE UNIVERSITY UNIVERSITY OF WISCONSIN	WILSON
MONT BLANC TUNNEL BETWEEN FRANCE AND ITALY	1981 ESTIMATED	5,000	150 TONS OF IRON INITIALLY (1-CM-THICK PLATES)	STREAMER TUBES BETWEEN PLATES	ITALIAN NATIONAL SYNCHROTRON LABORATORY AT FRASCATI UNIVERSITY OF MILAN UNIVERSITY OF TURIN	FIORINI
GRAND SASSO TUNNEL ITALY	UNDER CONSIDERATION	4,000	10,000 TONS OF IRON (3-MM PLATES)	FLASH CHAMBERS TRIGGERED BY STREAMER TUBES	ITALIAN NATIONAL SYNCHROTRON LABORATORY AT FRASCATI UNIVERSITY OF MILAN UNIVERSITY OF TURIN UNIVERSITY OF ROME	--
JAPAN	UNDER CONSIDERATION	2,700	3,400 TONS OF WATER	1,056 20-INCH PHOTOMULTIPLIER TUBES	UNIVERSITY OF TOKYO JAPANESE NATIONAL LABORATORY FOR HIGH-ENERGY PHYSICS UNIVERSITY OF TSUKUBA	--
FRÉJUS TUNNEL BETWEEN FRANCE AND ITALY	UNDER CONSIDERATION	4,500	1,500 TONS OF IRON (PLATES 3 TO 4 MM THICK)	FLASHTUBES AND STREAMER TUBES	ÉCOLE POLYTECHNIQUE UNIVERSITY OF PARIS AT ORSAY ÉCOLE DES MINES SACLAY	BARLOUTAUD
JAPAN	UNDER CONSIDERATION	2,700	600 TONS OF IRON	NEON FLASHTUBES	UNIVERSITY OF OSAKA UNIVERSITY OF TOKYO JAPANESE NATIONAL LABORATORY FOR HIGH-ENERGY PHYSICS	--
ARTYEMOVSK SALT MINE U.S.S.R.	UNDER CONSIDERATION	600	100 TONS OF LIQUID SCINTILLATOR	128 PHOTOMULTIPLIER TUBES	INSTITUTE FOR NUCLEAR RESEARCH, MOSCOW	--
DEEP OCEAN	?	4,500	10^6 TONS OF SEAWATER	MANY PHOTOMULTIPLIER TUBES	?	LEARNED

Table 2. Current and Projected Proton Decay Experiments

the depth in meters of water equivalent.) I should also mention the very different class of experiments on fossil tracks in mica discussed here by Bennett.

For the purposes of this summary, I have tried to understand what the various trade-offs are in these experiments. I probably will get everyone unhappy with this, but here is one theorist's attempt to summarize what are the relative advantages (of course, none of these experiments have any disadvantages) of the water Cerenkov counter experiments and the iron and concrete experiments with charged particle counters. Water is cheaper than anything else and also light goes further in water than charged particles go in iron or concrete so the water Cerenkov counters for a given amount of money can have a larger mass. On the other hand iron and concrete are denser than water which means you get a smaller volume for a given mass, and you don't have to have sump pumps and things like that, so the whole experiment can be more compact, and you can go deeper for a given amount of money or mass. The deepest experiments are all of the iron and concrete type. I don't know if that will always be true but, as shown by Table 2, it is true now. The water Cerenkov detectors have the advantage that it is fairly easy to tell the sense of the direction in which particles are going — you just have to time the arrival of the light at the different photomultiplier tubes. This capability would have been a great asset in interpreting the results of the Kolar gold field experiments, for instance. I understand that some of the dense detector experiments like the one in the Frejus Tunnel will have the capability of telling the sense of particle motion. The water Cerenkov detectors are excellent for photons, π^0's and electrons and may or may not, I'm not clear on that, be able to distinguish π^0's from photons and electrons. The iron and concrete detectors are excellent for high multiplicity events. They have no threshold like that in the Cerenkov detectors which can't see anything that travels slower than the speed of light in water. The water detectors have the advantage that there's less Fermi motion because they don't have heavy elements and that helps in analyzing events kinematically, but for the very reason that they don't have heavy elements, they don't get so many μ^- absorptions, while the fact that the iron concrete counters are composed of heavy elements means that they may be able to distinguish μ^+ from μ^- because μ^- gets absorbed and the μ^+ doesn't. That's very useful because it is so important to verify the general expectation that protons decay to antileptons but not leptons. A problem of all these experiments is that photons, positrons, and electrons are difficult to distinguish, though of course charge conservation and our knowledge of the proton charge would in some cases tell us that we have a positron and not an electron. It would be wonderful if magnets could be put into the iron detectors. I guess that's hopeless for water Cerenkov detectors. The capability of the water Cerenkov detectors to observe the electron from the decay of a muon produced in proton decay may allow determination of the muon longitudinal polarization and this would be very important as a means of distinguishing g_1 from g_2 and g_3 from g_4.

One more general remark. It's going to be very important to measure that the energy emitted in candidate proton decays is really about 940 MeV. For instance, you can play all sorts of games by imagining that some species of exotic particle is produced with the same abundance as baryons in the early universe, and decays with a lifetime of a few hundred million years, so that its abundance now is something like one part in 10^{22}. We would never see it in mass spectroscopy and yet it would accompany all ordinary matter. Its decay would appear like a proton decay with a lifetime of 10^{30} years. And the only way that I can think of — if I don't tell you what kind of exotic particle I am dreaming of — the only way I can think of to defeat that as a possible source of proton decay candidates is to really confirm that you are seeing about 940 MeV.

Until recently, the best lower bound on the proton lifetime was that set by the Witwatersrand experiment, about 10^{30} years. We now have a great improvement just beginning. There were two new results reported at this conference. The Kolar gold field experiment says the lifetime is now greater than or equal to 3×10^{30} years, and the Homestake experiment has also improved the previous lower bound. It is really good news that in just a few months, these new experiments have been able to set a limit on the proton lifetime which is three times better than we had before after many years of effort. This really is an extremely exciting period when the lifetime limit (or perhaps even the lifetime) is going to be so much better known. And when the larger detectors get turned on, of course, either proton decay will be seen or the lower bound on the lifetime is going to increase much more sharply.

A question: should I call the Kolar result a new lower bound on the lifetime or should I call it a value of the lifetime? I am not so foolish as to try to speculate on whether or not the events described here by Miyake are real proton decays. I am also not going to try to judge whether the few events that the Homestake experiment or the earlier Witwatersrand experiment reported as possible candidates were real events. They all are possible proton decays, and at this point we just don't know whether or not in fact the lifetime is 3×10^{30} years. If it is, then probably a few of the events of the Homestake or the Witwatersrand experiment were proton decays also. It seems to me though that it's not very important to say whether an individual event is or is not a proton decay. After all, if the proton lifetime is 3×10^{30} years then we probably each of us in our lifetime experience at least one proton decay in our bodies. I think the important thing is to be able to say that proton decay has been discovered, meaning that we now have some reasonable certainty that it occurs at a certain rate. We don't have that yet. The Kolar gold field experiment may succeed in discovering proton decay in that sense in the next few months if they go on finding these events. As an amateur cosmologist, I would say that what I would look for in the next few months of running is whether or not these events are going to satisfy the principles of isotropy and homogeneity. That is, are the events evenly distributed throughout the detector and is the orientation of the tracks isotropic with respect to the vertical direction? We will just have to see.

Since I am keeping an open mind about the results of current experiments, let me consider various alternative futures. One possibility is that proton decay will not be discovered: the proton decay lifetime will turn out to be greater than about 10^{34} years and baryon violation will never be seen in any other way either. Even so, many of us will go to our graves thinking that baryon number is probably violated, but it won't matter. There are then two possibilities with regard to the successful prediction of $\sin^2\theta$. One is that the correct grand unified theory if there is one is very different from any we know, so that $\sin^2\theta$ is given by some other formula, such as for instance the kind of formula that Hung described here. You would then just have to say that the present success in predicting $\sin^2\theta$ has no significance. Another possibility is that $\sin^2\theta$ is still correctly predicted by considerations of the unification of strong with electroweak interactions, and it is just the superheavy mass that came out wrong. Dimopoulos and Georgi discussed here how this could be true in the context of supersymmetry. If you fill out supermultiplets with particles that are very heavy in terms of ordinary energies but very light in terms of the superheavy mass scale, then the SU(3) octet of gluons has as a companion an SU(2) × U(1) singlet, SU(3) octet of fermions, and likewise for the electroweak gauge bosons. This would decrease the rate of change of all couplings with energy, so that the factor of $\exp(4\pi^2/11e^2)$ that I mentioned before would become $\exp(4\pi^2/9e^2)$. This is perhaps not spectacular but it would drive up the superheavy mass scale from 10^{15} GeV to 10^{18} GeV (or, according to Dimopoulos, 10^{17} GeV) which is more than enough to wipe out the possibility of observing proton decay in our lifetime, yet without appreciably affecting $\sin^2\theta$.

Alternative future number two is that baryon violation is discovered, but turns out to be weird. It might be a proton decay into a negative rather than a positive lepton or into three antileptons or three leptons. Another possibility is neutron-antineutron oscillation, which was discussed here by Green. Pati has considered all these different cases and has described here the different exotic particle mass scales that are needed. In order for these weird decays of protons or neutrons to occur, you need some kind of superheavy exotic particles with masses between 10^4 GeV and 10^{10} GeV. If they are heavier than that, you just won't see these effects. In such a theory, I presume $\sin^2\theta$ might come out right numerically, but the success of the particular formula which has been used so far to predict $\sin^2\theta$ would have to be regarded as an accident.

The third alternative future is that the proton decay <u>is</u> going to be discovered soon, which means that the lifetime would have to be 10^{30} to 10^{33} years, and that it turns out to be conventional proton decay into an antilepton plus mesons. Then you would tend to believe in the standard class of theories I mentioned earlier, but you still wouldn't know which is the right one. That would take years of effort. Is the correct theory one which has mostly scalar exchange? or one that has mostly vector exchange? This can be done as I said before by measuring things like antineutrino-positron ratios, fermion polarizations, etc., but it probably will not be until the eleventh workshop on grand unification that we can hope to get an answer to this.

Whichever one of these scenarios turns out to be correct, including the third one, we will still be faced with a variety of very deep problems. One of them, which was repeatedly mentioned today, is the hierarchy problem. Why are there scalar fields which trigger the spontaneous breakdown of SU(2) × U(1) and are so light compared to the grand unification scale and (even worse) compared to the Planck mass? As Yao showed here in describing the construction of effective field theories, this is a physical and not a mathematical difficulty. That is, there is no difficulty in saying that certain renormalized parameters like scalar masses have very small values. Veltman also discussed the fact that you can arrange for the cancellation of quadratic divergences in the low energy effective field theory and that this leads to interesting constraints. But even if we do all this, there is still a mystery: why is the scalar mass so light?

There are a number of alternatives which fill the literature. One is that in fact there are no elementary scalars, that the SU(2) × U(1) is broken dynamically. That is, the Goldstone bosons are composite particles like they are in superconductivity. Raby discussed this here. There are well known problems with this kind of theory, problems with producing the masses of the quarks and leptons and problems with including this kind of theory in any kind of grand unified theory.

The second way of approaching the hierarchy problem which was discussed here by Srednicki, Dimopoulos and Georgi is supersymmetry. One may hope that the scalars are light because they're members of supermultiplets and the other members of the supermultiplets are fermions which are light for a familiar reason, γ_5 invariance. If the supersymmetry is then broken at some scale, the scalars then get mass and maybe that has something to do with the breakdown of SU(2) × U(1) (although in fact in the model described by Srednicki that wasn't the case). This kind of theory, of course, has to face an enormous hurdle. Supersymmetry is not manifest in the world we know. It has to be unbroken all the way down to something like a few TeV in order to serve its function of preserving scalars from getting enormous masses, and yet it has to be broken where we live at a few GeV. That's very hard. Supersymmetric theories tend to resist spontaneous symmetry breaking. Srednicki discussed here the possibility that supersymmetries are broken dynamically; that is, their Goldstone bosons, the Goldstinos, are bound by strong forces. That sounds a little perverse because part of the reason for introducing supersymmetry is to allow light scalars so you don't have to think about dynamical symmetry breaking, but maybe that's the way the real world is. One very interesting aspect of the work Srednicki discussed is that the strong CP problem may be solved with an axion that couples too weakly to have been detected. Dimopoulos and Georgi considered another possibility, that the supersymmetry is broken in the Lagrangian, but softly, by terms which have lower dimensionality and therefore don't pollute the rest of the Lagrangian with a general breakdown of supersymmetry. Is this natural? Georgi very reasonably refers to something as technically natural if when you tune parameters you don't get infinities which require you to retune. To me something is natural if I can see where it comes from in terms of an exact symmetry of an underlying theory. So I'm not sure I would use the word natural as applied to the work he described the way he did. But the model of Dimopoulos and Georgi has many other attractive features and something like it may turn out to be right.

There's another deep problem, the problem of the cosmological constant, that was referred to here by Turner and Veltman. Basically, the problem is that in most theories the vacuum has a non-vanishing energy density ρ_0 (that is, the vacuum expectation value of the energy-momentum tensor is $\rho_0 g_{\mu\nu}$) and this turns up in the gravitational field equations as a term which looks just like the cosmological constant term which Einstein added in 1917 and then regretted ever after. Specifically, if the bare cosmological constant in the field equations is λ_B, then the effective cosmological constant which governs gravitational phenomena is

$$\lambda = \lambda_B + 8\pi G \rho_0 \tag{3}$$

Experimentally the value of λ/G is less than about 10^{-29} g/cm^3, or roughly 10^{-46} GeV^4 in Turner's way of putting it, while in ordinary field theories like QCD you expect ρ_0 to be of order 1 GeV^4 (whether or not there is any spontaneous symmetry breaking); in theories with exotic particles of mass 10^{15} GeV you might expect ρ_0 to be about 10^{60} GeV^4; and in quantum gravity ρ_0 would be expected to be about 10^{76} GeV^4. There is no mathematical problem here: just adjust λ_B so that it cancels $8\pi G \rho_0$ to 122 decimal places! But it's very disturbing that we have to arrange this cancellation in what appears like a thoroughly unnatural way.

I should add, contrary to one remark made here, that supersymmetry does not so far appear to be a help with this problem. If anything, it has made it worse because supersymmetry if it exists at all has to be spontaneously broken and if supersymmetry is spontaneously broken you can actually prove that ρ_0 is positive-definite.

In a sense, the problem of the cosmological constant is very similar to the problem of the light scalar mass because in both cases one is worried about a term in the Lagrangian which is super-renormalizable, i.e., which has dimensionality less than 4, and in both cases you have to throw away this term although there doesn't seem to be any reason to throw it away. The problem of the cosmological constant is so much more ubiquitous though, and so much more deadly in terms of numbers, that you might ask, since we have to worry about it anyway, why worry about the scalar mass? Perhaps it's premature to worry about the scalar mass and the hierarchy problem if we don't understand the cosmological constant at all.

Another problem which has been around for a long time is the problem of the multiplicity of generations and their various masses. In theories with elementary scalars this is a matter of a large number of free parameters which can be adjusted very easily, but which are mysterious. Veltman tried to get a handle on these by the "algebra of infinities" of his, the cancellation of the quadratic convergences already mentioned. Hill and Machacek here suggested fixed point conditions which are essentially the conditions that are seen at low energy if you demand that at high energy the Yukawa couplings become strong. This is interesting because if the Yukawa couplings do become strong, it suggests a new class of strong interactions at 10^{15} GeV which may have something to do with spontaneous breakdown of the grand unified theory. Maybe it's the grand unified theory that breaks down dynamically.

The problem of generations and masses also has to be confronted if you don't believe in elementary scalars, in which case it takes a somewhat different form. Various theorists have suggested extended hypercolor models, with or without tumbling. Raby surveyed these various ideas here. They lead to problems including large values for strangeness changing neutral current processes and dangerous pseudo-Goldstone bosons and they are also difficult to put into a grand unified theory.

There's yet another approach which has been much discussed in just the last year and was mentioned here by both Veltman and Pati. This is the idea that the reason we have these different generations is like the reason we have the different energy levels of hydrogen: the quarks and leptons are really composites. Until recently I thought this was a pretty implausible idea, because we know the quarks and leptons are pretty small — in fact, they're much smaller than their Compton wavelengths — and if they are composites, then they're the first composite particles in history which are small compared to their Compton wavelength. That requires a whole new kind of dynamics. There is one possibility as to how this could happen which has recently been widely studied. You can imagine something like a QCD-like force which has a characteristic scale of a TeV and which binds the free quarks into very small composites which we see as quarks and leptons, and yet these composites can be very much lighter than a TeV because of some unbroken chiral symmetry. The problem then is why quarks and leptons are not strictly massless. I will just mention that in this sort of model the bound states can get small masses from the ordinary strong and electroweak interactions, without any need for extended hypercolor, or anything like that.

Although there are a number of interesting ideas under study, I would say the whole situation regarding generations and masses is rather disappointing. One would have hoped that grand unified theories or dynamical symmetry breaking theories or composite models of quarks and leptons would have thrown great light on the problem of the generations and the problem of the mass values of the particles of different generations. So far they haven't, and this gives one a feeling of uneasiness about any attempt to regard the theories we are talking about now as truly fundamental.

These questions may be clarified in the next few years by very exciting experiments which are just on the horizon. (One should keep in mind that in order to do experimental physics it is not always necessary to dig a deep cavern in the ground. Sometimes you can actually do physics on the surface of the earth, with accelerators.) For example, pretty soon we hope the Z and maybe the W will be found; their masses will probably be measured pretty accurately. Marciano here gave values of their masses

$$m_W = 83.0 \pm 2.4 \text{ GeV} \qquad m_Z = 93.8 \pm 2.0 \text{ GeV}$$

which are calculated using the SU(2) × U(1) theory with radiative corrections and our best present information about $\sin^2\theta$ from neutral current experiments. It's going to be very exciting to see the experiments verify these predictions. These predictions incorporate one-loop radiative corrections, so you're really testing the electroweak gauge theory beyond the tree approximation in verifying these predictions at the 1 GeV level. In particular, in dynamical symmetry breaking theories we now understand how you can get the same values for the W and Z masses in tree approximation but the electroweak radiative corrections to the masses are presumably not what they are in the electroweak theory with elementary scalars. It would be very interesting to try to make some estimate even of just the order of magnitude of the uncertainty in the radiative corrections in a theory where you have dynamical symmetry breaking. (In a sense, Veltman described something like this, not with respect to the observed masses but with respect to the ρ parameter that we observe at low energies. The advantage of working with the actual W and Z masses is that they will eventually be measured much more accurately than the neutral current cross sections, and their interpretation does not rely on strong interaction theories like the parton model.)

As a by-product of the search for the W and Z, there may also be discovered narrow semi-weakly decaying scalar particles, the Higgs bosons. If there aren't any elementary scalars then we won't see any narrow Higgses, though they may show up as very broad enhancements at very high energy. However there will then be relatively light particles called pseudo-Goldstone bosons, somewhere in the neighborhood of 100 GeV or possibly much lighter. It may be hard to tell Higgs from pseudo-Goldstone bosons. It's going to be exciting because one or the other has to exist. I don't know of any theory that doesn't have either Higgs or pseudo-Goldstone bosons, so we have to find something.

Further, if you assume any sort of dynamical symmetry breaking, or if you believe in quarks and leptons as composites, either way there has to be a new class of extra-strong interactions, a whole new world of hyperhadronic physics with strong interactions that become strong at 1 TeV (give or take a factor of ten).

Another possibility which arises in many grand unified theories is that there may be mirror fermions at relatively low energy. That's true in theories based on groups like E_8 which is very attractive in many ways but leads to the prediction that there are right-handed doublets as well as left-handed doublets, maybe not at a few GeV but perhaps at a few hundred GeV. And finally I think the most exciting possibility of all is that when we get up to a few hundred GeV we may begin to see supermultiplets emerge out of the fog of particle taxonomy, that we may begin to see supersymmetry realized as an actual symmetry of nature.

I don't have much time to cover the last thing I wanted to talk about here which is cosmology. A general review was given here by Turner. I thought the most intriguing thing that he covered and that has happened in recent years in this general area is something that was just barely mentioned at the First Workshop: the development of the general idea that the phase transitions in which various gauge symmetries are broken may be first-order rather than second-order phase transitions. As is familiar from the first-order phase transition in which steam condenses into liquid water, it is then possible to have a period of supercooling in which you stay for a while in the wrong phase. This leads to all kinds of fascinating possibilities, such as that the heat finally released when the super-cooling is over may explain the enormous photon number of the universe: there are at least 10^{85} photons within the cosmic radius of curvature. (This bears on what Turner refers to as the flatness problem. That is, it is easy to show that the universe appears to be flat for a time which in units of the Planck time is something like the photon number within the radius of curvature of the universe to the 2/3 power. And so if there is a large photon number within the radius of the curvature of the universe then that explains why the universe appears to be flat for so many Planck times. Of course, it is possible that there is nothing here that needs to be explained — that in the large the universe is precisely flat, and that the photon number within the radius of curvature of the universe is therefore infinite.) First-order phase transitions may also help in explaining the observed homogeneity of the universe, and the absence of monopoles, which otherwise seems to be an endemic problem for grand unified theories.

The next year should be an exciting one. Either proton decay will be discovered or the limit on the lifetime will become much more constraining; the experimental indication of a neutrino mass (not discussed much here) may become clarified; and intermediate vector bosons may be found. Let's hope that theory will have something new to say about our outstanding problems — hierarchies, generations, etc. — that can match the enlightenment that will be provided by experiment.

WORKSHOP PROGRAM

Friday, April 24

Morning	M. Goldhaber, Chairman
R. H. Sands	Welcoming Comments
S. L. Glashow	Keynote Address: A Neutrino-Free Universe
P. Langacker	Theoretical Expectations for Proton Decay
E. Fiorini	Mont Blanc Detector
W. J. Marciano	SU(5) Predictions: Theory vs. Experiment
Afternoon	B. Kayser, Chairman
R. Barloutaud	Frejus Detector
M. Veltman	Infrared/Ultraviolet Connections
M. A. Shupe	Soudan Detector
Y.-P. Yao	Effective Lagrangians and Effective Parameters in Grand Unified Theories

Saturday, April 25

Morning	D. Cline, Chairman
J. C. Pati	Filling the Grand Plateau Between 10^2 and 10^{15} GeV
S. Miyake	Kolar Gold Field Detector
M. S. Turner	Grand Unification and the Fundamental Problems of Classical Cosmology
R. I. Steinberg	Pennsylvania Detector
Afternoon	G. Segre, Chairman
S. Raby	The TeV Picture
C. Wilson	Harvard-Purdue-Wisconsin Detector
C. T. Hill	Fixed Points Fermion Mass Predictions
M. Machacek	Fermion Mass Predictions: Another Look
D. Sinclair	Irvine-Michigan-Brookhaven Detector
C. L. Bennett	Limitations to Passive Proton Decay Experiments
J. G. Learned	Ultimate Limits to Proton Decay Detectors

Sunday, April 26

Morning	F. Reines, Chairman
K. Green	Review of $n\bar{n}$ Experiments
P. Q. Hung	Petite Unification of Quarks and Leptons
M. Srednicki	Supersymmetric Technicolor
S. Dimopoulos	Supersymmetric GUTs
H. Georgi	Supersymmetric GUTs, too
M. Yoshimura	Constraints on GUTs from Baryon Asymmetry
S. Weinberg	Conference Summary

LOCAL ORGANIZING COMMITTEE

R. Akhoury, M. Einhorn, D.R.T Jones, T.W. Jones, G.L. Kane, J.P. Leveille, J. LoSecco, D. Sinclair, T. Sterling, J. Stone, L. Sulak, Y. Tomozawa, D.G. Unger, J. Vander Velde, Y.-P. Yao.

ADVISORY COMMITTEE

R. Barloutaud, J.D. Bjorken, G. Cassidy, E. Fiorini, P. Frampton, J. Gaidos, M.K. Gaillard, F. Reines, K. Ruddick, F. Wilczek, A. Yildiz.

LIST OF PARTICIPANTS

AKHOURY, R.	University of Michigan
ALBRIGHT, Carl	DESY
ANTONIADIS, Ignatios	Ecole Polytechnique, Paris
AUBRECHT, Gordon	Ohio State University
BAIS, F. Alexander	University of Pennsylvania and of Utrecht
BALUNI, Varouzhan	Institute for Advanced Study, Princeton
BANKS, J.	Ohio State University
BARDEEN, William A.	Fermilab
BAREYRE, P.	CEN, Saclay, France
BARLOUTAUD, Roland	CEN, Saclay, France
BARS, Itzhak	Yale University
BENNETT, Charles	Princeton University
BINCER, Adam M.	University of Wisconsin
BIONTA, Richard	University of Michigan
BRANCO, Gustavo	Carnegie-Mellon University
BRATTON, Clyde W.	Cleveland State University/University of California, Irvine
BRYAN, Ronald	Texas A&M University
BURAS, Andrej	Fermilab
CAHILL, Kevin	University of New Mexico
CAMPBELL, B. A.	University of Toronto
CAPPS, Rick	Purdue University
CHANG, N. P.	City College of New York
CHENG, Hai	Purdue University
CHIANG, C. C.	University of Texas, Austin
CLARK, Tom	Purdue University
COHN, Hans O.	Oak Ridge National Laboratory
CORTEZ, B.	University of Michigan/Harvard University
DIMOPOULOS, Savas	Stanford University
EICHTEN, Estia	Harvard University
EINHORN, M. B.	University of Michigan
ELLIS, Steve	University of Washington
ERNWEIN, John	Fermilab
ERREDE, Steven	Ohio State University
FEDERBUSH, Paul	University of Michigan
FIELDS, Tom	Argonne National Laboratory

List of Participants, continued

FIORINI, Ettore	University of Milan
FISCHLER, M.	Fermilab
FISHBACH, Ephraim	Purdue University
FODA, Omar	Purdue University
FOGLEMAN, Guy	Indiana University
FOSTER, G. W.	University of Michigan/Harvard University
FRAMPTON, Paul	University of North Carolina
FRY, James	University of Chicago
GAIDOS, James	Purdue University
GAVELA, M. B.	LAPP, Annecy, France
GEORGI, Howard	Harvard University
GIRARDI, Georges	LAPP, Annecy, France
GLASHOW, Sheldon	Harvard University/Massachusetts Institute of Technology
GOLDHABER, Maurice	Brookhaven National Laboratory
GOLDMAN, Terry	Los Alamos National Laboratory
GOODMAN, Matt	Harvard University
GREEN, Keith	Rutherford-Appleton Laboratory
GREENBERG, J.	University of Michigan
GUNAYDIN, Murat	University of Bonn
HABER, Howard	University of Pennsylvania
HARRIS, Edward G.	University of Tennessee
HECHT, K. T.	University of Michigan
HEGYI, Dennis	University of Michigan
HELLER, Kenneth	University of Minnesota
HILL, Chris	Fermilab
HUNG, P. Q.	Lawrence Berkeley Laboratory
IAROCCI, Enzo	INFN, Frascati, Italy
JONES, D. R. T.	University of Michigan
JONES, Tegid W.	University of Michigan
JULLIAN, S.	LAL, Orsay, France
KANE, Gordy	University of Michigan
KARL, Gabriel	University of Guelph
KAYSER, Boris	NSF/Stanford Linear Accelerator Center
KROPP, William	University of California, Irvine
KUO, T. K.	Purdue University
LANE, Kenneth	Ohio State University
LANGACKER, Paul	University of Pennsylvania
LEARNED, John	University of Hawaii/University of California, Irvine
LEIBBRANDT, George	University of Guelph
LEVEILLE, Jacques	University of Michigan
LEVIT, Larry	LeCroy Research, New York
LI, Xiaoyuan	Virginia Polytechnic Institute
LICHTENBERG, D. B.	Indiana University
LINN, Leslie	University of Michigan
LONGO, Mike	University of Michigan
LO SECCO, John	University of Michigan
LOVE, Sherwin	Purdue University

List of Participants, continued

MA, Ernest	University of Hawaii/ University of Wisconsin
MACHACEK, Marie	Northeastern University
MAHANTHAPPA, K. T.	University of Colorado
MALOOF, David	Argonne National Laboratory
MANNHEIM, Philip	University of Connecticut
MARCIANO, William J.	Northwestern University
MATUTE, Ernesto A.	Purdue University
MAY, Edward	Argonne National Laboratory
McHENRY, Ronald	Purdue University
McMANUS, H.	Michigan State University
MESHKOV, Sydney	National Bureau of Standards
MEYER, Don	University of Michigan
MILTON, K. A.	Ohio State University
MIYAKE, Saburo	University of Tokyo
MORSE, R.	University of Wisconsin
MURTHY, P. V. Ramana	University of Michigan/Tata Institute
NAKAGAWA, Norio	Purdue University
NARASIMHAM, V.	Tata Institute, Bombay
NEGRET, Juan	Purdue University
NEWPORT, John	Purdue University
NIEVES, Jose	Carnegie-Mellon University
O'DONNELL, Patrick J.	University of Toronto
OAKES, Robert J.	Northwestern University
OLIVE, Keith	University of Chicago
ONG, Chong-Leong	University of California, San Diego
PAPAVASILIOU, Christos	University of Toronto
PATI, Jogesh	University of Maryland
PAULOT, M.	LAL, Orsay, France
PEASLEE, David	U. S. Department of Energy
PEREZ, Miguel A.	CIEA-IPN, Mexico
PERRY, Dave	Purdue University
PETERSON, Earl	University of Minnesota
PISARSKI, R.	Yale University
PRASAD, V. B.	Virginia Polytechnic Institute
PUHALA, Michael J.	Iowa State University
PUMPLIN, Jon	Michigan State University
RABY, Stuart	Stanford Linear Accelerator Center
RADFORD, Stanley F.	Wayne State University
RAJPOOT, Subash	Imperial College
REPKO, W.	Michigan State University
REINES, Fred	University of California, Irvine
ROBINETT, Richard W.	University of Minnesota
ROSEN, S. Peter	Purdue University
RUDDICK, Keith	University of Minnesota
SANDS, R. H.	University of Michigan
SAVIT, R.	University of Michigan
SCANIO, Joseph	University of Cincinnati
SCHULTZ, Jonas	University of California, Irvine
SEGRE, Gino	University of Pennsylvania
SHEPARD, Paul F.	University of Pittsburgh
SHER, Marc	University of California, Santa Cruz

List of Participants, continued

SHUMARD, E.	University of Michigan
SHUPE, Michael	University of Minnesota
SINCLAIR, Dan	University of Michigan
SIRLIN, Alberto	New York University
SMITH, Darrel	University of California, Irvine
SOCOLOVSKY, M.	CIEA-IPN, Mexico
SREDNICKI, Marc	Princeton University
STEINBERG, Richard	University of Pennsylvania
STERLING, Tom	University of Michigan
STONE, Jim	University of Michigan
STUMP, D.	Michigan State University
SULAK, Larry	University of Michigan
SURANYI, Peter	University of Cincinnati
TALLINI, Bruno	CEN, Saclay, France
TANAKA, Katsumi	Ohio State University
THURBER, James	Purdue University
TOMOZAWA, Yukio	University of Michigan
TURNER, Mike	University of Chicago
UNGER, David	University of Michigan
VAN DER VELDE, Jack	University of Michigan
VELTMAN, Martinus	University of Michigan/University of Utrecht
WADA, Walter	Ohio State University
WALKER, James K.	Fermilab
WEILER, Tom	Northeastern University
WEINBERG, Steven	Harvard University/University of Texas
WEISBERGER, W. I.	State University of New York, Stony Brook
WELDON, H. Arthur	University of Pennsylvania
WHEATER, John	Oxford University
WILLMAN, Robert	Purdue University
WILSON, Charles	Purdue University
WUEST, Craig	University of California, Irvine
YAO, Y.-P.	University of Michigan
YOUNG, Bin-Lin	Iowa State University
YOSHIMURA, Motohiko	KEK, Japan